Data Analytics in Power Markets

Qixin Chen · Hongye Guo · Kedi Zheng · Yi Wang

Data Analytics in Power Markets

科学出版社
Science Press

Springer

Qixin Chen
Department of Electrical Engineering
Tsinghua University
Beijing, China

Hongye Guo
Department of Electrical Engineering
Tsinghua University
Beijing, China

Kedi Zheng
Department of Electrical Engineering
Tsinghua University
Beijing, China

Yi Wang
Department of Electrical and Electronic Engineering
The University of Hong Kong
Hong Kong, China

ISBN 978-981-16-4974-5 ISBN 978-981-16-4975-2 (eBook)
https://doi.org/10.1007/978-981-16-4975-2

Jointly published with Science Press
The print edition is not for sale in China (Mainland). Customers from China (Mainland) please order the print book from: Science Press.

This Springer imprint is published by the registered company Springer Nature Singapore Pte Ltd.
The registered company address is: 152 Beach Road, #21-01/04 Gateway East, Singapore 189721, Singapore

To Our Alma Mater

—Tsinghua University

Preface

The deregulation of power sectors around the world allows for the entrance of competitors to buy and sell electricity while maximizing their own utility. The implementation of deregulated electricity markets builds on the principle of "bid-based, security-constrained economic dispatch". The market organizer clears the market by maximizing the overall welfare according to the submitted offers and bids of all market participants. The core problem for market participants is to achieve higher revenue or lower cost through optimal offering and bidding. Thus, the buyers need to perform accurate load forecasting to ensure a rational bid. Both the sellers and buyers need to perform electricity price forecasting as the price is used for market settlement. The core problem for market organizers is to improve the market efficiency and in some countries guide investments of generation and transmission assets. Thus, they also need to perform load forecasting (but from a higher level compared with market participants). In addition, they need to learn market participants' behaviors and conduct short-term and long-term market operation simulations. The daily operation of power markets creates a lot of valuable data, from which information can be extracted and used to support the aforementioned problems faced by market participants and organizers.

In recent years, the development of data analytic methods, especially machine learning techniques, has brought promising solutions to different industries. However, although a large volume of data related to power market operation is available, these data have not been fully utilized to support the decision-making in power markets yet. The reason behind this predicament is that the generation for market data (especially the price data) follows certain market rules and characteristics, and the market offer and bid data are unstructured and difficult to use.

The book aims to utilize available market data to support the decision-making of both market participants and market organizers in several key issues, with the support of data structuring and analytic methods. The book consists of three parts and thirteen chapters. It begins with an overview of worldwide electricity markets and an introduction to their organization and data disclosure policies. Then we focus on three major problems of data analytics in power markets: load modeling and forecasting, electricity price modeling and forecasting, and market bidding behavior analysis.

To help readers have a better understanding of what we have done, we would like to make a simple review of the thirteen chapters in the following.

Chapter 1 presents an introduction to power market data. An overview of electricity markets is given with examples of markets in the U.S., EU, and China. Then, the organization and data disclosure policies of electricity markets are introduced in detail.

Chapter 2 studies a probabilistic load forecasting (PLF) problem with smart meter data. It proposes a PLF framework with clustering and ensemble learning to leverage load information within the fine-grained smart meter data. Four different probabilistic ensemble learning methods are implemented and extended to rolling window-based online applications.

Chapter 3 provides a bad data identification and repairing method for load profiles. Singular Value Thresholding (SVT) algorithm is used to repair the missing data, and quantile regression with lag value is performed to detect local fluctuation. The impact of load data cleaning on load forecasting is also investigated.

Chapter 4 proposes a novel monthly electricity consumption forecasting method based on a seasonal vector correction model capturing the potential impacts from external economic factors. To select important variables, a self-adaptive screening method is integrated into the proposed forecasting framework.

Chapter 5 studies the problem of feature selection in PLF. It proposes a novel feature selection method based on sparse L_1-norm penalized quantile regression. An Alternating Direction Method of Multipliers (ADMM) algorithm is developed to boost the efficiency of model training in large-scale problems.

Chapter 6 investigates the fundamental distribution of the congestion part of LMPs in a high-dimensional Euclidean space using an unsupervised approach. It shows that the congestion part of LMPs is spanned by certain row vectors of the power transfer distribution factor (PTDF) matrix, and the subspace attributes of an LMP vector uniquely are found to reflect the instantaneous congestion status of all the transmission lines. It also proposes a framework for the identification of potential congestion status from LMP data.

Chapter 7 proposes an electricity price forecasting framework that trains individual forecasting models for components of locational marginal prices and stacks the composition forecasts with the direct forecasts to improve the overall forecasting accuracy. The characteristics of each component are analyzed and taken into consideration for feature engineering.

Chapter 8 proposes a method for quantifying the economic impact of price forecasting error for market participants. A general optimization problem of bidding in the electricity market is considered, and the forecasting error is modeled as a parameter for the optimization problem. Multiparametric linear programming (MPLP) is used to analyze the impact of forecasting error on the optimal bidding curves and revenues.

Chapter 9 proposes a financial transmission right (FTR) speculation framework based on LMP forecasting and stochastic programming. K-means clustering with network congestion patterns is used to select important nodes and reduce the problem size. Then quantile regression and copulas are used to forecast the nodal price scenarios.

Chapter 10 proposes a data-driven analysis framework for extracting typical bidding patterns from actual bidding data based on adaptive clustering. Considering the raw unstructured market bidding data, a data standardization processing method is also integrated to provide a foundation for bidding pattern extraction by addressing the bidding data particularities.

Chapter 11 proposes a forecasting framework for aggregated supply curve (ASC) based on a long short-term memory (LSTM) model and a paradigmatic data integration method. PCA-based dimension reduction is used for feature extraction.

Chapter 12 proposes a bottom-up data-driven power market simulation method based on individual offering strategy forecasting and realistic market clearing models. A deep neural network (DNN)-based model is used to learn the relationships between individual offering patterns and potential influencing factors.

Chapter 13 studies the problem of individual reward function (or objective function) identification for generation companies (GENCOs). The proposed data-driven identification framework is based on a standard Markov decision process and deep inverse reinforcement learning method.

To summarize, this book provides various applications of data analytic methods for several key problems in power markets. We hope this book can inspire readers to define new problems, develop novel methods, and achieve interesting results related to power markets.

Beijing, China Qixin Chen
Beijing, China Hongye Guo
Beijing, China Kedi Zheng
Hong Kong, China Yi Wang
September 2021

Acknowledgements

This book made a summary of our recent research about the application of data analytic methods in power markets. These works were carried out in the Energy Intelligence Laboratory (EILAB), Department of Electrical Engineering, Tsinghua University, Beijing, China.

Many people contributed to this book in various ways. The authors are indebted to Prof. Mohammad Shahidehpour from Illinois Institute of Technology, Prof. Le Xie from Texas A&M University, Prof. Gabriela Hug from ETH Zürich, Prof. Qing Xia, Prof. Chongqing Kang, Dr. Ning Zhang, Mr. Yuxuan Gu, Mr. Dahua Gan, Mr. Xichen Fang, and Mr. Xiao Cai from Tsinghua University, who have contributed materials to this book.

We also thank Mr. Yinxiao Li, Mr. Cheng Feng, Ms. Kexin Li, Ms. Xichen Tian, Mr. Qinghu Tang, Mr. Dawei Wang, Ms. Tianyu Zhang, and Mr. Ruike Lyu at Tsinghua University for their assistance in pointing out typos and checking the whole book.

In addition, we acknowledge the innovative works contributed by others in this increasingly important area especially through IEEE Power & Energy Society Working Group on Load Aggregator and Distribution Market and appreciate the staff at Springer for their assistance and help in the preparation of this book.

This book is supported in part by the Major Smart Grid Joint Project of the National Natural Science Foundation of China and State Grid (U2066205) and in part by the Guangdong Province Key Area R&D Project (2019B111109002). The authors really appreciate their supports.

Qixin Chen
Hongye Guo
Kedi Zheng
Yi Wang

Contents

Chapter 1
Introduction to Power Market Data

1.1 Overview of Electricity Markets

Electricity is widely used as a kind of efficient and convenient energy. The trading of electricity not only has typical characteristics of bulk commodity trading but also must meet certain constraints of physical laws. Hogan pointed out that a successful electricity market should be based on the principle of "bid-based, security-constrained economic dispatch (SCED)" [1]. At present, the electricity spot markets in the world (including markets in the U.S., EU, Australia, and other countries, as well as the pilot market in the trial operation stage in China) basically follow this principle. "Bid-based" refers to that market participants need to declare their own cost or utility of generating or purchasing electricity, so as to form the supply and demand curves of the market. The SCED model is usually a linear programming model whose objective function is the maximization of social welfare. The balance of energy, transmission capacity, as well as upper and lower limits of generation are included in the constraints of SCED. The optimal solution determines the cleared capacity of the market participants, and the shadow prices of the constraints are used to calculate the locational marginal prices (LMP), representing the marginal cost of the power system caused by the incremental demand at each node.

Figure 1.1 shows the map of wholesale electricity markets organized by regional transmission operators (RTOs) and independent system operators (ISOs) in the U.S. and Canada. The deregulation of electricity markets in the U.S. began in the 1990s. At present, there are seven ISO/RTO markets in the U.S., including PJM [3], CAISO [4], NYISO [5], etc., most of which are under the regulation of the Federal Energy Regulatory Commission (FERC) [6], except for the ERCOT [7] in Texas. Take PJM, the earliest deregulated market in the U.S., as an example [8]. PJM is located in the northeast of the U.S., and its main management area covers Pennsylvania, New Jersey, and Maryland. In 1997, PJM opened its first bid-based energy market. Later that year, the FERC approved PJM as the nation's first fully functioning ISO. Then, from 1998 to 2001, PJM completed the implementation of LMP mechanism, financial transmission right (FTR), day-ahead market, and ancillary services market of load frequency regulation. PJM became the nation's first fully functioning RTO in 2002. After 2002, PJM focuses on the integration of other small regional transmission networks and expansion of the wholesale market scale. Currently, the wholesale

Q. Chen et al., *Data Analytics in Power Markets*,
https://doi.org/10.1007/978-981-16-4975-2_1

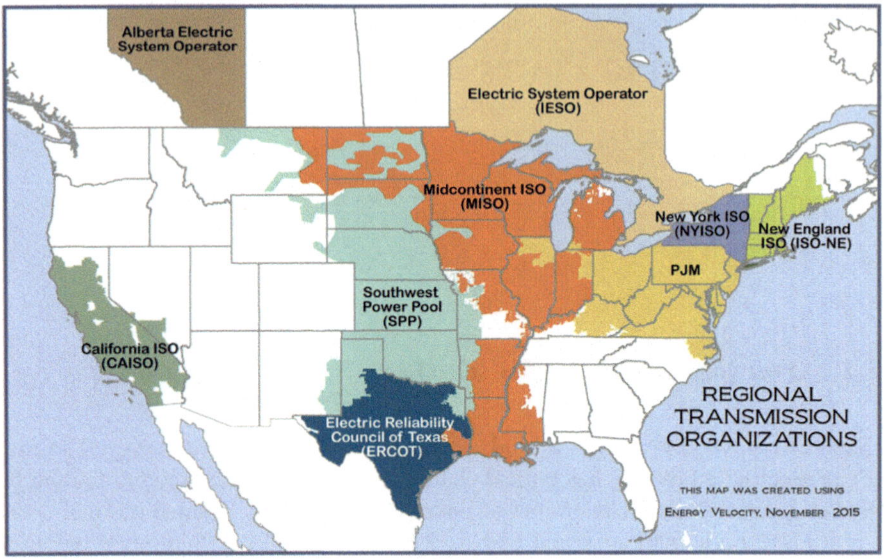

Fig. 1.1 Wholesale electricity markets operated by ISOs and RTOs in U.S. and Canada [2]

markets in the U.S. adopt a two-settlement (i.e., day-ahead and real-time settlement) system and use the LMP calculated based on SCED models for payment.

The goal of electricity market deregulation in EU is to establish an integrated electricity market for the whole union. As EU has many member countries, its implementation of cross-border market coupling is divided into several steps. In the first step, national markets were created, starting from the partial market opening in 1999. As the next step, interconnectors were forced to open to market-based capacity allocation from 2003. Then, market coupling was established and expanded after 2005 [10]. By 2019, the EU power system had covered 36 countries and 43 transmission system operators (TSO), and the European power exchange center EPEX SPOT [11] is responsible for cross-border transactions as well as day-ahead and intra-day coupling markets. The EU power market has completed unified day-ahead operation of 25 countries and intra-day operation of 14 countries. Compared with the U.S. electricity markets, a significant difference of the EU market is the adoption of zonal price instead of the nodal price scheme, which usually requires sufficient transmission capacity in each zone to avoid congestion and ensure a uniform zonal price. If congestion occurs, the local TSO needs to redispatch the units in the zone and make certain compensation. Figure 1.2 shows the bidding zone configuration of EU in September 2020.

The recent deregulation of China's power sector started with the release of the ninth document entitled *Several Opinions of the CPC Central Committee and the State Council on Further Deepening the Reform of the Electric Power System* in March 2015. At present, the construction and operation of medium and long-term

Fig. 1.2 The bidding zone configuration in Europe in September 2020 [9]

markets have been completed in all provinces of China. The construction of spot markets has almost been finished in eight pilot provinces and will be expanded to other provinces in the future. Currently, the spot markets have entered the trial operation stage in these pilot provinces. The map in Fig. 1.3 shows the eight pilot provinces for electricity spot markets in China. Although there are differences in the detailed rules of these pilot spot markets, the principle in their organization, transaction and settlement is similar. The data analytic methods in power markets should be conducted based on the aforementioned principle of "bid-based SCED" regardless the specific market form and location.

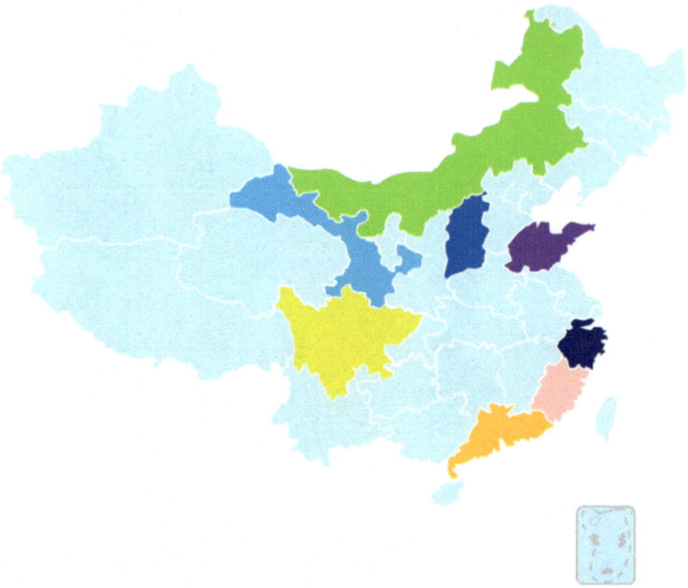

Fig. 1.3 Eight pilot provinces for electricity spot market in China

1.2 Organization and Data Disclosure of Electricity Market

Figure 1.4 shows the information flow in the clearing procedure of electricity spot markets, including the collection of information (state estimation, market bids & offers, load forecasting, etc.), the construction and solving of security-constrained unit commitment (SCUC) and SCED models, and the release of clearing results. Taking PJM's day-ahead market as an example, market participants need to submit their bid and offer data to PJM before 11:00 on the day before the trading day. The market clearing is conducted from 11:00 to 13:30, and the results will be released before 13:30.

The cumulation of market participants' bid & offer curves forms the objective function in the SCED model. To ensure global optimality of the solution, market operators usually require that the offer curves of generators are convex, and the bid curves of demand are concave. Additionally, offer curves are usually submitted to the market operator in the form of (Price, Quantity) blocks for convenience.

The spot market is closely coupled with the operation state of power systems. Due to the complexity and security requirements of power systems, market organizers only disclose part of the market information to support the decision-making of participants and market regulation. There are many electricity markets of different types (e.g., energy market, auxiliary service market, capacity market) and operation modes (e.g., centralized and optimized operation of electricity, physical execution of medium and long-term contracts) in the world with different levels of data disclosure, i.e.,

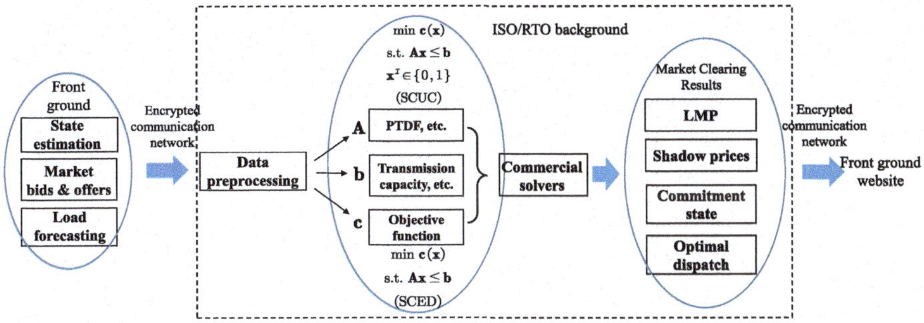

Fig. 1.4 General information flow in the clearing procedure of electricity spot markets

full disclosure, anonymous disclosure, and non-disclosure. Generally, the data types published by the energy markets can mainly be divided into five categories, including: transaction data, price data, supply and demand data, operation data, and forecast data.

1.2.1 Transaction Data

The transaction data are the bid & offer data submitted by the market participants and their aggregate data. Specifically, the bid & offer data submitted by market participants include: individual supply offer data, individual demand bid data and virtual bidding data. The aggregate data include the supply offer and demand bid data at the system level.

1.2.1.1 Individual Supply Offer Data

The individual supply offer data are submitted by market participants, including traditional power sources such as coal, gas, nuclear energy and hydroelectricity, intermittent renewable energy sources such as wind and solar, and new market subjects such as energy storage systems and dispatchable load.

Each item of data generally includes the segmented offer information of (Price, Quantity) pair format, unit economic operation constraints (e.g., start-up and shutdown costs), operation constraints (e.g., maximum and minimum output in economic and emergency mode, ramping capacity).

Due to the fact that individual offer data involve the privacy of market participants, there are three types of policies for the disclosure of individual offer data: (1) full disclosure, i.e., completely publishing all the data; (2) anonymous disclosure, i.e., publishing the data with a pseudo name for each individual which is usually changed every year; (3) non-disclosure, i.e., considering the individual data as private and confidential information.

Table 1.1 Information disclosure policy of individual supply offer data in the main electricity markets

Electricity market	Disclosure policy	Disclosure time
PJM (U.S.) [3]	Anonymous disclosure	4 months after the operation day
CAISO (U.S.) [4]	Anonymous disclosure	3 months after the operation day
MISO (U.S.) [12]	Anonymous disclosure	3 months after the operation day
ISONE (U.S.) [13]	Anonymous disclosure	3 months after the operation day
AEMO (Australia) [14]	Full disclosure	1 day after the operation day
NZEM (New Zealand) [15]	Full disclosure	2 weeks after the operation day
AESO (Canada) [16]	Full disclosure	2 months after the operation day
Nord Pool (Europe) [17]	Non disclosure	/

The information disclosure policy of individual supply offer data of some energy markets in the world is shown in Table 1.1.

1.2.1.2 Individual Demand Bid Data

The individual demand bid data are submitted by each consumer. At present, two forms of individual demand bids are used in most power markets: (1) submitting the segmented demand curve in (Price, Quantity) pair format, which reflects the price sensitivity of consumers; (2) submitting only the power required by load without any corresponding price.

The individual demand bid data also involve the privacy of market participants, the information disclosure policies are different by markets, and can also be categorized into three types: full disclosure (e.g., Australian electricity market), anonymous disclosure (e.g., PJM), and non-disclosure (e.g., Nord Pool).

1.2.1.3 System Aggregate Supply Offer Data

For each time interval, all the individual supply offer data in the market are aggregated to form the system supply offer data. The data directly reflect the overall supply situation of the market, and the system marginal price can be obtained by the intersection of the system supply and demand curve.

The information disclosure policies of system supply offer data are different: some power markets (e.g., PJM, NYISO, ERCOT, and Singapore electricity market) will process and publish the system supply offer data for each interval; Other markets (e.g., MISO, and Australian electricity market) do not provide such data in a unified form, and researchers need to obtain such data with the published individual supply offer data by themselves.

1.2.1.4 System Aggregate Demand Bid Data

For each interval, all the individual demand bid data in the market are aggregated to form the system demand bid data. The data directly reflect the overall demand situation of the market. There are different forms of system demand bid data by markets, which can be categorized into two forms: A monotonically decreasing demand curve or a single demand value. The latter one is the system demand value, which is available in all markets.

1.2.1.5 Virtual Bidding Data

Some electricity markets allow financial subjects without any generation or consumption assets participating in the market, and their transaction is called virtual bidding. Virtual bidding has no physical delivery, and the electricity bought (or sold) by any financial entity in the day-ahead market will be automatically sold (or bought) in the real-time market to close the position.

1.2.2 Price Data

The price data in the energy markets include LMP, aggregate price, and congestion price data.

1.2.2.1 LMP Data

In markets settled with nodal prices, LMP data are published by market operators in time, including the day-ahead hourly LMP, real-time hourly LMP, and fine-grained (e.g., 5-minute) LMP. The data granularity is the same as market clearing frequency.

1.2.2.2 Zonal Price Data

Markets settled with nodal prices will also publish the zonal price based on node aggregation. For example, PJM provides zonal and hub prices for the settlement of some market participants.

In markets settled with zonal prices (e.g., Australian electricity market), a zonal price is the weighted average price of a certain region and is considered the main price index.

1.2.2.3 System Marginal Prices

Some markets would publish the system marginal price, which is often used when there is sufficient transmission capacity in the market.

1.2.3 Supply and Demand Data

1.2.3.1 Supply Data

System available capacity: Usually updated on a daily basis, including total installed capacity (calculated based on the capacity market), upper operating limit in economic operation or under emergency, etc.

Available capacity by fuel type: The system available capacity is partitioned according to the fuel type, e.g., coal, gas, wind, and solar.

System outage capacity: The system capacity that needs daily maintenance according to the scheduled outages.

1.2.3.2 Demand Data

System demand capacity: The total demand capacity in the system according to the demand bids from consumers.

Zonal demand capacity: The zonal demand capacity in the system according to the demand bids from consumers.

1.2.4 System Operation Data

1.2.4.1 Individual Operation Data

Individual dispatching information: The dispatched generation power of each unit. The information disclosure policies are different by markets due to privacy concerns. Some markets do not disclose the information (e.g., PJM), while other markets disclose the information (e.g., Australian power market).

Individual outage information: The actual outage situation of each unit. The disclosure policy of such data at each market is usually the same as that of the individual dispatching information.

Nodal load information: The actual load by each user. The disclosure policy of such data is usually the same as that of the individual dispatching information.

1.2.4.2 System Operation Data

System generation and load: The generation and load data of the whole system and each zone.

Renewable energy source generation: The renewable energy source generation data of the system and each zone, usually released by renewable energy type.

System power flow: The power flow data and the congestion condition for each transmission line.

Inter-provincial and interregional power exchange: The power exchange data for inter-provincial and interregional lines.

1.2.5 Forecast Data

Forecast data are key indicators reflecting the incoming operation status in the markets, which are given by market operators using big data techniques. Forecast data help market participants in generation planning, optimal bidding, and offering. The data include load forecasting, electricity price forecasting, and renewable energy generation forecasting results.

Load forecast data: The load forecast data of the system and load zones in different time scales, e.g. 5-minute ahead, 30-minute ahead, 1-hour ahead, 1-day ahead, and 7-day ahead.

Electricity price forecast data: The electricity price forecast data of the system, zones, and pricing nodes in different time scales.

Renewable energy generation forecast data: The renewable energy generation forecast data of the system and zones by resource type in different time scales.

1.2.6 Confidential Data

Some data are usually confidential in almost all power markets, such as detailed system topology information, admittance of transmission lines, and accurate geographic information of each node in the system.

1.3 Conclusions

In this chapter, a brief overview of the electricity markets in the world is given. The principle behind the spot market is introduced, and the history and recent progress on market deregulation in U.S., EU, and China are presented. Then, the general organization and data disclosure policy of electricity spot markets are introduced, with a discussion on different categories of market data.

References

1. Hogan, William W. 2014. Electricity market design and efficient pricing: Applications for New England and beyond. *The Electricity Journal* 27 (7): 23–49.
2. Federal Energy Regulatory Commission. Regional transmission organization map. 2015. https://www.ferc.gov/sites/default/files/2020-05/elec-ovr-rto-map.pdf.
3. Pennsylvania-New Jersey-Maryland (PJM) Interconnection. About PJM. 2021. https://www.pjm.com/about-pjm.
4. California ISO. About us. 2021. http://www.caiso.com/about/Pages/default.aspx.
5. New York Independent System Operator. What we do. 2021. https://www.nyiso.com/what-we-do.
6. Federal Energy Regulatory Commission. About FERC. 2021. https://www.ferc.gov/about/what-ferc.
7. Electric Reliability Council of Texas. About ERCOT. 2021. http://www.ercot.com/about.
8. Pennsylvania-New Jersey-Maryland (PJM) Interconnection. PJM history. 2020. https://www.pjm.com/about-pjm/who-we-are/pjm-history.aspx.
9. Florence School of Regulation. Electricity markets in the EU. 2020. https://fsr.eui.eu/electricity-markets-in-the-eu/.
10. Supponen, Matti, Qing Xia, Qixin Chen and Nan Jiang. 2020. Electricity markets and systems in the EU and China. Technical report. https://ec.europa.eu/energy/sites/ener/files/electricity_markets_report_en.pdf.
11. EPEX SPOT. History. 2021. https://www.epexspot.com/en/10years#history.
12. Midcontinent Independent System Operator. About MISO. 2021. https://www.misoenergy.org/about.
13. ISO New England. About us. 2021. https://www.iso-ne.com/about.
14. Australian Energy Market Operator. About AEMO. 2021. https://aemo.com.au/en/about.
15. Wholesale Information Trading System. Dashboard. 2021. https://www.electricityinfo.co.nz.
16. Alberta Electric System Operator. About the AESO. 2021. https://www.aeso.ca/aeso/about-the-aeso.
17. Nord Pool. About us. 2021. https://www.nordpoolgroup.com/About-us.

Part I
Load Modeling and Forecasting

Chapter 2
Load Forecasting with Smart Meter Data

2.1 Introduction

In recent years, the area of load forecasting has experienced a resurgence due to the prevalence of smart meters. These measuring devices offer more fine-grained consumption information and are able to boost the performance of load forecasting [1].

Load forecasting research are generally divided into individual and aggregation load forecasting. In individual forecasting, the power consumption of each consumer is predicted separately. Investigation for this subarea has been done by [2], among others. Commonly applied forecasting techniques include linear regression, autoregressive models, support vector regression (SVR), and artificial neural networks (ANN) [3]. Individual load forecasting turns out to be difficult by virtue of the particular stochastic and volatile behavior. The second subarea is constituted by aggregation load forecasting, where the total consumption of all consumers in a region of interest is predicted. Traditional approaches try to do this directly using historical data at the aggregation level. In contrast, smart meters pave the way for an alternative approach. In [4], a bottom-up approach was proposed where the consumers are first predicted individually and then aggregated. In order to improve procedural efficiency, a clustering-based approach was suggested in [5]. The procedure comprises two steps. In the first step, the consumers are grouped according to their load patterns; in the second step, the different groups are predicted separately and eventually aggregated to a final forecast. It was found that the clustering-based aggregation forecasting displays higher accuracy than the traditional direct approach. Similar works such as [6] confirmed these findings. On this basis, an ensemble approach was proposed in [7] to further improve forecasting accuracy by combining the advantageous effects of clustering-based forecasting with an averaging method based on a convex optimization problem.

In order to account for the uncertainties in loads, probabilistic load forecasting (PLF) has been introduced, e.g. [8]. Several ensemble learning approaches were studied to combine different probabilistic forecasts [9]. Similar to deterministic load forecasting, these PLF methods were implemented using historical aggregated load

Q. Chen et al., *Data Analytics in Power Markets*,
https://doi.org/10.1007/978-981-16-4975-2_2

data. An intuitive question is: *how can we leverage the fine-grained smart meter data to enhance the PLF performance just like the practices in deterministic load forecasting* [5–7]? To the best of our knowledge, this question has not been fully investigated yet.

To fill this gap, this chapter proposes a novel framework for probabilistic aggregated load forecasting based on consumer clustering and probabilistic ensemble learning. The framework includes three major stages. The first stage is to cluster the consumers into different groups, such that a limited number of underlying cluster patterns can be leveraged. The second stage is to conduct forecasting for each group and obtain the aggregated forecasts. The third stage is to combine these aggregated forecasts with the help of ensemble learning. The ensemble methods of interest are quantile regression averaging (QRA), factor quantile regression averaging (FQRA), LASSO quantile regression averaging (LQRA), and quantile gradient boosting regression (QGBR). In addition, a window-based approach is studied to update the weights/parameters of the ensemble model.

This chapter makes the following contributions:

1. Propose a framework for PLF based on consumer clustering and probabilistic ensemble learning that can leverage information within the fine-grained smart meter data.
2. Investigate and compare four different probabilistic ensemble learning approaches, including QRA, FQRA, LQRA, and QGBR.
3. Extend the offline ensemble to the rolling window-based ensemble such that the weights/parameters can be updated to integrate more new information.

The remainder of this chapter is organized as follows. Section 2.2 provides a solution framework for PLF with smart meter data. Section 2.3 introduces four probabilistic ensemble learning methods and a rolling window-based method. Section 2.4 conducts case studies to verify the effectiveness of the proposed method. Finally, Sect. 2.5 draws the conclusions.

2.2 Framework

Assume that the aggregated load profile to be predicted has M subprofiles represented in a matrix $\mathbf{L}_{M \times T}$, which contains the load values $l_{m,t}$ for consumers $m = 1, .., M$ at times $t = 1, ..., T$. The aggregated load at time t is $l_{ag,t} = \sum_{m=1}^{M} l_{m,t}$. Traditional forecasting methods are trained only based on the time series $l_{ag,t}$ and other relevant factors such as day type, whether data, etc. This section gives a framework to make full use of the information within $\mathbf{L}_{M \times T}$ to produce probabilistic forecasts with higher accuracy.

The framework can be divided into preprocessing and three main stages, as shown in Fig. 2.1. The basic step is preprocessing the data which involves cleaning the data from corrupted or missing load measurements and partitioning the data into training $\mathbf{D}_{\text{Train}}$, ensemble $\mathbf{D}_{\text{Ensemble}}$, and test set $\mathbf{D}_{\text{Test}}$, respectively. Figure 2.2 visualizes the

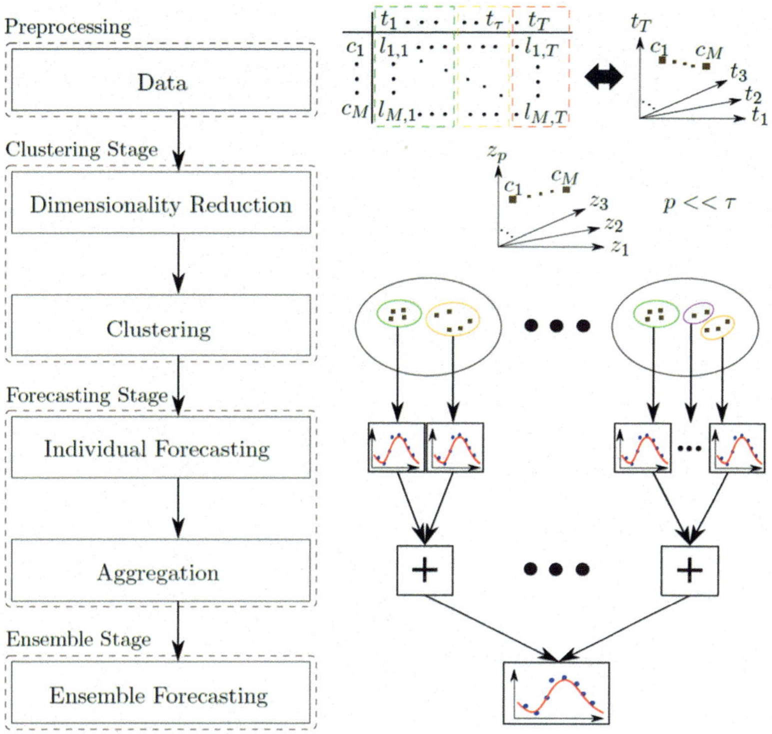

Fig. 2.1 Illustration of the data pipeline

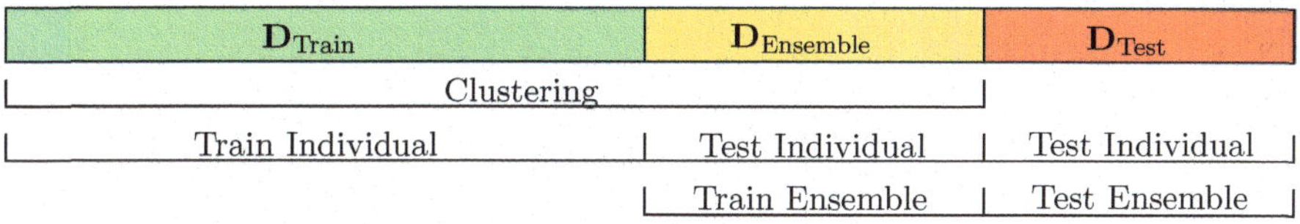

Fig. 2.2 Data partitioning in Train Ensemble Test set

partitioning into these three sub datasets and illustrates which parts of the algorithm use which sub datasets.

The first stage (**clustering stage**) corresponds to partitioning the consumers into different groups using clustering algorithms. The clustering of the consumers is performed on the data $\mathbf{D}_{\text{Train}} \cup \mathbf{D}_{\text{Ensemble}}$.

Since the time series load profile is high-dimensional, principal component analysis (PCA), a commonly used dimension reduction method, is employed first to reduce redundant information in the dimension of time and tackle "the curse of dimensionality" when implementing clustering algorithms. In PCA, it is important to know which principal components to retain. In general, there is no formal procedure but rather rules or guidelines. A possibility is to plot the cumulative explained variance for the principal components. This allows targeting a certain percentage of total explained variance to keep [10].

Then, the data obtained by PCA that still includes one profile per customer are grouped into clusters to leverage commonalities among consumers. For that purpose, the methods k-means, spectral clustering, agglomerative clustering, and Gaussian mixture model (GMM) are investigated. For each of these methods, the optimal number of clusters is determined with the help of clustering metrics. All the clustering algorithms are seeded using a fixed random seed in order to get reproducible clustering results. K-means is initialized with the k-means++ initialization strategy. The spectral clustering is performed with an RBF kernel and a label assignment strategy named discretization [11]. The agglomerative clustering is done with the euclidean distance measure and Ward's minimum variance linkage criterion. Ward's criterion generally promotes more equal-sized clusters. For the GMM, the co-variance type parameters are not restricted in order to allow for arbitrary cluster shapes.

The second stage (**forecasting stage**) comprises training individual forecasting models and making predictions for each cluster. Hence, a regression model is first trained for each cluster of consumers using the respective data in $\mathbf{D}_{\text{Train}}$. Then, the aggregated load of each cluster is predicted on $\mathbf{D}_{\text{Ensemble}}$. For later comparison, the prediction using the same regression model is also made on $\mathbf{D}_{\text{Test}}$. After obtaining the forecast of each group, i.e. cluster, of consumers, these forecasts are aggregated to form the final point forecast. The three different regression methods, SVR, random forest regression (RFR), and GBR are used for the individual forecasting task. The hyper-parameters of different models are optimized using cross-validation on dataset $\mathbf{D}_{\text{Train}}$.

For each clustering algorithm in the first stage and each regression method in the second stage, a point forecast of the aggregated load is produced. Since we consider four clustering methods in the first stage and three regression methods in the second stage, a total of $4 \times 3 = 12$ forecasts can be obtained. However, it should be noted that the proposed framework is not limited to these clustering and regression methods.

The third stage (**ensemble stage**) investigates different probabilistic ensemble learning methods to combine the point forecasts obtain in the first and second stages and then produce the final probabilistic forecasts. Amongst the considered methods are QRA, FQRA, LQRA, and QGBR, which are introduced in the following section. The ensemble methods are trained on $\mathbf{D}_{\text{Ensemble}}$ with features containing the forecasts of the individual regression methods.

2.3 Ensemble Learning for Probabilistic Forecasting

Quantile regression is a commonly used and effective regression method for probabilistic forecasting. Quantiles are statistical location parameters to divide the probability distribution into different intervals. For a random variable Y, a real number y_q is called the q-th quantile if

$$\Pr(Y \leq y_q) \geq q \quad \text{and} \quad \Pr(y_q \leq Y) \geq 1 - q \tag{2.1}$$

where $\Pr(\cdot)$ denotes the probability function. Intuitively, an observation is above y_p with probability q and below y_p with probability $1 - q$. The quantile is calculated in terms of an optimization problem where the quantile loss ρ is defined as

$$\rho_q(u) = \begin{cases} qu, & \text{if} u \geq 0 \\ (q-1)u, & \text{if } u < 0 \end{cases} \tag{2.2}$$

and is of the form of a tilted absolute value function. The q-th sample quantile is obtained by

$$\hat{y}_q = \arg\min_{y_q} \sum_{i=1}^{n} \rho_q(y_i - y_q) \tag{2.3}$$

where i and n denote the index and total number of the training data samples. This is used in quantile regression, where the conditional quantile function Q describes the regression problem for a data set $D = \{(\mathbf{x}_1, y_1), \ldots, (\mathbf{x}_i, y_i), \ldots, (\mathbf{x}_n, y_n)\}$ as follows

$$Q_y(q|\mathbf{x}_i) = \mathbf{x}_i \mathbf{w}_q \tag{2.4}$$

The corresponding quantile weights are found through

$$\hat{\mathbf{w}}_q = \arg\min_{\mathbf{w}_q} \sum_{i=1}^{n} \rho_q(y_i - \mathbf{x}_i \mathbf{w}_q) \tag{2.5}$$

2.3.1 Quantile Regression Averaging

For traditional quantile regression, the inputs $\mathbf{x}_i$ are relevant factors such as historical load and temperature. QRA, a special form of quantile regression, is a kind of model averaging method [12]. It aims to produce a series of quantiles by combining K point forecasts $\hat{y}_1$, ..., $\hat{y}_K$ obtained in the first two stages described in Section II where K is equal to 12 in this chapter. In this case, the regressors $\mathbf{x}_i$ are the point forecasts and a constant offset

$$\mathbf{x}_i = [1, \hat{y}_{1,i}, \ldots, \hat{y}_{K,i}], \qquad i \in [1, \ldots, n] \tag{2.6}$$

Thus, QRA takes advantage of the development in the point load forecasting and transforms several point forecasts into a series of quantiles.

The quantiles can be used to construct a Prediction Interval (PI) with corresponding Prediction Interval Nominal Confidence (PINC) given as

$$
\begin{aligned}
PI &= [y_{q_l}, y_{q_u}] \\
PINC[\%] &= P(y_{q_l} \le Y \le y_{q_u}) = 100\% \cdot (1-\alpha) \\
\alpha &= q_u - q_l, \qquad \alpha \in [0, 1]
\end{aligned} \tag{2.7}
$$

where y_{q_l} and y_{q_u} are lower and upper bound of the interval, respectively. The PINC gives a measure of how likely it is for a new observation to be within the range $[y_{q_l}, y_{q_u}]$.

2.3.2 *Factor Quantile Regression Averaging*

Similar to linear regression, if the regressors $\mathbf{x}_i$ of the QRA model exhibit high multicollinearity, there will be an over-fitting risk. This is particularly the case when the K individual models are performing well individually and therefore have a high correlation. From this perspective, it might not be optimal to include all of the available models in the regression [13].

An approach to solve this issue is called factor quantile regression averaging (FQRA) [14]. It first extracts the relevant information (principal components) $\mathbf{Z}$ in all the models $\mathbf{X} = [\mathbf{x}_1^\top, \ldots, \mathbf{x}_i^\top, \ldots, \mathbf{x}_n^\top]^\top$ by performing PCA. PCA projects each high-dimensional data $\mathbf{x}_i$ onto only the first few principal components to obtain lower-dimensional data $\mathbf{z}_i$. Then, the extracted principal components $\mathbf{z}_i$ are denoted "factors" and used as regressors for quantile regression:

$$
\hat{\mathbf{w}}_q = \arg\min_{\mathbf{w}_q} \sum_{i=1}^{n} \rho_q(y_i - \mathbf{z}_i \mathbf{w}_q) \tag{2.8}
$$

The main difference between QRA and FQRA is the input feature for quantile regression: the former is the point original point forecasts $\mathbf{x}_i$; the latter is the reduced data $\mathbf{z}_i$ by PCA.

2.3.3 *LASSO Quantile Regression Averaging*

Another approach to reduce the over-fitting risk is to use a regularization method just as for LASSO. By adding the L_1-norm penalty term to the loss function, we determine the estimated weights through

$$
\hat{\mathbf{w}}_q = \arg\min_{\mathbf{w}_q} \sum_{i=1}^{n} \rho_q(y_i - \mathbf{x}_i \mathbf{w}_q) + \lambda \|\mathbf{w}_q\|_1 \tag{2.9}
$$

where the regularization parameter λ acts as a regulator to encourage sparse weights. The method combining LASSO and quantile regression averaging is therefore named LQRA in this chapter [15].

2.3.4 Quantile Gradient Boosting Regression Tree

The aforementioned three ensemble learning approaches are based on linear regression, which are "weak learners." Boosting refers to a family of algorithms that are capable of converting so-called "weak learners" into "strong learners" [16]. The main idea is to combine M learners that perform slightly better than random guessing into a single strong learner. In a regression setting, one approach is to make use of regression trees. Since the input of the QGBR is the K point forecasts, tree-based methods partition the K-dimensional feature space into P regions R_1, ..., R_P and fit a model in each one of these. If the fitted region models are assumed to be constants c_p, we can write the overall tree function T as

$$T(\mathbf{x}, \Theta) = \sum_{p=1}^{P} c_p \mathbb{1}\left(\mathbf{x} \in R_p\right) \tag{2.10}$$

where $\mathbf{x} = [\hat{y}_1, \ldots, \hat{y}_K]$ is the input feature vector of the QGBRT model; $\mathbb{1}(\cdot)$ is the indicator function. The region and constant parameters $\Theta = \{R_p, c_p\}$ are found by minimizing the empirical risk with a loss function given by

$$\hat{\Theta} = \arg\min_{\Theta} \sum_{p=1}^{P} \sum_{\mathbf{x}_i \in R_p} l(y_i, c_p) \tag{2.11}$$

The boosted tree model can be represented as the sum of M trees in

$$f_M(\mathbf{x}) = \sum_{m=1}^{M} T(\mathbf{x}, \Theta_m) \tag{2.12}$$

This is generally a difficult combinatorial optimization problem and is done in a forward stagewise additive manner. One idea is to sequentially add new parameters to the tree model based on estimated regions and region constants. The iterative procedure is divided into three steps. First, one finds the direction of steepest descent r_{im} and fits the current regression tree model f_{m-1} to determine the Regions R_{pm}. In the second step, the optimal constants c_{pm} are calculated. Finally, the overall tree model is updated by adding the calculated constants to the current model f_{m-1}. Theoretically, gradient tree boosting allows minimizing an arbitrary differentiable loss function, which makes it applicable in a variety of fields. If the loss function is squared error loss, the gradient boosting regression is a deterministic regression model. If the loss

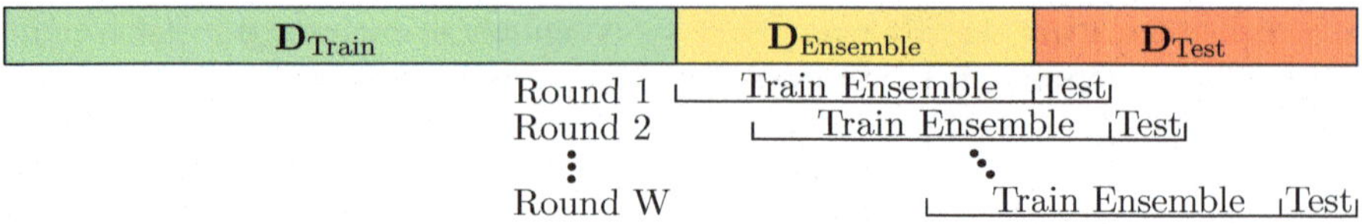

Fig. 2.3 Illustration of the rolling window-based forecasting

function is a quantile loss function, it provides probabilistic forecasting, just like QRA. We denote this model as QGBR.

2.3.5 Rolling Window-Based Forecasting

The weights $\mathbf{w}_q$ in the above probabilistic ensemble learning model are determined on $\mathbf{D}_{\text{Ensemble}}$ and stay unchanged in the test phase on $\mathbf{D}_{\text{Test}}$. To further improve the forecasting performance, a rolling window forecasting is studied, where the weights $\mathbf{w}_q$ are updated regularly by integrating the new information in $\mathbf{D}_{\text{Test}}$. Hereby the joint data set $\mathbf{D}_{\text{Ensemble}} \cup \mathbf{D}_{\text{Test}}$ is taken as train-test set. The approach consists of starting with $\mathbf{D}_{\text{Ensemble}}$ as a training set for the ensemble methods and testing their performance on the following week in $\mathbf{D}_{\text{Test}}$. This train-test window is then repeatedly shifted in a forward fashion by one week as illustrated in Fig. 2.3. The final rolling window performance is then obtained by averaging the error metrics on $\mathbf{D}_{\text{Test}}$.

2.4 Case Study

This section conducts case studies on the residential load data from the Smart Metering Customer Behaviour Trials initiated by the Commission for Energy Regulation (CER) [17]. The dataset contains electricity consumption data of over 5000 Irish residential consumers and small and medium enterprises. The consumption itself was recorded during half-hour intervals between July 20, 2009 and December 26, 2010.

2.4.1 Experimental Setups

In the data preprocessing stage, the consumers with a large number of missing or zero values were excluded. This left a total of 5237 consumers for the case study. The whole data set was then partitioned in $\mathbf{D}_{\text{Train}}$, $\mathbf{D}_{\text{Ensemble}}$ and $\mathbf{D}_{\text{Test}}$ of size 50%, 25%, and 25%, respectively.

In the clustering stage, to keep at least 75% total explained variance, 7 principal components are retained in PCA. The optimal numbers of clusters for different clustering algorithms are selected by comprehensively considering the Silhouette Coefficient, Calinski-Harabasz Index, and Davies-Bouldin Index.

In the forecasting stage, the input features of all point forecasting models are the same as those used in [7]. Specifically, for the SVR, the features were standardized before training. The penalty and deviation hyperparameters in SVR are estimated according to [18]. Regarding the tree-based regression methods, no feature scaling is needed since they are invariant to monotonic transformations of the regressors [19]. Since the main focus of this chapter is not on the individual regression model development, both the RFR and GBR are trained with standard hyperparameters.

In the ensemble stage, the parameters of the quantile regression ensemble are optimized using cross-validation. For EQRA, the number of principal components is set to 11; for LQRA, the regularization parameter λ is set to 1000; for QGBR, the number of trees M is set to 1000.

2.4.2 Evaluation Criteria

A metric for evaluating probabilistic forecasts is the Pinball Loss (PBL), which is also the loss minimized in quantile regression and is given by

$$PBL = \begin{cases} q(y_t - \hat{y}_{t,q}), & y_t \geq \hat{y}_{t,q} \\ (q-1)(y_t - \hat{y}_{t,q}), & y_t < \hat{y}_{t,q} \end{cases} \tag{2.13}$$

where q is the target quantile. The PBL of a PI is the averaged PBL over all quantiles $q = 0.01, ..., 0.99$. A lower score relates therefore to a better PI.

The reliability of a PI can be measured with the Average Coverage Error (ACE). It is defined as the difference between PI coverage probability and nominal confidence given as

$$ACE = \frac{1}{T}\sum_{t=1}^{T} \mathbb{1}(y_t \in [L_t, U_t]) - (1-\alpha) \tag{2.14}$$

Thus, the absolute ACE of a good model should be as close to zero as possible. Since higher reliability can be obtained by simply widening the PI, the sharpness is also of importance. This is assessed through the Winkler Score (WKS) written as

$$WKS = \begin{cases} \delta_t, & L_t \leq y_t \leq U_t \\ \delta_t + 2(L_t - y_t)/\alpha, & y_t < L_t \\ \delta_t + 2(y_t - U_t)/\alpha, & y_t > U_t \end{cases} \tag{2.15}$$

Table 2.1 Error metric comparison for all ensemble methods with a prediction interval of 90%

Ensemble method	Error metrics	Offline ensemble	Benchmark 1	Rolling window-based ensemble	Benchmark 2
QRA	ACE	−1.73%	−1.85%	−0.56%	−0.92%
	PBL	45.82	50.19	42.28	46.52
	WKS	788.62	846.89	728.13	791.78
FQRA	ACE	−1.80%	−1.85%	**−0.45%**	−0.92%
	PBL	45.82	50.19	**42.26**	46.52
	WKS	787.26	846.89	**727.24**	791.77
LQRA	ACE	**−1.71%**	−1.83%	−0.63%	−0.98%
	PBL	**45.84**	50.2	42.26	46.53
	WKS	**785.77**	845.7	724.74	791.55
QGBR	ACE	−17.29%	−17.59%	−5.46%	−5.44%
	PBL	96.71	93.87	44.25	48.32
	WKS	3082.76	2825.13	818.69	889.86

where, $\delta_t = U_t - L_t$ is the interval width given by the PI's lower and upper bound L_t and U_t. The WKS penalizes if the true value lies outside of the constructed PI. A lower score indicates a PI of higher quality.

2.4.3 Experimental Results

Table 2.1 gives the performance of different probabilistic ensemble learning methods including two naive benchmarks in terms of ACE, PBL, and WKS. The two naive benchmarks are obtained by directly forecasting the total loads without dimension reduction and clustering. The rest of the pipeline is done in exactly the same way as for the other ensemble methods. That is to say, all the consumers are treated as one group, i.e. one aggregated load, and is predicted as such. Consequently, the information of the fine-grained smart meter data is not used. Benchmark 2 updates the weights in a rolling window-based approach, while Benchmark 1 does not.

By comparing the columns Offline Ensemble and Benchmark 1, it can be found that except for QGBR, all ensembles display superior performance compared to Benchmark 1 across all metrics. It verifies that clustering-based forecasting is generally able to provide better performance than the direct benchmark approach for probabilistic forecasting by exploring more information in the fine-grained smart meter data. QRA, FQRA, and LQRA have similar performance, among which LQRA is the best with respect to (w.r.t.) ACE and WKS; QRA and FQRA are the best w.r.t. PBL. It should be noted that there might be a conflict between ACE and WKS. For example, a forecast may have high reliability, i.e., a good ACE performance, but

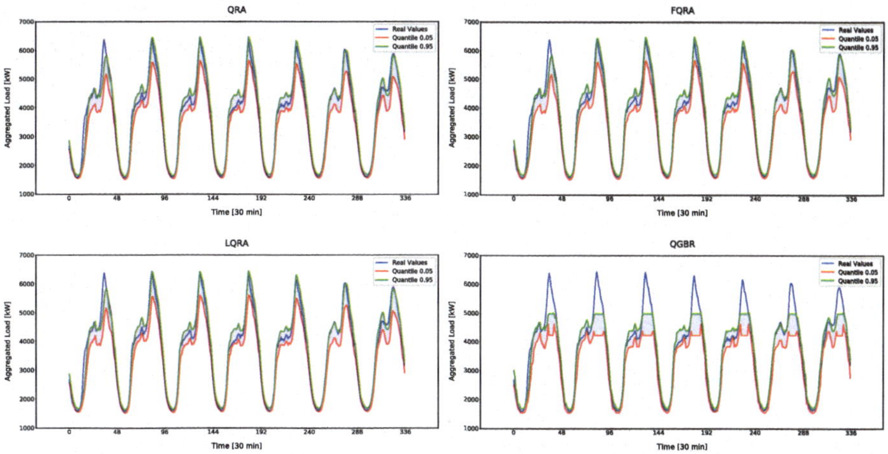

Fig. 2.4 Probabilistic load forecasts over a representative week from the test set

poor sharpness and resolution, i.e., a poor WKS performance. Depending on the use case, one might prefer a lower WKS in this dilemma if the goal is to capture as many consumption observations as possible. On the other hand, the objective might be to give a more precise estimation of the prediction interval. In this case, one might prefer a lower ACE.

Another interesting observation is the poor performance of QGBR. The overfitting of QGBR might nullify the advantageous effect of aggregated load forecasting since it is a very strong learner. Figure 2.4 shows the predicted interval over a representative week from the test set based on different combining methods. It can be found that the predicted interval by QGBR cannot capture the peaks during this week since these peak values have not appeared in the training dataset. The strong learner, QGBR, is hard to be generalized to new scenarios.

Figure 2.5 presents the reliability diagram for four different quantile ensembles. For comparison, the perfect reliability curve is additionally indicated. It shows that the reliability curves of QRA, FQRA, and LQRA are much closer to the perfect curve, i.e., very similar and with high reliability. In contrast, QGBR has substantially lower reliability, especially for higher quantiles. It again shows the worse performance of QGBR.

By comparing the columns Offline Ensemble and Rolling Window-based Ensemble, it is clear that all ensembles considerably improve their metrics in the rolling window-based forecasting. This can be explained by the sliding training window with more recent training data up to the test week. The ACE of QRA, FQRA, and LQRA improves from an average of −1.74% to an average of −0.54%; the PBL improves from an average of 45.83 to an average of 42.27; the WKS improves from an average of 787.22 to an average of 726.70. Large improvements can also be observed by comparing the columns Benchmark 1 and Benchmark 2. These significant improvements show the importance of making full use of the most recent load data.

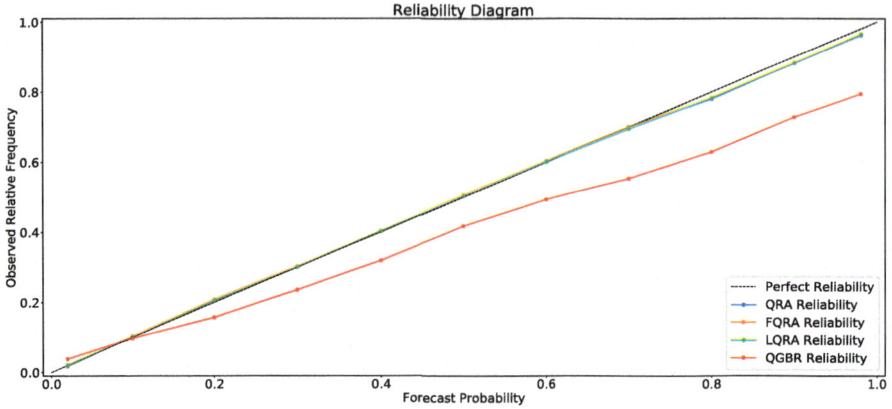

Fig. 2.5 Reliability diagram for all quantile ensembles

2.5 Conclusions

This chapter proposes a novel framework for PLF with fine-grained smart meter data. Four probabilistic ensemble learning models are studied. Case studies on an open dataset show that by making full use of the fine-grained smart meter data, the forecasting performance can be improved with weak but effective linear regression. Since the point forecasts are produced by complex procedures and strong regression models, it is better to apply weak learners in the ensemble stage to reduce the over-fitting risk. Besides, rolling window-based methods can update the parameters of the combined model by integrating the information with the most recent data, and thus further improve the forecasting performance.

References

1. Wang, Yi., Qixin Chen, Tao Hong, and Chongqing Kang. 2018. Review of smart meter data analytics: Applications, methodologies, and challenges. *IEEE Transactions on Smart Grid* 10 (3): 3125–3148.
2. Yu, PChun-Nam, Piotr Mirowski, and Tin Kam Ho. 2017. A sparse coding approach to household electricity demand forecasting in smart grids. *IEEE Transactions on Smart Grid* 8 (2): 738–748.
3. Edwards, Richard, Joshua New, and Lynne Parker. 2012. Predicting future hourly residential electrical consumption: A machine learning case study. *Energy and Buildings* 49: 591–603.
4. Stephen, Bruce, Xiaoqing Tang, Poppy R Harvey, Stuart Galloway, and Kyle I Jennett. 2017. Incorporating practice theory in sub-profile models for short term aggregated residential load forecasting. *IEEE Transactions on Smart Grid* 8 (4): 1591–1598.
5. Quilumba, Franklin L, Wei-Jen Lee, Heng Huang, David Y Wang, and Robert L Szabados. 2015. Using smart meter data to improve the accuracy of intraday load forecasting considering customer behavior similarities. *IEEE Transactions on Smart Grid* 6 (2): 911–918.

6. Sevlian, Raffi, and Ram Rajagopal. 2018. A scaling law for short term load forecasting on varying levels of aggregation. *International Journal of Electrical Power & Energy Systems* 98: 350–361.
7. Wang, Yi., Qixin Chen, Mingyang Sun, Chongqing Kang, and Qing Xia. 2018. An ensemble forecasting method for the aggregated load with subprofiles. *IEEE Transactions on Smart Grid* 9 (4): 3906–3908.
8. Hong, Tao, and Shu Fan. 2016. Probabilistic electric load forecasting: A tutorial review. *International Journal of Forecasting* 32 (3): 914–938.
9. Wang, Yi, Ning Zhang, Yushi Tan, Tao Hong, Daniel S Kirschen, and Chongqing Kang. 2018. Combining probabilistic load forecasts. *IEEE Transactions on Smart Grid* 10 (4): 3664–3674.
10. Kent, John T. 1983. Information gain and a general measure of correlation. *Biometrika* 70 (1): 163–173.
11. Stella X Yu, and Jianbo Shi. 2003. Multiclass spectral clustering. In *Proceedings Ninth IEEE International Conference on Computer Vision*, 313–319.
12. Liu, Bidong, Jakub Nowotarski, Tao Hong, and Rafał Weron. 2017. Probabilistic load forecasting via quantile regression averaging on sister forecasts. *IEEE Transactions on Smart Grid* 8 (2): 730–737.
13. Nowotarski, Jakub, and Rafał Weron. 2015. Computing electricity spot price prediction intervals using quantile regression and forecast averaging. *Computational Statistics* 30 (3): 791–803.
14. Maciejowska, Katarzyna, Jakub Nowotarski, and Rafał Weron. 2016. Probabilistic forecasting of electricity spot prices using factor quantile regression averaging. *International Journal of Forecasting* 32 (3): 957–965.
15. Li, Youjuan, and Ji. Zhu. 2008. L1-norm quantile regression. *Journal of Computational and Graphical Statistics* 17 (1): 163–185.
16. Schapire, Robert E. 1990. The strength of weak learnability. *Machine Learning* 5 (2): 197–227.
17. Irish Social Science Data Archive (2012). Commision for Energy Regulation Smart Metering Project-Electricity Customer Behaviour Trial, 2009–2010 [dataset]. www.ucd.ie/issda/CER-electricity.
18. Cherkassky, Vladimir, and Yunqian Ma. 2004. Practical selection of SVM parameters and noise estimation for svm regression. *Neural Networks* 17 (1): 113–126.
19. LeBlanc, Michael, and John Crowley. 1995. *A review of tree-based prognostic models*, 113–124. Springer US, Boston, MA.

Chapter 3
Load Data Cleaning and Forecasting

3.1 Introduction

Bad data including missing data and anomaly data may be introduced to load profiles in the process of data acquisition, transmission, and storage deliberately or accidentally due to cyber attack and equipment failure [1]. The bad data may cause bias for data analytics such as forecasting. Thus, identifying and repairing bad data are important to guarantee the unbiasedness and effectiveness of the data analytics models [2, 3].

A substantial of effort has been devoted to bad data identification. The research on load profile bad data identification and repairing can be divided into two categories: state estimation-based and time series-based methods. Power system state estimation is one of the key functions to obtain a real-time network model by analyzing redundant measurements including voltage magnitude and phase, active and reactive injections, and solving a (non)linear least-squares problem [4].

The state estimation mainly focuses on a single time period. For low-voltage networks, redundant measurements such as reactive injections, voltage magnitudes, and phases are not easy to obtain.

Time series-based data cleaning methods utilize historical load data itself to detect and impute bad data. The bad data or corrupted data are classified into locally corrupted data and globally corrupted data in [5]. The B-Spline smoothing and Kernel smoothing-based method is proposed to automate the data cleaning process and a man-machine dialogue is designed to further improve the performance. The low-rank matrix technique is used to exploit the spatio-temporal correlation of load profiles in [6]. A distributed principal components pursuit algorithm is proposed to make the cleaning method applicable for networked devices. A two-stage bus load data cleaning strategy is proposed in [7] to improve the accuracy of bus load forecasting. In the first stage, bad data patterns including consecutive points and sudden rising up or falling down are identified by a certain rule and Chebyshev's Inequality; in the second stage, the feature curve is extracted using Fourier transform and the bad data are restored based on the feature curve.

Q. Chen et al., *Data Analytics in Power Markets*,
https://doi.org/10.1007/978-981-16-4975-2_3

A segmented clustering method is proposed for load estimation in [8]. Specifically, k-means is performed on segmented load profiles to obtain typical load profiles and then the missing data or future data are estimated according to the typical load profile. Different distance measurements including Canberra, Manhattan, Euclidean, and Pearson correlation distances have been used.

A periodic auto-regression with exogenous variables (PARX) model is established in [9] to model the electricity consumption behaviors in different calendar hours and temperatures. With Gaussian distribution consumption, the anomaly data can be detected according to the regression residues. A Lambda architecture is proposed for the detection system implementation.

The anomaly detection methods for other time series can also be introduced to load profiles. Long short term memory (LSTM) networks, an efficient time series model, is used in [10] for anomaly detection.

Identification of the types of bad data can help a lot for data repairing. Generally, there are two types of bad data in time series: additive bad data and innovative bad data [11]. The former is "real" and caused by an isolated event; the latter is "false" and caused by system errors. Specifically, for load profiles, the bad data are classified into three types which are caused by signal transmission, system operation, and electricity consumption respectively in [7]. Based on the state estimation based bad data identification results, the bad data are classified into five patterns in [12] which are the bad data caused by scaling, time delay, a sudden failure of a measuring device, continuous failure of a measuring device, or communication system, and other types. Cyber attacks such as energy theft is another bad data source for load profiles in the information era. Six types of malicious data are generated in [13] to simulate the energy theft behavior. On this basis, the other four types of malicious data are provided in [14] for energy theft detection research.

There are various types of bad data that are too complex to detect using only one method. The majority of existing works applied only one method such as regression model and low-rank matrix to detect some specific bad data. This chapter proposed a comprehensive framework to tackle different kinds of bad data. The possible types of bad data are firstly analyzed. On this basis, the low-rank matrix technique is applied to complete missing data and tackle with sudden changes of load profiles; then, quantile regression is applied to detect the small but anomaly fluctuations. Load forecasting is performed on the dataset with and without data cleaning to investigate the accuracy improvement.

This chapter makes the following contributions:

1. Proposing a novel data-driven bad data identification and repairing method for load profiles by integrating the low-rank matrix analysis and quantile regression to exploit the interday and intraday temporal correlations;
2. Investigating the point and probabilistic forecasting performance improvement brought by bad data identification and repairing.

The rest of this chapter is organized as follows: Sect. 3.2 discusses the characteristic of load profiles and introduces different types of bad data which are the basis of the proposed bad data identification and repairing method; Sect. 3.3 provides the framework and details of the proposed method; Sect. 3.4 provides the evaluation criteria to quantify the performance of the proposed method; Sect. 3.5 conducts case studies on two open datasets to verify the effectiveness of the proposed method; Sect. 3.6 draws the conclusions.

3.2 Characteristics of Load Profiles

In this section, we will explore the basic characteristics of load profiles based on the load data from Fujian Province, China. The dataset used here contains three-year load profiles from January 1, 2003, to December 31, 2005. Then, various typical bad data are introduced in the following.

3.2.1 Low-Rank Property of Load Profiles

We rearrange these load profile as a matrix where each row represents a daily load profile and each column represents one time period. Thus, the load profile can be rearranged to a $N \times 96$ matrix $\mathbf{L}$. Then, we roughly divide $\mathbf{L}$ in to several $n \times 96$ matrices $\mathbf{L}_j$ (load matrix of each year is divided into four quarter matrices, $n < 96$). In this way, the impact caused by season can be reduced. Singular-value decomposition (SVD) [15] is performed on each load matrix $\mathbf{L}_j$:

$$\mathbf{L}_j = \mathbf{U}_j \mathbf{D}_j \mathbf{V}_j^* \tag{3.1}$$

where $\mathbf{U}_j$ is a $n \times n$ unitary orthogonal matrix; $\mathbf{D}_j$ is a diagonal $n \times 96$ matrix with non-negative real numbers on the diagonal; $\mathbf{V}_j$ is a 96×96 unitary orthogonal matrix; $\mathbf{V}_j^*$ is the conjugate transpose of $\mathbf{V}_j$.

The element of the diagonal matrix $\mathbf{D}_j$, $\mathbf{D}_j(t,t)$ represents the variability of load profile in the t-th transformed space. Then, we normalize the singular-values:

$$\sigma_j(t) = \frac{\mathbf{D}_j(t,t)}{\sum_{t=1}^{n} \mathbf{D}_j(t,t)}, \quad t = 1, 2, 3, \cdots, n \tag{3.2}$$

Hence the summation of the normalized singular-values is equal to 1 and we can sort them in descending order and get the cumulative variability. Figure 3.1 shows the cumulative variability of all $\mathbf{L}_j$ with different numbers of singular values.

It can be seen that the first singular value represents at least 80% of the total variance which means that there exist strong correlations among the loads at different hours of the intraday and different interdays. Among all the $\mathbf{L}_j$, 95% variance can be represented by only 57 singular values, respectively. Even though these singular values cannot represent the 100% variance, they can form good approximations of original load profiles.

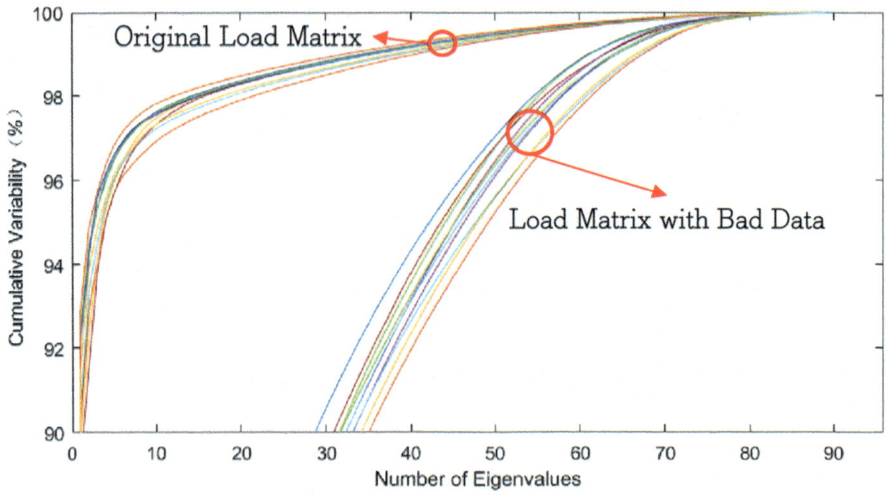

Fig. 3.1 Cumulative variability of each substation load profile

If the summation of the first T singular values $\sigma_{j,T} = \sum_{t=1}^{T} \sigma_j(t)$ is greater than a certain value α, then we call this observation as the approximate low-rank property of load profiles.

3.2.2 Bad Data in Load Profiles

As stated in the introduction section, anomaly data can be introduced in the process of data collection, transmission, and storage. Sudden changes of external conditions such as sports events can also introduce anomaly usage patterns.

Figure 3.2 shows four typical types of bad data selected from the real dataset. The first one is anomaly spikes. These spikes are as low as zeros or even negative; they may also be much higher than the normal value. The second one is consecutive constant points which may be caused by communication failure. The third one is local anomaly pattern which is very different from the load patterns at the same time period in recent days. The last one is null value which is a very common type of bad data.

Load forecasting usually includes three phases: data normalization, model training, and model testing. These kinds of bad data may undermine the basis of the forecasting model and influence the forecasts when using these bad data as inputs. For example, the anomaly spikes have a significant impact on load data normalization if the data are normalized according to the minimum and maximum loads; many forecasting models cannot deal with the input data with null value. Not only load forecasting, but also many other applications motivate us to identify and repair the bad data.

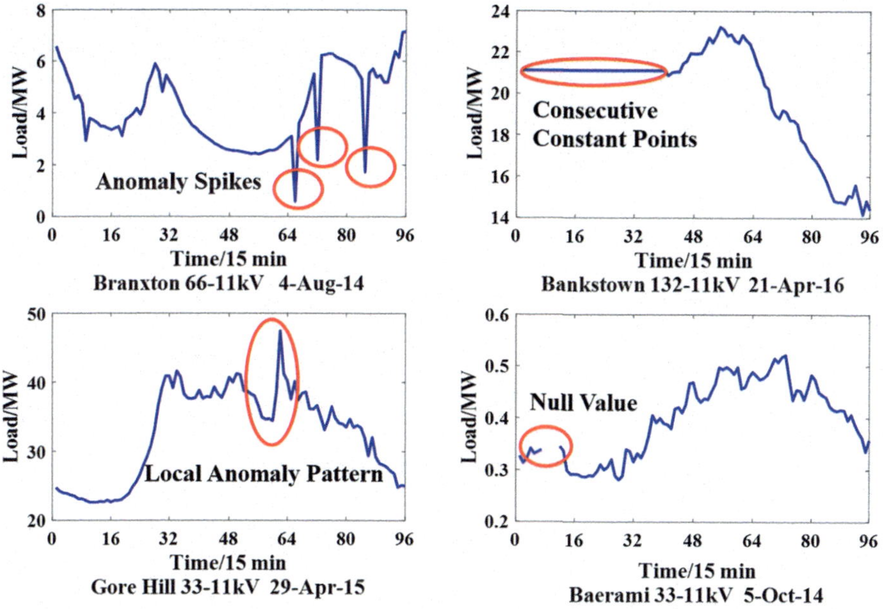

Fig. 3.2 Typical bad data in load profiles

3.3 Methodology

In this section, we first introduce the framework of our proposed data cleaning method and then turn to the implementation details.

3.3.1 *Framework*

The proposed method consists of two stages. First, low-rank matrix completion is used to explore the long-term correlations, detect some obvious bad data, and complete the null values. Second, a quantile regression model is integrated into the RF model (QRF) to describe the short-term volatility.

Specifically, the whole procedures of the proposed method are shown in Algorithm 3.1. Only the load profiles that have approximate low-rank property are processed by the low-rank matrix completion technique SVT. Otherwise, the load profiles are directly processed by QRF. The implementation details of singular value thresholding (SVT) based low-rank matrix completion $f_{SVT}(\mathbf{L}_i)$, QRF $f_{QL}(\mathbf{L}_i)$, and load forecasting test are introduced in the following subsections.

Since the following statements are for load data of one specific quarter, we use $\mathbf{L}$ instead of $\mathbf{L}_j$ for simplicity.

Algorithm 3.1 Load Data Cleaning

Require: Original data load **L**, the number of low-voltage loads N; the threshold for low-ranks matrix implementation α.
for $i = 1 : N$ **do**
  Divide **L** into twelve $n \times 96$ matrices $\mathbf{L}_j$
  Conduct SVD on $\mathbf{L}_j$ according to (3.1);
  Calculate the sum of the first T normalized singular-values $\sigma_{j,T}$;
  if $\sigma_{j,T} > \alpha$ **then**
    Perform SVT algorithm on load data $\mathbf{L}_j = f_{SVT}(\mathbf{L}_j)$;
  end if
  Perform Quantile RF algorithm on load data $\mathbf{L}_j = f_{QL}(\mathbf{L}_j)$;
  Conduct load forecasting tests on the original and cleaned load data;
  Calculate and record the performance criteria;
end for

3.3.2 Singular Value Thresholding (SVT)

The bad data includes missing data and anomaly data. The missing data are first simply completed by the average of the loads of the last week and next week at the same time period:

$$L_{d,t} = \frac{1}{2}(L_{d+7,t} + L_{d-7,t}) \tag{3.3}$$

Then, the measured load data can be viewed as the summation of real (normal) load data **A** and measurement errors (anomaly data) **E**:

$$\mathbf{L} = \mathbf{A} + \mathbf{E} \tag{3.4}$$

We formulate the anomaly detection problem as an optimization problem according to the approximate low-rank characteristic:

$$\begin{aligned} \min \quad & \|\mathbf{A}\|_* + \lambda|\mathbf{E}|_1 \\ s.t. \quad & \mathbf{L} = \mathbf{A} + \mathbf{E} \end{aligned} \tag{3.5}$$

where $\|\cdot\|_*$ denotes the nuclear norm which is the summation of all the singular values; minimizing the nuclear norm is an approximation of minimizing the rank of matrix. $|\mathbf{E}|_1$ presents the sparsity of the error; λ is the relative weight of the sparsity of the error. A grater λ results in a higher sparsity of the error and vice versa.

The above optimization problem is convex and can be reformulated as a semidefinite programming problem. It can be solved with interior-point methods but it is very time-consuming and is not applicable for large problems. We use the singular value thresholding method to solve this problem. The singular value shrinkage operator is first defined:

$$\begin{aligned} S_\tau(\mathbf{L}) &= \mathbf{U}S_\tau(\mathbf{D})\mathbf{V}^*, \\ S_\tau(\mathbf{D}) &= diag(\{d_i - \tau\}_+) \qquad i = 1, 2, 3, ..., n \end{aligned} \tag{3.6}$$

where $\mathbf{D}$ is the diagonal matrix of $\mathbf{L}$ obtained by SVD; d_i is the singular value of the i th;$\{d_i - \tau\}_+ = \max(0, d_i - \tau)$. The singular value shrinkage operator simply shrinks the singular values below τ towards zeros. Then, the SVT method is implemented by a series of iterations:

$$\begin{aligned}\mathbf{A}^k &= S_\tau(\mathbf{Y}^{k-1}) \\ \mathbf{E}^k &= sign(\mathbf{Y}^{k-1}) \times \max(0, abs(\mathbf{Y}^{k-1}) - \lambda\tau) \\ \mathbf{Y}^k &= \mathbf{Y}^{k-1} + \delta(\mathbf{L} - \mathbf{A}^k - \mathbf{E}^k)\end{aligned} \tag{3.7}$$

where k is the number of iteration; δ denotes the learning step-size and it is valued in (0, 1). For each iteration, we only need to compute SVD and perform elementary matrix operations which are computationally efficient.

It can be proven that the iteration is equal to solving the following optimization problem [16]:

$$\begin{aligned}\min \quad & \tau(\|\mathbf{A}\|_* + \lambda|\mathbf{E}|_1) + \|\mathbf{A}\|_F^2 \\ s.t. \quad & \mathbf{L} = \mathbf{A} + \mathbf{E}\end{aligned} \tag{3.8}$$

Obviously, when $\tau \to \infty$, the solution to the optimization problem (3.8) converges to that of (3.5). Thus, we choose a large value of the threshold τ. It should also be noted that a larger value of τ causes slower convergence of the iterations.

We use the relative reconstruction error as the convergence index e if it is lower than a very small value ϵ:

$$e = \frac{\|\mathbf{L} - \mathbf{A} - \mathbf{E}\|_F^2}{\|\mathbf{L}\|_F^2} \tag{3.9}$$

Thus, the proposed SVT algorithm can be neatly presented by Algorithm 3.2.

Algorithm 3.2 Singular Value Thresholding

Require: Measured load data $\mathbf{L}$, the relative weight of error sparsity λ, learning step-size δ, the maximum number of iterations K, convergence threshold ϵ.

Set $\mathbf{Y} = \mathbf{0}$, $\mathbf{A} = \mathbf{0}$, $\mathbf{E} = \mathbf{0}$, $k = 0$;
while $k \leq N \&\& \; e > \epsilon$ **do**
  $[\mathbf{U}, \mathbf{D}, \mathbf{V}] = \text{SVD}(\mathbf{Y})$;
  $\mathbf{A} = S_\tau(\mathbf{Y})$;
  $\mathbf{E} = sign(\mathbf{Y}) \times \max(0, abs(\mathbf{Y}) - \lambda\tau)$;
  $\mathbf{Y} = \mathbf{Y} + \delta(\mathbf{L} - \mathbf{A} - \mathbf{E}^k)$;
  $e = \|\mathbf{L} - \mathbf{A} - \mathbf{E}\|_F^2 / \|\mathbf{L}\|_F^2$;
  $k = k + 1$;
end while

3.3.3 Quantile RF Regression

This subsection uses quantile RF regression to further detect anomaly data after the SVT method. The main idea is to use the strength of quantile regression: it can judge whether the data are anomalous by the predefined quantiles. Random forest (RF) is an algorithm that uses multiple decision trees to train and predict samples. Each decision tree samples the original training data set while training. N_{RF} indicates the number of trees to be grown, and M_{RF} indicates the number of variables randomly selected at each division. Therefore, for each decision tree $T(W_{n,q,n_{RF}})$, the parameter $W_{n,q}$ includes the random parameters $W_{n,q,n_{RF}}$. The parameters for each tree can be obtained from the M_{RF} training data. The loss function of the whole network is the total pinball loss:

$$\min \frac{1}{T} \sum_{t=1}^{N} P_{\tau}(L_t, \hat{q}_t^{(\tau)}) \tag{3.10}$$

where L_t denotes the measured load at time t; $\hat{q}_t^{(\tau)}$ denotes the estimated τ quantile at time t; P_{τ} denotes the pinball loss for the τ quantile:

$$P_{\tau}(L_t, \hat{q}_t^{(\tau)}) = \begin{cases} (1-\tau)(\hat{q}_t^{(\tau)} - L_t)\hat{q}_t^{(\tau)} \geq L_t \\ \tau(L_t - \hat{q}_t^{(\tau)})\hat{q}_t^{(\tau)} < L_t. \end{cases} \tag{3.11}$$

Random forests are integrated through the results of multiple decision trees

$$\hat{y}_{t,n_{RF}} = T(W_{n,q,n_{RF}, X_{n,t}}) \tag{3.12}$$

Then the forecast result is:

$$\hat{y}_t = \frac{1}{N_{RF}} \sum_{n_{RF}=1}^{N_{RF}} \hat{y}_{t,n_{RF}} \tag{3.13}$$

The higher the N_{RF}, the longer the time required for the operation, but the corresponding prediction works better. The value used in this chapter is 200.

We need to set a threshold β to identify the anomaly load data: if the data do not fall in a β-length interval, then the data are identified as anomaly ones and they will be replaced by the 0.5-quantile value. Thus, three quantile regression models are trained to obtain the estimated quantiles: $\tau = \beta/2$, $\tau = 0.5$, and $\tau = 1 - \beta/2$. The quantile RF algorithm is summarized in Algorithm 3.3.

Algorithm 3.3 QRF Regression based cleaning

Require: Load data **L**; the threshold for QRF β.
Data normalization for **L**;
Form the input $\mathbf{X}_t$ and output L_t of QRF;
Train the $\beta/2$, 0.5, and $(1-\beta/2)$ QRF models;
for $t = 1 : T$ **do**
  if $L_t < \hat{q}_t^{(\beta/2)}$ or $L_t > \hat{q}_t^{(1-\beta/2)}$ **then**
    $L_t = \hat{q}_t^{(0.5)}$;
  end if
end for

3.3.4 *Load Forecasting*

The load data processed by SVT and quantile RF are then used as historical load data to forecast future load. We conduct both point load forecasting and probabilistic load forecasting to investigate the impact of data cleaning on forecasting performance. The forecasting models are trained on the original dataset, bad dataset, and clean dataset, respectively.

3.4 Evaluation Criteria

In this section, several evaluation criteria are proposed from the perspective of data cleaning and load forecasting performance.

3.4.1 *Data Cleaning-Based Criteria*

We denote the real data and the repaired data as L and $\hat{L}$, respectively. A perfect data cleaning work exactly reverts bad data to good data. The Root Mean Square Error (RMSE) is used to quantify the distance between the reverted data and the actual good data.

$$\text{RMSE} = \|L - \hat{L}\|_2 \tag{3.14}$$

3.4.2 *Load Forecasting-Based Criteria*

One main application of data cleaning is to improve the performance of load forecasting. We use both point forecasting and probabilistic forecasting to test how much the performance can be improved by data cleaning. Thus, the improvement of forecasting accuracy is also used to quantify the performance of our work. RMSE and mean absolute error (MAE) are used to evaluate the accuracy of point forecasting, and pinball loss is used to evaluate the accuracy of quantile forecasting.

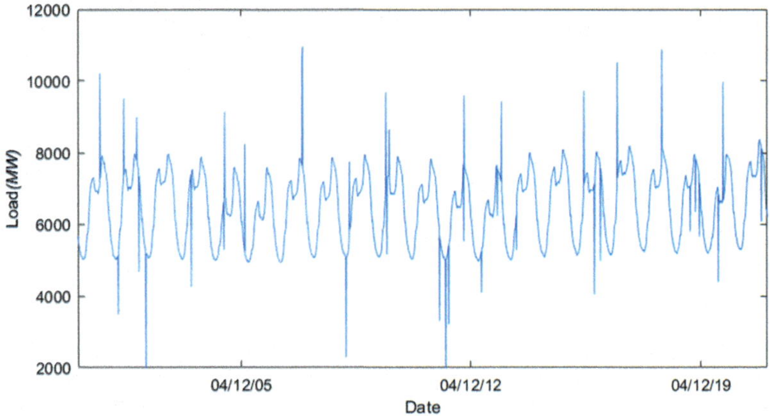

Fig. 3.3 Overview of bad data

3.5 Case Study

The case studies are conducted on the data set from Fujian Province, China which includes the load profiles from 2003 to 2005. Since the error load value of the continuous constant is easily recognized by other methods, this type of error is not considered. In this chapter, a positive or negative error of 0.1 to 0.3 times the maximum value of the original data set is added to the original data set with a probability of 0.02, and the original data are erased with a probability of 0.005 to create a data set with bad data. An example of a part of the original data set with bad data is shown in Fig. 3.3.

3.5.1 Result of Data Cleaning

First, we use the characteristics of low-ranking data to initially identify error data points in the load matrix. Before data cleaning, the RMSE of the bad data set is 1.35×10^4 MW.

Figure 3.4 shows that after Step I, most of the bad data are identified, but the data are still far from the true values. At this time, the RMSE of the data set is 1.08×10^4 MW.

Then, QRF is trained based on the results of the data after the low-ranking process. QRF forecast gives the results of the quantiles: $\tau = \beta/2$, $\tau = 0.5$, and $\tau = 1 - \beta/2$. The β value in this chapter is 0.98.

After going through Step II, almost all of the bad data (>98%) are returned to the true value (Fig. 3.5). Although the method in this chapter will cause some data to be misidentified, the corrected value is still close to the real value and has little effect on the load forecast. At this time, the RMSE of the data set is 6.27×10^3 MW.

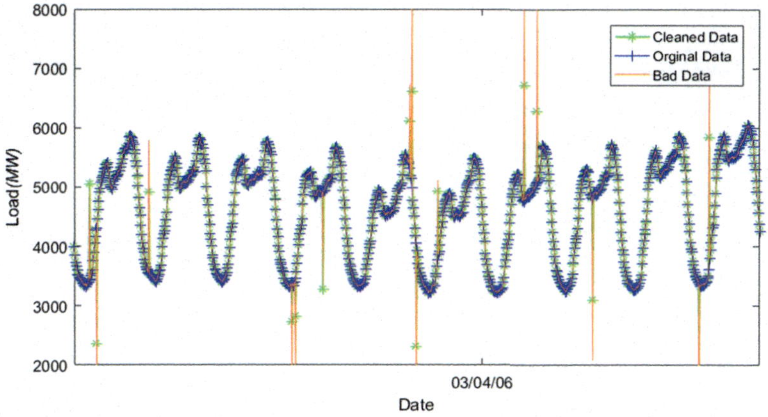

Fig. 3.4 Data cleaned after step 1

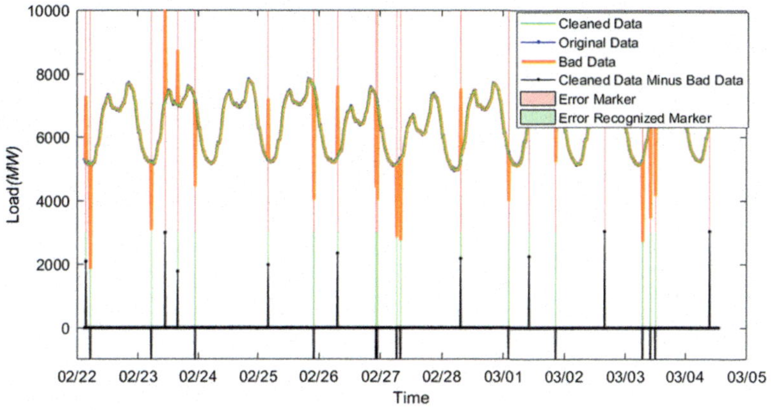

Fig. 3.5 Data cleaned after step 2

3.5.2 Day Ahead Point Forecast

LightGBM is chosen to perform point forecasts. The results of the point prediction are shown in Fig. 3.6.

Figure 3.6 shows the point forecasts based on bad data, cleaned data, and the original data. It is shown that the performance of the forecast based on cleaned data is closed to that based on the actual data, while the forecast based on the bad data has some abnormal data values.

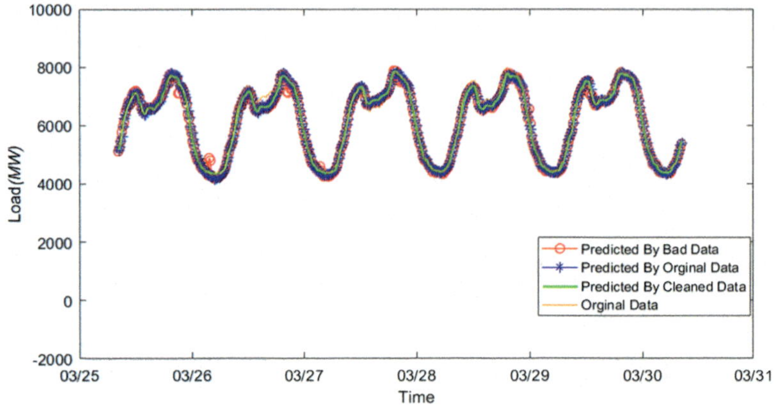

Fig. 3.6 Point forecast of LightGBM

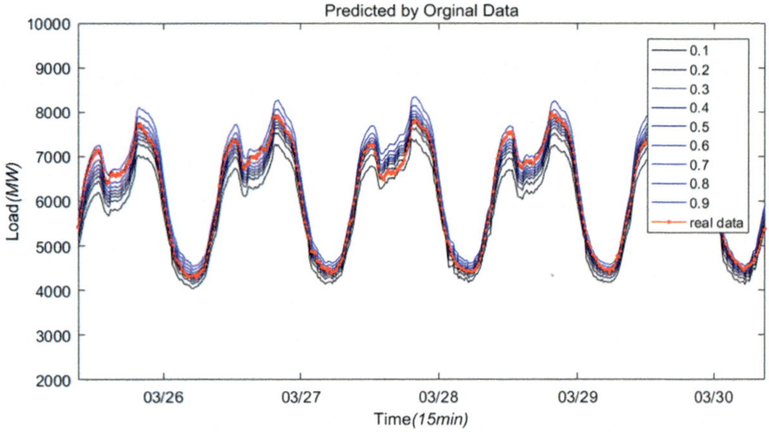

Fig. 3.7 Quantile forecast base on original data

3.5.3 Day Ahead Probabilistic Forecast

Figures 3.7, 3.8 and 3.9 presents the performance of probabilistic forecasting. Similar to point forecasts, some abnormal jump values appear in the probabilistic forecasts. In addition, due to the interference of bad data, the values of each quantile appear to be relatively dispersed in the prediction results based on bad data. We take the β of 0.6, 0.7, 0.8, 0.9, and 0.98 separately to determine the effect of data intervals on data recognition. The error evaluation indicators of load forecasting with different values of β are shown in the Table 3.1. As can be seen from the Table 3.1, whether

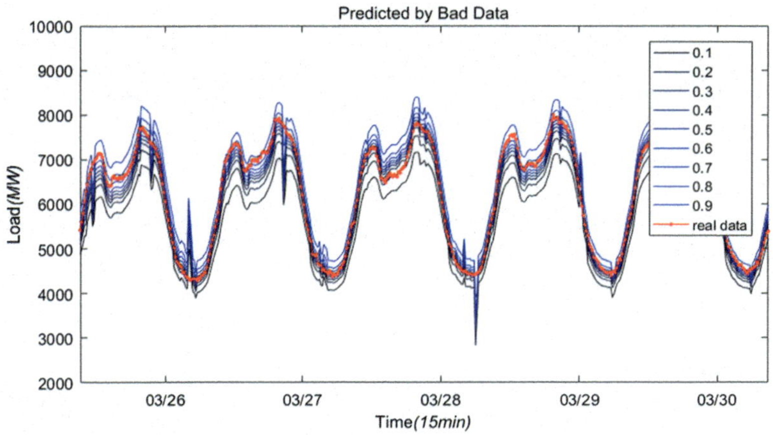

Fig. 3.8 Quantile forecast base on bad data

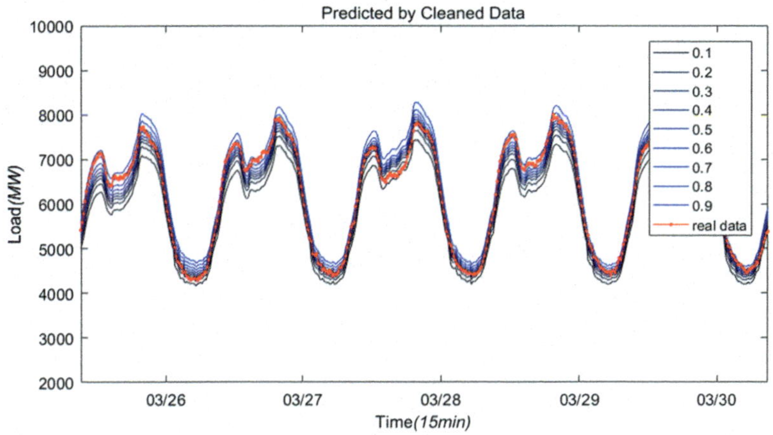

Fig. 3.9 Quantile forecast base on cleaned data

Table 3.1 Accuracy improvement after error correction

Loss	Original	Bad	$\beta = 0.98$	$\beta = 0.9$	$\beta = 0.8$	$\beta = 0.6$
MAE	2.31	2.5	2.4	2.42	2.44	2.45
RMSE	3.42	3.7	3.52	3.56	3.59	3.64
Pinball	73.48	86.76	77.98	77.67	77.45	78.01

using point forecasting or quantile forecasting, the accuracy of the bad data set will be lower than the original data set, and after the data cleaning, the prediction result is improved. The larger the β, the better the effect of error recovery is.

3.6 Conclusions

This chapter proposes a load profile cleaning method that combines the low-rank property of the matrix with the quantile regression random forests. The whole framework relies on the low-rank property of the matrix to initially identify outliers in the data, and quantile prediction to further identify and restore data. Tests in the data set show that this method effectively identifies and restores the anomaly points in the load data and improves the accuracy of the load forecasting.

References

1. Wang, Yi., Qixin Chen, Tao Hong, and Chongqing Kang. 2019. Review of smart meter data analytics: Applications, methodologies, and challenges. *IEEE Transactions on Smart Grid* 10 (3): 3125–3148.
2. Wang, Yi, Ning Zhang, Qixin Chen, D. S. Kirschen, P. Li, and Qing Xia. 2018. Data-driven probabilistic net load forecasting with high penetration of behind-the-meter PV. *IEEE Transactions on Power Systems* 33 (3): 3255–3264.
3. Wang, Yi., Ning Zhang, Chongqing Kang, Miao Miao, Rui Shi, and Qing Xia. 2018. An efficient approach to power system uncertainty analysis with high-dimensional dependencies. *IEEE Transactions on Power Systems* 33 (3): 2984–2994.
4. Monticelli, A. 2000. Electric power system state estimation. *Proceedings of the IEEE* 88 (2): 262–282.
5. Chen, Jiyi, Wenyuan Li, Adriel Lau, Jiguo Cao, and Ke. Wang. 2010. Automated load curve data cleansing in power systems. *IEEE Transactions on Smart Grid* 1 (2): 213–221.
6. Mateos, Gonzalo, and Georgios B Giannakis. 2013. Load curve data cleansing and imputation via sparsity and low rank. *IEEE Transactions on Smart Grid* 4 (4): 2347–2355.
7. Chen, Xinyu, Chongqing Kang, Xing Tong, Qing Xia, and Junfeng Yang. 2014. Improving the accuracy of bus load forecasting by a two-stage bad data identification method. *IEEE Transactions on Power Systems* 29 (4): 1634–1641.
8. Al-Wakeel, Ali, Wu. Jianzhong, and Nick Jenkins. 2016. State estimation of medium voltage distribution networks using smart meter measurements. *Applied Energy* 184: 207–218.
9. Liu, Xiufeng, Nadeem Iftikhar, Per Sieverts Nielsen, and Alfred Heller. 2016. Online anomaly energy consumption detection using lambda architecture. In *International Conference on Big Data Analytics and Knowledge Discovery*, 193–209. Springer.
10. Malhotra, Pankaj, Lovekesh Vig, Gautam Shroff, and Puneet Agarwal. 2015. Long short term memory networks for anomaly detection in time series. In *23rd European Symposium on Artificial Neural Networks, Computational Intelligence and Machine Learning*, 89–94.
11. Choy, Kokyo. 2001. Outlier detection for stationary time series. *Journal of Statistical Planning and Inference* 99 (2): 111–127.
12. Sheng, Tongtian, Wu. Wenchuan, Hongbin Sun, Qinglai Guo, et al. 2018. Abductive identification of bad data: Methodology and field test. *IET Generation, Transmission & Distribution* 12 (1): 150–159.
13. Jokar, Paria, Nasim Arianpoo, and Victor CM Leung. 2016. Electricity theft detection in ami using customers' consumption patterns. *IEEE Transactions on Smart Grid* 7 (1): 216–226.
14. Zanetti, Marcelo, Edgard Jamhour, Marcelo Pellenz, Manoel Penna, Voldi Zambenedetti, and Ivan Chueiri. 2019. A tunable fraud detection system for advanced metering infrastructure using short-lived patterns. *IEEE Transactions on Smart grid* 10 (1): 830–840.

15. Golub, Gene H, and Christian Reinsch. 1970. Singular value decomposition and least squares solutions. *Numerische mathematik* 14 (5): 403–420.
16. Cai, Jian-Feng, Emmanuel J Candès, and Zuowei Shen. 2010. A singular value thresholding algorithm for matrix completion. *SIAM Journal on Optimization* 20 (4): 1956–1982.

Chapter 4
Monthly Electricity Consumption Forecasting

4.1 Introduction

Electricity consumption forecasting is a significant issue in the field of energy economics, which can be classified into different categories. According to the determinacy of outputs, it is classified as deterministic forecasting [1–3] and probabilistic forecasting [4–6]. According to relative time scales, it is classified as long-term, mid-term and short-term. Mid-term electricity consumption forecasting always refers to the monthly time scale, which is important in many business and government decision-making processes. On the one hand, economic statistics and government policies, which are closely related to electricity consumption forecasting, are always performed and analyzed by the month. On the other hand, decisions on equipment maintenance, fuel trading, and bilateral transactions of electricity are generally performed months ahead. These decisions could be greatly affected by mid-term electricity consumption forecasting [7].

As a fundamental industry of the national economic system, the monthly forecasting of electricity consumption is inevitably influenced by external economic factors (EEFs) [8, 9], which, in the past, were usually represented by several common macroeconomic indexes.

In recent years, economic growth has greatly fluctuated around the world, mostly slowing down in many countries and regions. Facing new situations, the conventional mid-term electricity consumption forecasting methods can no longer precisely characterize the influence of EEFs [10, 11].

EEFs include detailed industrial data, commodity trade data, and price index data, with as many as hundreds of types, released on a monthly or seasonal basis. Theoretically, with sufficient EEF data, the influential factors on electricity consumption can be clearly identified, and, therefore, the prediction will become more accurate. However, conventional electricity consumption forecasting models are unable to process a large number of EEFs. First, too many input data would introduce the "curse of dimensionality" into the forecasting models. Second, although there are many types of EEFs, the lengths of most EEF data are relatively short, considering the monthly or seasonally released frequency. It is difficult to solve the parameters of the forecasting model by simply taking large amounts of EEFs as inputs without sufficient data length.

Q. Chen et al., *Data Analytics in Power Markets*,
https://doi.org/10.1007/978-981-16-4975-2_4

Currently, some papers have performed research on mid-term electricity consumption forecasting methods considering various EEFs. The existing forecasting methods can be generally classified into two categories. The first category focuses on statistical techniques. The major feature is to consider the relevancy of electricity consumption and EEFs and to build parametric models, which include the time series method [12–15] and the multivariate regression method [1, 16–20]. The second category focuses on intelligent algorithms, which include artificial neural networks [3, 21–23], support vector machines (SVM) [2, 24–26] and so on [27]. These algorithms consider the past electricity consumption and EEFs as inputs and current electricity consumption as outputs.

More specifically, the first category analyzes the relationships between electricity consumption and EEFs using multivariable regression or correlation analyses. Because the impact of some EEFs on electricity consumption might last for months, a precise analysis of the relationship between electricity consumption and EEFs needs to consider both all EEFs at the same time and trends implicit in the historical data. However, most of the existing parametric forecasting methods did not pay sufficient attention to time lag effects of EEFs. Forecasting is mostly performed based on the synchronous data of EEFs [1, 16]. Although some researchers have considered time lag effects of EEFs, a determined period is always subjectively given to represent the time lag effects before model training [3]. Moreover, the quantity of the EEF considered in parametric methods is usually constrained by the chosen methods. Considering the hundreds of input data types, conventional parametric forecasting methods combining several regression functions do not have the ability to self-adaptively choose appropriate data as inputs. Some papers have attempted to make a "filtrate" job in advance and then select a small number of appropriate EEFs for electricity consumption forecasting [2, 17]. This measure indicates that these forecasting methods have limitation on handling so large economic datasets.

The second category, which is mostly based on AI techniques, uses machine learning or pattern recognition abilities to organize the input factors adaptively. However, one weakness of AI techniques is that the entire computational process is mostly a "black bo", which is not as understandable as the parametric methods [28]. Therefore, it is extremely difficult to understand the process and the interaction of how inputs influence the outputs. From an economic perspective, it would be important to understand the relationships between mid-term electricity consumption and the corresponding EEF. Moreover, the application of AI techniques involves a large number of training data. Even though the available data volume is increasing rapidly, considering the hundreds of factor types, it would still take a long time for the requirements of conventional AI methods' training data amount to be met.

In summary, because the set of EEFs has greatly increased, conventional parametric models might be too simple to capture the abundant information hidden within the data, and AI models do not solve the aforementioned contradiction between data quantity and data length. Therefore, a new monthly electricity consumption forecasting method that can effectively and credibly integrate large amounts of EEF data and thus self-adaptively improve the forecasting accuracy is needed.

To handle the contradiction between data quantity and data length, the vector error correction model (VECM) [29] has been broadly applied in macro-econometrics in recent years [10, 30]. Several studies have used VECM to describe the relationship between electricity consumption and EEFs [10, 11, 30, 31] and performed forecasting of electricity consumption. However, the types of input data for these works were limited, and the forecasting periods were usually long, such as annually or quarterly. VECM is essentially a type of parameter model; thus, the data length requirements are relaxed relative to the commonly used AI models. By building up the influence matrix and the coefficient matrixes, VECM is able to represent the long-run (historical trends) and the short-run (recent changes) equilibrium relationships between input factors and identify the time lag effects in the most relevant sense. As a fundamental industry of the national economic system, electricity consumption can be integrated as a regular factor in the VECM; thus, hidden correlated information between electricity consumption and EEFs can be clearly identified and integrated into the forecasting model of electricity consumption. Moreover, to handle the monthly data more accurately, a seasonal decomposition method, X-12-ARIMA [32], can also be incorporated into the proposed forecasting framework before performing the VECM.

The authors of this chapter have proposed a novel electricity consumption forecasting method using VECM and X-12-ARIMA [33]. The forecasting method considers the relationships between mid-term electricity consumption and the corresponding EEF. However, the quantity of its input data types is relatively few and it doesn't have the ability to self-adaptively choose factors from a large data set. Thus, the input factors of the forecasting method are randomly chosen, which makes the forecasting performances of the forecasting method relatively unstable. In other words, the forecasting accuracy is quite dependent on the choice of input EEFs. It won't perform well if the proper EEFs haven't been selected.

To handle the problem of the previous forecasting method, a variable selection process is required before forecasting to identify the most effective factors. Conventional variable selection methods implement similarity measures [1] regression analysis [17], spline models [34], memetic algorithms [2], maximum relevance minimum redundancy feature selection methods [35], among others. These methods usually focus on the relevance calculation of inputs without considering the contributions of historical forecasting errors from the inputs. In this chapter, a self-adaptive screening (SAS) method for variable selection is proposed and integrated into the proposed forecasting framework, which makes the previous forecasting method change a lot in the model structure. The SAS method makes use of training based on historical data to identify the factors with the best historical performance by error analysis. These factors might naturally become the most influential factors on electricity consumption in the subsequent "rea" forecasting, when the long-run equilibrium relationships of these factors in seasonal VECM forecasting (SVECM) are not significantly changed. For the factors with significant impacts on the short-run equilibrium relationships, the SAS method is then also able to identify them. With the accumulation and updates of the historical electricity consumption forecasting and the SVECM results, the SAS process can therefore self-adaptively identify the changed influential factors when the external economic situation has changed.

The contributions of this chapter are proposed:

1. A novel forecasting method, SVECM, is proposed, which is capable of capturing the potential impacts on electricity consumption from a "bigge" set of EEFs and considering the time lag effects of different factors.
2. An original variable-selection method, SAS, is proposed, which is capable of handling the contradiction between data type and data length, and self-adaptively identifying the most possible influential EEF as forecasting inputs.
3. A novel monthly electricity consumption forecasting framework, namely SAS-SVCEM, is proposed, which is composed of SAS and SVECM, whose benefits in forecasting and variable-selection are also integrated.

The feasibility and precision performance of SAS-SVECM are examined using the latest data from China's economic system and the electricity industry. A comparative study of the forecasting accuracies of the proposed method and the conventional forecasting methods is also included.

The remainder of the chapter is organized as follows. The framework and flowchart of the proposed method are presented in Sect. 4.2. In Sect. 4.3, we show how we treat the collected data to set up the foundation of the forecasting data. In Sect. 4.4, the SVECM is described and formulated. It is followed by the SAS method in Sect. 4.5. Experimental results are then presented and discussed in Sect. 4.6. Finally, we conclude the work of the chapter and propose future works.

4.2 Framework

The proposed SAS-SVECM forecasting framework comprises three major procedures: (1) Data Collection and Treatment, (2) SVECM Forecasting, (3) SAS method. The SAS method comprises three sub-procedures: (a) Influential EEF Identification, (b) Influential EEF Grouping, (c) Forecasting Performance Evaluation Considering Different EEF Groups.

The flowchart of the proposed forecasting framework, SAS-SVECM, is shown in Fig. 4.1.

Details of the procedures are discussed below.

4.2.1 Data Collection and Treatment

Before forecasting with SVCEM, a large number of EEF data should be collected as potential input variables. To prevent spurious regression of input factors, an augmented Dickey-Fuller (ADF) test [36] is applied. Co-integration tests [29] are then run between electricity consumption and various EEFs. EEFs that fail the test are removed.

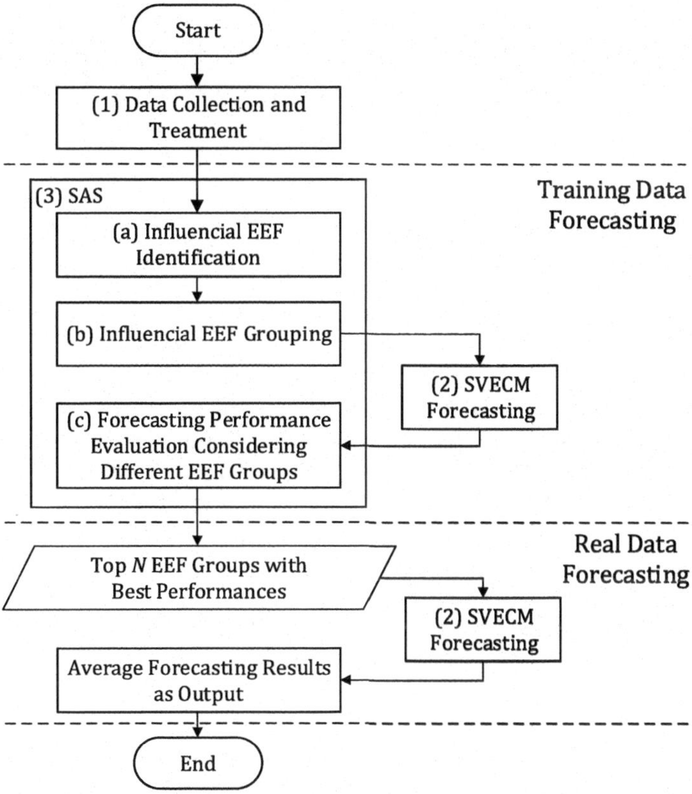

Fig. 4.1 Flowchart of the SAS-SVECM Framework Proposed in this chapter

After the tests for the input data, X-12-ARIMA is used to seasonally adjust the input variable time series, after which each variable is decomposed into four time-series components: trend components (TC), seasonal components (SC), holiday components (HC), and irregular components (IC). A multiplication model connects them together.

4.2.2 SVECM Forecasting

Considering the characteristics of the four time series components, different forecasting strategies should be applied. Because TC represents long-run interaction and IC represents short-run interaction between the EEF and the electricity consumption, TC and IC are integrated—considered as a multiplication, TIC—and forecasted by the VECM. Using the identified effective EEF and the historical TIC data of electricity consumption as the inputs, VECM can forecast the future TIC of electricity consumption as the outputs with the consideration of time lag effects. However, for SC and

HC, because they are self-correlated and unrelated to the EEF, the two components are each forecasted by implementing time series extrapolation. The four forecasted components are then integrated, as a reverse action of seasonal decomposition, also by a multiplication model.

4.2.3 Self-adaptive Screening

Owing to contrary data types and the data length, the SAS takes three steps to identify the most influential EEF set for forecasting. In the first step, (a) Influential EEF Identification, two types of EEF are identified as influential EEFs. The first type of EEF comes from historical forecasting, which brings positive impacts to improve the forecasting performance when they are considered in the forecasting model. The second type indicates those that impose significant impacts on the short-run equilibrium relationships of VECM. Because the influential EEF might comprise a large "poo" of factors, in the second step of SAS, (b) Influential EEF Grouping, they are selected and combined as smaller input factor groups (IFGs) as forecasting inputs ergodicity for SVECM training. Each IFG corresponds to a specific training forecasting result. For the last procedure, (c) Forecasting Performance Evaluation Considering Different EEF Groups, the training forecasting results and different corresponding EEF Groups are sorted in ascending order according to the forecasting errors. Thus, the top "N" EEF Groups with the best training forecasting performance can be identified. The identified EEF Groups are integrated into the subsequent real data forecasting procedure. Finally, the forecasting results corresponding to the identified EEF Groups are averaged as the final forecasting result.

4.2.4 Novelty and Characteristics of SAS-SVECM

Different from the existing researches, the proposed method, SAS-SVECM, builds a totally new forecasting framework to combine a parametric model and a self-adaptive model together, with the acquisition of their both benefits, to forecast more accurately.

In detail, with the help of SVECM and SAS, the proposed forecasting framework is not only able to understand the forecasting calculation process, but also able to analyze the impact of various EEFs on electricity consumption. In addition, it can deal with the contradiction between data type and data length, choosing the appropriate IFG as the optimal input data.

4.3 Data Collection and Treatment

The data collection and treatment procedure is performed before forecasting.

4.3.1 Data Collection and Tests

The existing EEF can be broadly classified into three categories: detailed industrial data, commodity trade data, and price index data. The first category includes industry, investment in fixed assets, real estate, basic conditions of transport, postal and telecommunication data, among others. The second category includes domestic trade, foreign trade, and economic cooperation. The last category includes various price index data and is the indicator of the economy. Each subcategory mentioned above also contains more detailed data; consequently, the collection of data might contain as many as hundreds of types.

An ADF test [36] and the co-integration test [29] are then performed between electricity consumption and various EEFs.

4.3.2 Seasonal Adjustments Based on X-12-ARIMA

The X-12-ARIMA seasonal adjustments algorithm has previously been described in [32]. After X-12-ARIMA, the time series is decomposed into four components: TC, SC, HC and IC. For the convenience of comparison with all EEFs and electricity consumption, a multiplication model for combining four components is applied:

$$Y_t = T_t \cdot S_t \cdot I_t \cdot H_t \tag{4.1}$$

where T_t represents the TC in the tth month, having the same dimension as the multiplication Y_t. The other three components, S_t, H_t, andI_t, indicate SC, HC and IC in the tth month, respectively. In the following process, four components will be treated in different ways.

4.4 SVECM Forecasting

4.4.1 VECM Forecasting

4.4.1.1 Vector Error Correction Model

Time series of exogenous variables can be proposed as [29].

$$\mathbf{F}_t = \left[f_{1,t}, \ldots, f_{i,t}, \ldots, f_{n,t}\right]^{\mathrm{T}} \tag{4.2}$$

where $f_{i,t}$ represents the ith EEF's time series. In particular, $f_{1,t}$ represents the electricity consumption, and the other $f_{i,t}$ represents EEF.

We then use linear regression analysis on $\mathbf{F}_t$, which will turn into the Vector Auto-regression model (VAR) as follows:

$$\mathbf{F}_t = \mathbf{A}_0 + \mathbf{A}_1\mathbf{F}_{t-1} + \cdots + \mathbf{A}_k\mathbf{F}_{t-k} + \cdots + \mathbf{A}_p\mathbf{F}_{t-p} + \mathbf{U}_t \tag{4.3}$$

where $\mathbf{A}_k$ is an $n \times n$ order matrix of the model; $\mathbf{A}_0$ is a constant term of the polynomial; $\mathbf{F}_t$ is the time series of EEF; $\mathbf{F}_{t-k}$ is k orders later than $\mathbf{F}_t$; p is the time lag period of the model; and $\mathbf{U}_t$ represents the disturbance vector with n orders.

Regarding $\mathbf{F}_{t-1}$ as the base value, we use difference transformation towards variables in (4.3). We can then obtain

$$\begin{cases} \mathbf{F}_t = \mathbf{F}_{t-1} + \Delta\mathbf{F}_t \\ \mathbf{F}_{t-k} = \mathbf{F}_{t-1} - \sum\limits_{j=1}^{k-1} \Delta\mathbf{F}_{t-j}, \quad k = 2, \ldots, p \end{cases} \tag{4.4}$$

By plugging (4.4) into (4.3) and then applying co-integration transformations, we are able to obtain the first-order differential form of (4.3), which is

$$\Delta\mathbf{F}_t = \Pi\mathbf{F}_{t-1} + \sum_{k=1}^{p-1} \Gamma_k \Delta\mathbf{F}_{t-k} + \mathbf{U}_t \tag{4.5}$$

where Π is considered as the influence matrix and Γ_k is considered as the coefficient matrix:

$$\begin{cases} \Pi = \sum\limits_{k=1}^{p} \mathbf{A}_k - \mathbf{I} \\ \Gamma_k = -\sum\limits_{j=k+1}^{p} \mathbf{A}_j, \quad k = 1, 2, \cdots, p-1 \end{cases} \tag{4.6}$$

where $\mathbf{I}$ is a unit matrix. Matrix Π represents long-run mutual effects between various exogenous variables, with order $m \times m$. Matrix Γ_k represents short-run relevant disturbances between various EEFs, with order $m \times m$.

Definition. A series with no deterministic component that has a stationary, invertible, ARMA representation after differencing d times, is said to be integrated of order d, denoted $x_t \sim I(d)$ [29].

Because of $f_{i,t} \sim I(1)$, we can obtain $\Delta f_{i,t} \sim I(0)$. We know that each parameter in (4.5), except $\Pi\mathbf{F}_{t-1}$, is stationary. Thus, only if $\Pi\mathbf{F}_{t-1}$ is a stationary time series will the whole equation be stationary. In other words, the co-integration characteristic of the whole equation depends on matrix Π.

We assume that the rank of matrix Π is $r \in (0, n)$. It can then be decomposed into a multiplication product of two $m \times r$ rank matrixes α and β.

$$\Pi = \alpha \cdot \beta^{\top} \tag{4.7}$$

where α is the adjustment coefficients, and β is the co-integrating vectors. Their ranks are both r. We then plug (4.7) into (4.5) and can obtain

$$\Delta \mathbf{F}_t = \alpha \beta^{\top} \cdot \mathbf{F}_{t-1} + \sum_{i=1}^{p-1} \Gamma_i \Delta \mathbf{F}_{t-i} + \mathbf{U}_t \tag{4.8}$$

With advanced transformations, the equation above can be turned into the normal form of the vector error correction model:

$$\Delta \mathbf{F}_t = \alpha \cdot \mathbf{vecm}_{t-1} + \sum_{i=1}^{p-1} \Gamma_i \Delta \mathbf{F}_{t-i} + \mathbf{U}_t \tag{4.9}$$

where $\mathbf{vecm}_{t-1} = \beta^{\top} \cdot \mathbf{F}_{t-1}$, $\mathbf{F}_{t-1}$ represents the each EEF's deviation degree of the overall trend; α represents the recovery rate of EEF, which deviates from the overall trend; and Γ_k represents the EEF's short-run relevant disturbance in the period k on the other periods.

The multiplication of TC and IC of EEF, TIC, represents the long-run developing trend of each time series

$$C_t = T_t \cdot I_t \tag{4.10}$$

where C_t denotes TIC. Considering the mutual relations towards electricity consumption and EEF, we select TIC as the forecasting target. For simplicity, EEF-TIC denotes the TIC of EEF, and EC-TIC, EC-SC and EC-HC represent TIC, SC and HC of electricity consumption, respectively.

4.4.1.2 Determination of Time Lag Period *p*

The upper limit of the time period, in which EEFs and electricity consumption have strong mutual relations, is the time lag period p of the VECM. Theoretically, the higher p is, the greater the capacity of VECM to reveal the connections between EEFs. With higher p, there are more parameters of the VECM in the requirement of the calculation, and the freedom degree of VECM is lower.

In common statistical methods, the Akaike information criterion (AIC) and Schwartz-Bayes criterion (SBC) are used for the model order selection [37]. Because the forecasting error is the most significant indicator directly related to forecast, the parameters in the forecasting model should be chosen with the consideration of AIC and SBC results and forecasting error results.

4.4.1.3 VECM Parameters Calculation

We set up the VECM based on the number of EEFs. TIC of electricity consumption and EEFs are considered as historical training data, which are displayed as matrixes of $\Delta\mathbf{F}_t$, $\mathbf{F}_{t-1}$ and $\Delta\mathbf{F}_{t-k}$. We then use $\Delta\mathbf{F}_t$, $\mathbf{F}_{t-1}$ and $\Delta\mathbf{F}_{t-k}$ as historical data in Eq. (4.5) to calculate matrixes Π and Γ_k. α and β can be obtained by the decomposition of matrix Π.

4.4.1.4 Forecasting Results Output

After the settlement of the VECM, we calculate the increment of month t, $\Delta\mathbf{F}_t$, which we select to forecast from month $t-1$. We then use (4.11) to calculate the final forecasting results.

$$\mathbf{F}_t = \Delta\mathbf{F}_t + \mathbf{F}_{t-1} \tag{4.11}$$

4.4.2 *Time Series Extrapolation Forecasting*

In the consideration of the similarity of the EC-SC each year and the high correlation between time and EC-SC, the time series extrapolation method is selected to forecast EC-SC. S_t denotes EC-SC in year t:

$$\mathbf{S}_t = \left[s_{t,1}, \ldots, s_{t,i}, \ldots, s_{t,12}\right] \tag{4.12}$$

where $s_{i,t}$ is EC-SC in the ith month of the tth year. Thus, the EC-SC forecasting equation is

$$\bar{\mathbf{S}}_t = \frac{1}{2}\left(\mathbf{S}_{t-1} + \mathbf{S}_{t-2}\right) \tag{4.13}$$

where $\bar{\mathbf{S}}_t$ is the forecasting EC-SC series in the tth year. Regarding each component:

$$\bar{s}_{t,i} = \frac{1}{2}\left(s_{t-1,i} + s_{t-2,i}\right),\ i = 1, 2, \ldots, 12 \tag{4.14}$$

EC-HC, which has the same characteristic as EC-SC, could be calculated in the same way.

4.5 Self-adaptive Screening

Figure 4.2 depicts the flowchart of the SAS method.

4.5.1 Influential EEF Identification

Because of the excessive data types and the limited data length, not all EEFs should be considered as the inputs of the forecasting model. Thus, influential EEFs should be picked out of the ocean of EEFs.

Two types of influential EEFs are identified in this section. First, EEFs with long-run effects on electricity consumption are identified. To achieve this aim, historical forecasting IFGs with the least error and the corresponding EEF inputs are identified as influential EEFs. Second, a short-run influence analysis between EEFs and electricity consumption is carried out, and the EEFs with significant impacts on the short-run equilibrium relationships are also identified.

The two parts of EEFs consist of the set of influential EEFs. This step aims to identify as many influential EEFs as possible to avoid oversight. Thus, the number of EEFs in this set, N_1, could be relatively large.

4.5.2 Influential EEF Grouping

Although influential EEFs have been identified, the dataset is still too large for VECM, with the limited data length. Because the number of parameters included within VECM is restricted, we have

$$n_{\text{para}} = m \cdot m + m \cdot m \cdot (p - 1) = pm^2 \tag{4.15}$$

where n_{para} represents the number of parameters in VECM to be calculated; m represents the number of input factors and the order of matrixes Π and Γ_k; and p represents the time lag order of the VECM. Because we use the ordinary least squares method (OLS) to calculate n_{para}, the number of training data time spans, n_{training}, should be greater than n_{para}.

For example, if 100 EEFs are chosen as VECM inputs with the time lag order 6, the data length of EEFs should be more than 60,000 months, which is impossible. Actually, restricted to the data length, l, the number of input EEF could not exceed

$$m < \sqrt{l/p} \tag{4.16}$$

Thus, it is reasonable to choose limited EEFs in the forecasting model at one time. With the changes of the economy, the types of the most influential EEF would change.

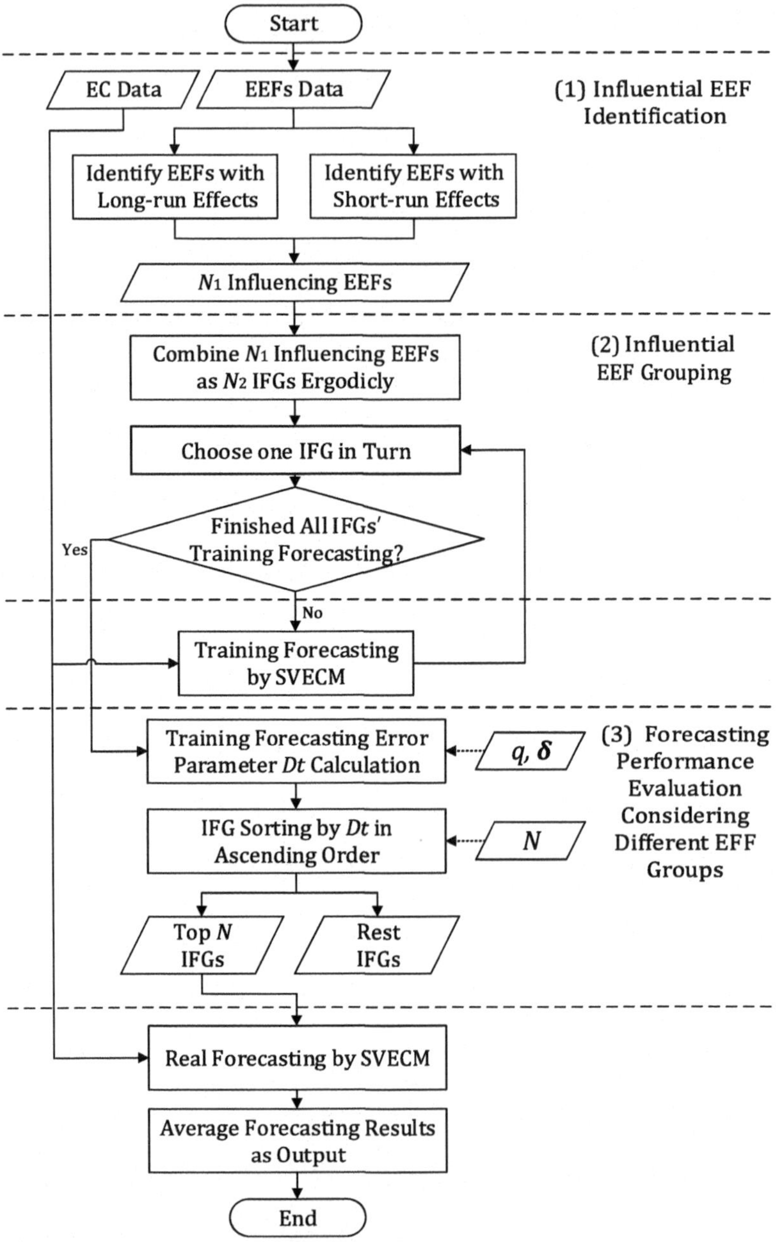

Fig. 4.2 Flowchart of the self-adaptive screening method, SAS

However, the structural economic changes usually take place smoothly; the EEF with the best performance would not change significantly in adjacent months. Thus, the most effective EEF in the past months' forecasting could be accurately identified based on the lowest mean absolute percentage error (MAPE) of the training months.

$$E_r = \frac{1}{T_0}\sum_{i=1}^{T_0}\left|\frac{E_i - \hat{E}_i}{E_i}\right| \tag{4.17}$$

where E_r represents the MAPE of electricity consumption's forecasting error; E_i and $\hat{E}_i$ represent the actual and predicted electricity consumption, respectively; and T_0 represents the time span of the forecasting.

A small number of EEFs can then be combined as IFG for future monthly electricity consumption training forecasting. Considering the length of EEF data, the number of EEFs in the IFG should be less than $m_{\max}$, according to (4.16). In this case, we could obtain no more than N_2 IFGs. We have

$$N_2 = \sum_{i=1}^{m_{\max}} C_{N_1}^{m_{\max}-i+1} \tag{4.18}$$

For each IFG, corresponding training forecasting results could be obtained by implementing the SVECM.

4.5.3 Forecasting Performance Evaluation Considering Different EEF Groups

According to the corresponding training forecasting errors in the previous section, the IFGs would be sorted in ascending order, and the top few IFGs with best forecasting performance would be selected. The training forecasting error parameter is determined as follows:

$$D_t = \frac{1}{q}\sum_{i=1}^{q}\delta_i R_{t-i} \tag{4.19}$$

where D_t represents the training forecasting error parameter of the tth month; q represents the training month of self-adaptive screening; R_{t-i}represents the training forecasting error in the ith month before; and δ_i represents the attenuation coefficient of R_{t-i}. In this study, we set $\delta_i = 0.8^i$ to weaken the influence of months from the distant past.

To avoid the contingency of EEF selection, SAS selects the top N IFGs based on the ascending order of D_t for real electricity consumption forecasting by SVECM. The top N IFGs are likely to include the most influential EEFs in recent months.

Because the forecasting performance of the top IFG varies with time, the final result of real forecasting is more appropriate to be used as the average of the top N IFGs' real forecasting results. The average procedure considers all N IFGs' results and reduces the contingency in the forecasted IFG selection.

4.6 Case Study

4.6.1 Basic Data and Tests

4.6.1.1 Basic Data Collection

In addition to monthly electricity consumption data in China, this chapter collected a large number of monthly macroeconomic data of China (2000–2014) from China's National Bureau of Statistics, which are shown in Table 4.1.

Some of their values are presented in Fig. 4.3.

It should be noted that all electricity consumption and EEF data are normalized by each of their data in the last month, which makes it possible for all data to be shown in the same figure. It is obvious that there are various fluctuations in every time series of electricity consumption and EEFs. All normalized values have the characteristic of a growth trend, which would make the time series nonstationary.

Table 4.1 Corresponding table of macroeconomic data types

Abbreviation	Full name	Description
EXP	Gross export	The current value (k $)
IMP	Gross import	The current value (k $)
COP	Crude oil price	Platts monthly average in WTI market ($/barrel)
PPI	Producer price index	Growth rate based on 2010 (%)
CPI	Consumer price index	Growth rate based on 2010 (%)
RRR	Required reserves rate	Monthly data (%)
RPI	Retail price index	Growth rate based on 2010 (%)
EXR	Dollar exchange rate	Exchange rate of RMB and Dollar
IPI	Industrial production index	Growth rate based on 2010 (%)
MSP	M2 supply	The current value (10^8 RMB)
FAI	Fixed-asset investment	The current value (10^8 RMB)
RSCG	Retail sales of consumer goods	The current value (10^8 RMB)
REI	Real estate investment	The current value (10^8 RMB)
VFR	Volume of railway freight	The current value (10^4 t)

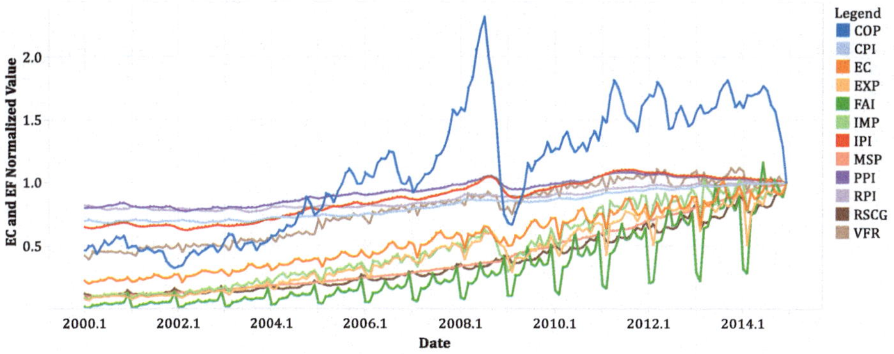

Fig. 4.3 Electricity consumption and EEF normalized values from 2000 to 2014 in China

4.6.1.2 Basic Data Tests

To prevent spurious regression of the model, the ADF tests and Johansen co-integration tests are performed before forecasting. According to the results, only 3 factors–COP, CPI, and RPI–are stationary at the 5% level. As stated previously, the factors should not be simply connected without any modification. Two Johansen co-integration tests are then performed. According to the tests, the combination of electricity consumption and EEFs has passed the Johansen co-integration tests as well.

4.6.1.3 Basic Data Seasonal Adjustments

The data from the first month in the sample (January 2000) is selected to act as the base of this type of price index data. Thus, the fixed base price index data can be determined. Seasonal adjustments based on X-12-ARIMA are then applied to the data, which have seasonal peaks in their time series. Specifically, some price index EEFs, including CPI, PPI, RPI, and IPI, do not have seasonal peaks, which means that there is no seasonal adjustment demand for them. The EC-TC and EEF-TC are shown in Fig. 4.4.

It is obvious that the time series of electricity consumption and EEFs after seasonal adjustments become smoother than before. In addition, the impact of the worldwide economic crisis of 2008 can easily be observed in the figure. HC, IC, and SC of electricity consumption are presented in Fig. 4.5.

There are apparent periodicities in the time series of EC-SC and EC-HC. It is clear from the line of EC-HC that holiday effects exist only from January to March. EC-IC seemingly has the characteristic of white noise, which could not be specifically classified. However, there are short-term relationships between electricity consumption

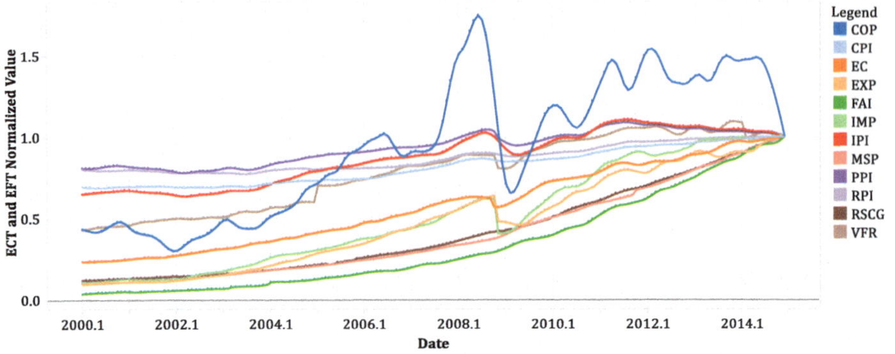

Fig. 4.4 EC-TC and EEF-TC normalized values from 2000 to 2014 in China

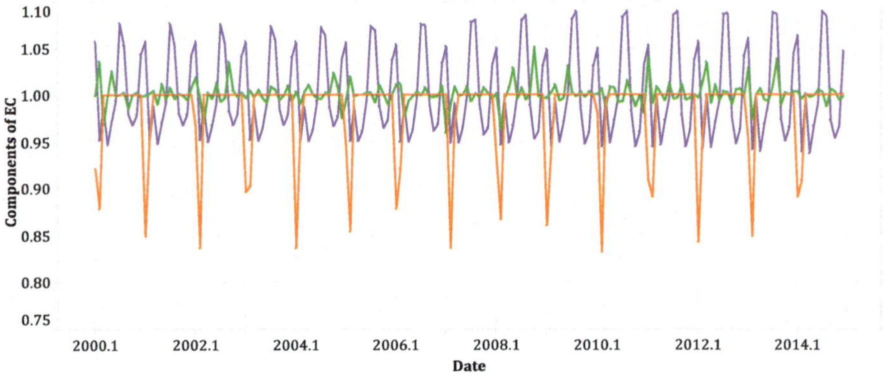

Fig. 4.5 Holiday, irregular and seasonal components of electricity consumption

and EEFs buried deeply in their irregular components. Thus, our forecasting object of VECM is determined as EC-TIC, the multiplication of EC-TC and EC-IC, with the purpose of maximal use of related information between electricity consumption and EEFs.

4.6.2 Electricity Consumption Forecasting Performance Without SAS

In this section, we only use SVECM to forecast electricity consumption, without the help of SAS. In SVECM, EC-TC and EC-IC are forecasted based on VECM, whereas EC-SC and EC-HC use the forecasting method of time series extrapolation.

4.6.2.1 SVECM Forecasting

Without the EEF selection of SAS, we could select only 15 different IFGs randomly to compare their performance in SVECM. The member number of IFG varies from 3 to 6. According to the test results, the maximum lag order of VECM is 5, and in this situation, the corresponding rank of matrix Π is 1. In this section, we use lag order $p = 5$ and rank $r = 1$ as example SVECM parameters. The forecasting results of EC-TIC are shown in Fig. 4.6.

Taking 2014 as an example, as shown in Fig. 4.6, the broken lines denote forecasting EC-TIC of 15 different IFGs. The EC-TIC forecasting MAPEs and max errors of 15 different IFGs are shown in Table 4.2. It is clear from the figure that the forecasting performance of different IFGs varies in a relatively large scale.

Among all IFGs, IFG 11, with the largest EC-TIC forecasting MAPE of 1.84%, contains EXP, IMP, COP, CPI, IPI, and FAI, with the maximum forecasting error of 3.74%. IFG 13 has the smallest MAPE of 0.98%, containing EXP, PPI, CPI, RPI, IPI, and VFR, with the maximum forecasting error of 2.32%.

4.6.2.2 Time Series Extrapolation Forecasting

Considering the strong time correlation of EC-SC and EC-HC, we select time series extrapolation to forecast them. The MAPE of forecasting EC-SC is 0.39%, which shows that the forecasting method performs well. The forecasted EC-HC MAPEs in January and February are 0.09% and 0.24%, respectively, which proves the validity and applicability of seasonal adjustments and the forecasting method.

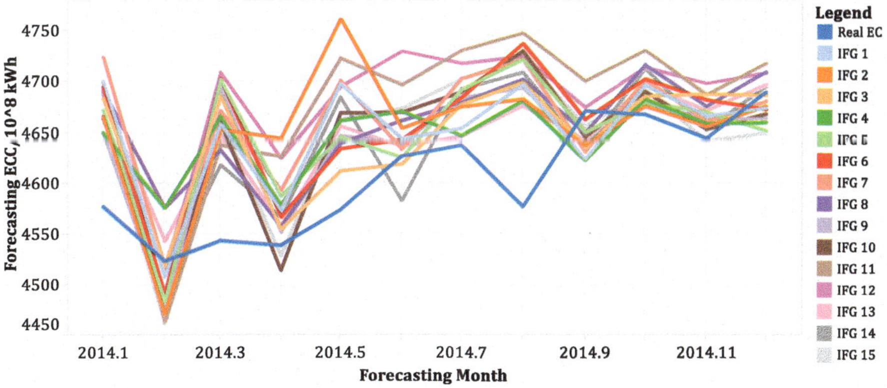

Fig. 4.6 EC-TIC forecasting results of 15 different IFGs

Table 4.2 EC-TIC forecasting results of 15 different IFGs

IFG No.	IFG member	MAPE (%)	Max error (%)
1	EXP, IMP, FAI	1.24	2.72
2	IMP, COP, FAI	1.43	4.12
3	RPI, IPI, FAI	1.05	3.13
4	EXP, IMP, COP, RSCG	1.16	2.69
5	COP, CPI, MSP, FAI	1.30	3.49
6	PPI, MSP, FAI, RSCG	1.31	3.51
7	EXP, IMP, CPI, IPI, FAI	1.42	3.22
8	EXP, PPI, IPI, FAI, VFR	1.21	2.74
9	COP, CPI, IPI, FAI, RSCG	1.14	3.31
10	CPI, MSP, FAI, RSCG, VFR	1.25	3.36
11	EXP, EMP, COP, CPI, IPI, FAI	1.84	3.74
12	EXP, COP, PPI, CPI, RPI, FAI	1.72	3.64
13	EXP, COP, CPI, IPI, FAI, VFR	0.98	2.32
14	IMP, PPI, IPI, FAI, RSCG, VFR	1.21	2.90
15	COP, IPI, MSP, FAI, RSCG, VFR	1.30	3.22

4.6.2.3 Analysis of Electric Consumption Forecasting Results

By using function (4.1), it is convenient for us to calculate the forecasting results and MAPE of the selected 15 EEF IFGs in 2014, which are shown in Fig. 4.7.

As shown in Fig. 4.7, the broken lines denote forecasting electricity consumption of 15 different IFGs. The electricity consumption forecasting MAPEs and maximum errors of 15 different IFGs are shown in Table 4.3. Among all EEF IFGs, the largest

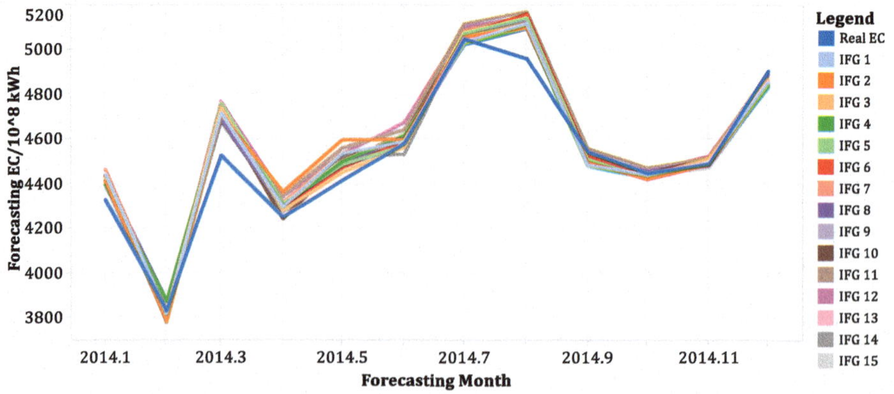

Fig. 4.7 Electricity consumption forecasting results of 15 different IFGs

Table 4.3 Electricity consumption forecasting results of 15 different IFGs

IFG No.	IFG member	MAPE (%)	Max error (%)
1	EXP, IMP, FAI	1.45	4.16
2	IMP, COP, FAI	1.67	4.09
3	RPI, IPI, FAI	1.28	4.76
4	EXP, IMP, COP, RSCG	1.45	4.31
5	COP, CPI, MSP, FAI	1.54	5.13
6	PPI, MSP, FAI, RSCG	1.44	5.14
7	EXP, IMP, CPI, IPI, FAI	1.56	4.44
8	EXP, PPI, IPI, FAI, VFR	1.25	3.58
9	COP, CPI, IPI, FAI, RSCG	1.36	4.94
10	CPI, MSP, FAI, RSCG, VFR	1.40	4.19
11	EXP, EMP, COP, CPI, IPI, FAI	1.81	4.24
12	EXP, COP, PPI, CPI, RPI, FAI	1.77	5.28
13	EXP, COP, CPI, IPI, FAI, VFR	1.19	3.91
14	IMP, PPI, IPI, FAI, RSCG, VFR	1.37	3.40
15	COP, IPI, MSP, FAI, RSCG, VFR	1.50	4.24

MAPE is 1.81% for IFG 11, whereas the smallest MAPE is 1.19% for IFG 13. For the maximum single month error, the largest is 3.40% for IFG 14, whereas the maximum error for IFG 11 is 4.24%.

It is clear from the forecasting results that EC forecasting MAPEs of different IFGs vary in a relatively large scale. We can observe that the forecasting errors might differ by approximately one-third. Thus, it would be necessary to select the appropriate IFGs as forecasting inputs with the help of SAS.

4.6.3 EC Forecasting Performance with SAS

In this section, the SAS proposed in Sect. 4.5 is applied to enhance the EC forecasting accuracy. After the model training by historical data, we determined that the best parameters for VECM are lag order $p = 3$ and matrix rank $r = 3$, and the best parameters for the training forecasting in the SAS are training month $q = 6$ and the top $N = 50$ IFGs. The forecasting performance comparison between SVECM with and without SAS is presented in Sect. 4.6.3.1. The forecasting performance of different order p and rank r are presented in Sect. 4.6.3.2, and the forecasting performance of different training parameters for training forecasting, month q and top N IFGs are presented in Sect. 4.6.3.3.

4.6.3.1 Forecasting Performance Comparison Between SVECM with and Without SAS

By using SVECM with SAS, we forecast the electricity consumption in 2014. The forecasting results are listed in Table 4.4 with some typical forecasting results from Sect. 4.6.2.3. To draw an appropriate comparison, we select IFGs with the best and worst forecasting performance, which are IFG 13 and IFG 11, respectively. We also integrated the average forecasting performance into the comparison as the reference substance.

The numbers in the table are centesimal. Based on the forecasting results, Fig. 4.8 demonstrates the forecasting performance of different methods, where blue and purple dashed lines represent the worst and best forecasting performance, respectively, in each month of 15 random IFGs from Sect. 4.6.2. The green line in the figure denotes SAS-SVECM forecasting performance.

Specifically, in 2014, with the same SVECM parameters $p = 5$ and $r = 1$, the forecasting error of SVECM without SAS fluctuates violently, which we can see

Table 4.4 SVECM forecasting performance with and without SAS

Methods/Month		Jan.	Feb.	Mar.	Apr.	May	Jun.	Jul.	Aug.	Sep.	Oct.	Nov.	Dec.	MAPE
Without SAS	Best	1.56	0.16	3.26	0.14	0.81	0.03	0.14	2.77	0.23	0.08	0.00	0.11	0.7743
	Worst	3.19	1.41	5.28	2.69	4.09	2.05	1.34	4.24	1.36	0.65	0.81	1.56	2.3889
With SAS		2.21	0.16	3.14	0.96	1.72	0.38	0.14	2.95	1.16	0.21	0.49	0.81	1.1941

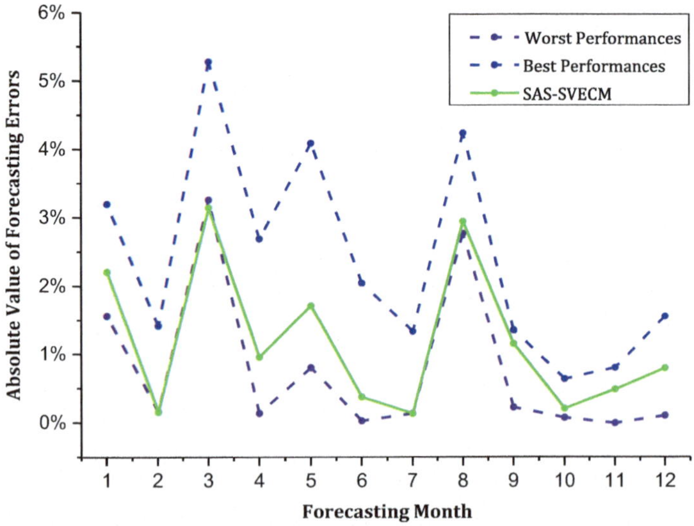

Fig. 4.8 SVECM forecasting performance with and without SAS

in Tables 4.3, 4.4 and Fig. 4.8, according to different IFG selection. The MAPE of SVECM with SAS is 1.19% according to self-adaptively screening the IFG selection. Figure 4.8 clearly shows that SAS-SVECM has similar performances to the best performances of each month, which demonstrates the efficiency of SAS and the forecasting accuracy of SAS-SVECM.

4.6.3.2 Sensitivity Analysis with Different Lag Order and Rank

To compare the forecasting performance of different VECM parameters, we applied lag order p and rank r of VECM in the forecasting model. The electricity consumption forecasting results from 2012 to 2014 are shown in Fig. 4.9.

As Fig. 4.9 shows, SVECM with different p and r achieves different performance in monthly electricity consumption forecasting, where lines denote real electricity consumption and predicted electricity consumption and circles denote relative forecasting errors. Although the electricity consumption prediction accuracy with various SVECM parameters differ slightly, they are basically able to remain consistent with the actual electricity consumption curve and achieve better forecasting performance than the method using only SVECM in Sect. 4.6.2.

Regarding the forecasting tests from 2012 to 2014, the months with relatively large forecasting errors generally appear in January and February, when the Chinese Spring Festival is celebrated, and July and August, when summer temperatures influence monthly electricity consumption by the increased use of air conditioners. For a more direct observation on SVECM with different parameters, the MAPE of electricity consumption forecasting in three testing years is listed in Table 4.5.

In the table, bold numbers indicate the smallest forecasting error among all SVECM parameter selections in each year, and * indicates the number used as the

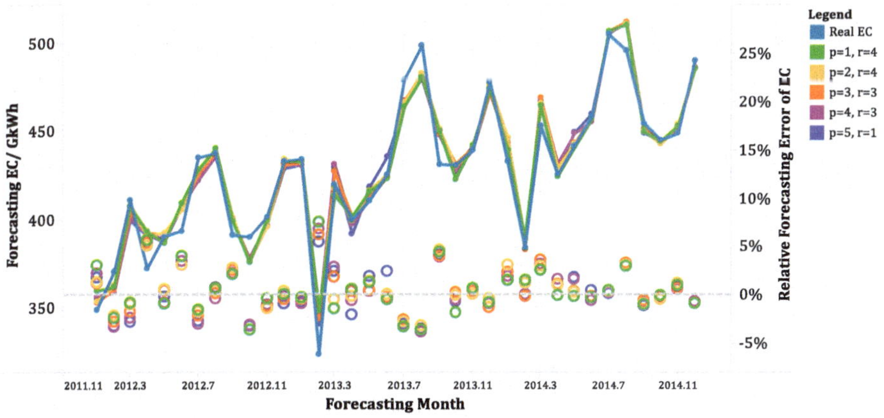

Fig. 4.9 The monthly electricity consumption forecasting comparison of VECM with different p and r parameters

Table 4.5 MAPE comparison of SVECM with different p and r parameters

Year	MAPE with different VECM parameters				
	$p = 1\ r = 4$	$p = 2\ r = 4$	$p = 3\ r = 3$	$p = 4\ r = 3$	$p = 5\ r = 1$
2012	2.10%	**1.94%**	2.07%	2.28%	2.28%
2013	2.18%	1.78%	**1.76%**	2.01%	2.31%
2014	**1.05%**	1.28%	1.15%	1.19%	1.19%*
Average	1.78%	1.67%	**1.66%**	1.83%	1.93%

comparison with the MAPE in Sect. 4.6.3.1. As shown in Table 4.5, although the forecasting results of the SVECM parameter selection, $p = 3$ and $r = 3$, do not achieve the highest accuracy in each testing year, they achieve the best forecast performance with the average MAPE and the most stable forecasting performance.

4.6.3.3 Sensitivity Analysis with Different SAS Training Settings

In this section, we choose lag order $p = 3$ and rank $r = 3$. To compare electricity consumption forecasting performance with different SAS training parameters, we applied different training months q and top N combinations to forecast monthly electricity consumption from 2012 to 2014.

As Fig. 4.10 shows, SAS with different q and N achieves different performance in monthly electricity consumption forecasting, where lines denote real electricity consumption and predicted electricity consumption and circles denote relative forecasting errors. The forecasting results of SAS-SVECM with different SAS parameters have slight differences, and the several months with large forecasting errors can be

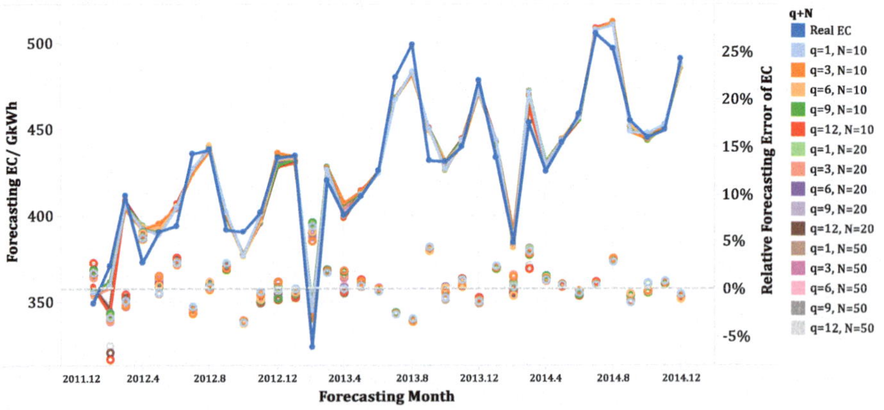

Fig. 4.10 The monthly electricity consumption forecasting comparison of SAS with different q and N parameters

Table 4.6 MAPE comparison of SAS with different q and N parameters

Top N Comb.	Year	MAPE with different training month numbers				
		1 month (%)	3 months (%)	6 months (%)	9 months (%)	12 months (%)
10	2012	1.93	2.10	2.15	2.10	2.58
	2013	1.85	1.83	1.69	2.05	1.98
	2014	1.24	1.27	1.22	1.20	1.10
	Average	**1.67**	**1.73**	**1.69**	**1.78**	**1.88**
20	2012	1.96	2.07	2.15	2.08	2.52
	2013	1.89	1.80	1.71	2.08	1.93
	2014	1.30	1.27	1.20	1.22	1.13
	Average	**1.72**	**1.71**	**1.69**	**1.79**	**1.86**
50	2012	2.01	2.08	2.07	2.06	2.41
	2013	1.92	1.82	1.76	2.04	1.91
	2014	1.31	1.29	1.15	1.21	1.14
	Average	**1.75**	**1.73**	**1.66**	**1.77**	**1.82**

explained in the same manner as in Sect. 4.6.3.1. For a more direct observation on SAS with different parameters, the MAPE of electricity consumption forecasting in three testing years is listed in Table 4.6.

As shown in Table 4.6, although the chosen selection of SAS parameter forecasting results does not achieve the highest accuracy in each testing year, it has the best forecasting performance with an average MAPE in three years of 1.66%. More importantly, this combination of SAS parameters has relatively stable forecasting performance in each year without forecasting error peaks and valleys, which have a greater significance in practice. In different years, the overall forecasting performance of SAS-SVECM is different, facing structural changes and growth pattern adjustments in the economy during these years. It is visible from the prediction accuracies that SAS-SVECM has sufficient ability to track economic alterations.

4.6.4 SAS-SVECM Forecasting Comparisons with Other Forecasting Methods

To determine the forecast performance of the optimized SAS-SVECM, we use several normal forecasting methods and the only SVECM to predict monthly electricity consumption from 2012 to 2014 in China. The forecasting results are shown in Table 4.7.

Table 4.7 MAPE of different forecasting methods

Forecasting method/MAPE	2012 (%)	2013 (%)	2014 (%)	Average (%)
Time series based on X-12-ARIMA	5.17	5.59	2.47	4.39
Growth equation method	4.48	3.65	3.96	4.03
Artificial neural network method	3.24	3.18	2.03	2.82
SVM with X-12-ARIMA	1.97	2.72	1.67	2.12
SVR with X-12-ARIMA	1.86	2.24	1.35	1.89
SAS-SVECM	**2.07**	**1.76**	**1.15**	**1.66**

The input EEFs of the first four forecasting methods include EXP, IMP, COP, CPI, PPI, and VRF, which are monthly data from 2000 to 2014 in China. The forecasting parameters and results of the first four methods can be obtained in the paper [33].

The first two forecasting methods place more emphasis on the electricity consumption time series, lacking the reaction of the variation in the economic situation, and they are unable to predict electricity consumption in times of economic change. Although the artificial neural network is able to take advantage of EEFs, it is difficult to forecast precisely given the lack of data quantity. SVM has become a popular model for monthly electricity consumption forecasting in recent years, which has relatively high accuracy forecasting performances with the given specific input EEFs and the help of X-12-ARIMA. Despite SVM has better performance than SAS-SVECM in 2012, its stability is relatively weak. As a generalized multi-dimensional SVM that uses a cost function with a hyperspherical intensive zone [2], SVR also performs well. It has the best forecasting performance in the first year among all methods. Compared to SVM, it has better performances in the last three years. Even though, its results in the last three years are not as good as SAS-SVECM proposed in this chapter. The table demonstrates the superiority and adaptability of the SAS-SVECM forecasting method.

Above all, the last three methods, which are SVM, SVR, and SAS-SVECM have the best forecasting performances among all. Analyzing the error composition of the three methods, we can see their forecasting errors of most of the months are really low and the obvious errors usually exist in the winter and summer, which can be seen in Table 4.8.

Table 4.8 Monthly forecasting MAPEs of the three methods in 2014 (%)

Month	1	2	3	4	5	6	7	8	9	10	11	12
SVM	3.75	1.72	2.44	1.51	2.10	2.42	2.21	1.20	0.92	0.94	0.54	0.33
SVR	2.23	1.15	2.56	1.24	1.89	1.64	0.89	1.14	1.07	1.42	0.56	0.42
SAS-SVECM	2.01	0.16	3.14	0.96	1.72	0.38	0.14	2.65	1.16	0.21	0.49	0.81

It can be concluded that besides the EEFs' influences, the temperatures in winter and summer also put a lot of impacts on electricity consumption. In general, with the ability to identify the most influential EEFs and select the best IFGs for different economic situations, SAS-SVECM has more stable forecasting performances and highly forecasting accuracy in total.

Despite the advantages of SAS-SVECM, it should be noted that the whole calculation of this relatively complex model will take a long time, which is dependent on the input datasets. In this chapter, it usually takes several hours to have a forecasting result, which should be improved as our next task.

4.7 Conclusions

This chapter proposes a novel monthly electricity consumption forecasting framework, SAS-SVECM, which comprises three procedures—X-12-ARIMA seasonal decomposition, VECM, and the self-adaptive screening method—to enhance the forecasting accuracy and adaptability to the changing economic environments and the expanding EEF dataset. The identified influencing EEFs and time lag effects between EEFs and electricity consumption are integrated into the SVECM. The proposed method SAS is able to address the problem of data types and data length by identifying the most influential IFGs as inputs for SVECM. Furthermore, an empirical analysis based on real electricity consumption and EEF data of China from 2000 to 2014 is performed. The results verify that the SAS-SVECM achieves better forecasting performance than the conventional forecasting methods. In summary, the proposed SAS-SVECM enhances the accuracy of monthly electricity consumption forecasting, accomplishes more stable forecasting performance, and gains the ability to select the most influential EEFs as forecasting inputs self-adaptively.

In the future, more data types should be further integrated into the model to enhance the model's performance. The different forecasting horizons should be considered. The forecasting adaptation of SAS-SVECM on other variables could be tested. Moreover, the method to enhance the calculation efficiency should be studied as well.

References

1. Shao, Zhen, Fei Gao, Qiang Zhang, and Shan-Lin. Yang. 2015. Multivariate statistical and similarity measure based semiparametric modeling of the probability distribution: A novel approach to the case study of mid-long term electricity consumption forecasting in China. *Applied Energy* 156: 502–518.
2. Zhongyi, Hu., Yukun Bao, Raymond Chiong, and Tao Xiong. 2015. Mid-term interval load forecasting using multi-output support vector regression with a memetic algorithm for feature selection. *Energy* 84: 419–431.

3. Ardakani, F.J., and M.M. Ardehali. 2014. Long-term electrical energy consumption forecasting for developing and developed economies based on different optimized models and historical data types. *Energy* 65: 452–461.
4. Hong, Tao, Jason Wilson, and Jingrui Xie. 2013. Long term probabilistic load forecasting and normalization with hourly information. *IEEE Transactions on Smart Grid* 5 (1): 456–462.
5. Hyndman, Rob J and Shu Fan. 2009. Density forecasting for long-term peak electricity demand. *IEEE Transactions on Power Systems* 25 (2): 1142–1153.
6. Hong, Tao, Pierre Pinson, Shu Fan, Hamidreza Zareipour, Alberto Troccoli, and Rob J Hyndman. 2016. Probabilistic energy forecasting: Global energy forecasting competition 2014 and beyond. *International Journal of Forecasting* 32 (3): 896–913.
7. Chen, Bo-Juen., Ming-Wei. Chang, et al. 2004. Load forecasting using support vector machines: A study on eunite competition 2001. *IEEE Transactions on Power Systems* 19 (4): 1821–1830.
8. Karanfil, Fatih, and Yuanjing Li. 2015. Electricity consumption and economic growth: exploring panel-specific differences. *Energy Policy* 82: 264–277.
9. Zhang, Chi, Kaile Zhou, Shanlin Yang, and Zhen Shao. 2017. On electricity consumption and economic growth in China. *Renewable and Sustainable Energy Reviews* 76: 353–368.
10. Al-mulali, Usama, Hassan Gholipour Fereidouni, and Janice YM Lee. 2014. Electricity consumption from renewable and non-renewable sources and economic growth. Evidence from latin American countries. *Renewable and Sustainable Energy Reviews* 30: 290–298.
11. Muhammad Shahbaz and Hooi Hooi Lean. 2012. The dynamics of electricity consumption and economic growth: A revisit study of their causality in Pakistan. *Energy* 39 (1): 146–153.
12. Croonenbroeck, Carsten, and Daniel Ambach. 2015. A selection of time series models for short-to medium-term wind power forecasting. *Journal of Wind Engineering and Industrial Aerodynamics* 136: 201–210.
13. Garcia-Ascanio, Carolina, and Carlos Maté. 2010. Electric power demand forecasting using interval time series: A comparison between var and imlp. *Energy Policy* 38 (2): 715–725.
14. Wang, Shaojian, Chuanglin Fang, Xingliang Guan, Bo. Pang, and Haitao Ma. 2014. Urbanisation, energy consumption, and carbon dioxide emissions in China: A panel data analysis of China's provinces. *Applied Energy* 136: 738–749.
15. Vaghefi, A, Mohsen A Jafari, Emmanuel Bisse, Yan Lu, and Jack Brouwer. 2014. Modeling and forecasting of cooling and electricity load demand. *Applied Energy* 136: 186–196.
16. De Felice, Matteo, Andrea Alessandri, and Franco Catalano. 2015. Seasonal climate forecasts for medium-term electricity demand forecasting. *Applied Energy* 137: 435–444.
17. Vu, Dao Hoang, Kashem M Muttaqi, and AP Agalgaonkar. 2015. A variance inflation factor and backward elimination based robust regression model for forecasting monthly electricity demand using climatic variables. *Applied Energy* 140: 385–394.
18. Sanstad, Alan H, Stuart McMenamin, Andrew Sukenik, Galen L Barbose, and Charles A Goldman. 2014. Modeling an aggressive energy-efficiency scenario in long-range load forecasting for electric power transmission planning. *Applied Energy* 128: 265–276.
19. Mavromatidis, Lazaros Elias, Anna Bykalyuk, and Hervé Lequay. 2013. Development of polynomial regression models for composite dynamic envelopes' thermal performance forecasting. *Applied Energy* 104: 379–391.
20. Hor, Ching-Lai, Simon J Watson, and Shanti Majithia. 2005. Analyzing the impact of weather variables on monthly electricity demand. *IEEE Transactions on Power Systems* 20 (4): 2078–2085.
21. Bunnoon, Pituk, Kusumal Chalermyanont, and Chusak Limsakul. 2012. Mid-term load forecasting: Level suitably of wavelet and neural network based on factor selection. *Energy Procedia* 14: 438–444.
22. Maia, André Luis Santiago, and Francisco de AT de Carvalho. 2011. Holt's exponential smoothing and neural network models for forecasting interval-valued time series. *International Journal of Forecasting* 27 (3): 740–759.
23. Tsekouras, George J, Nikos D Hatziargyriou and Evangelos N Dialynas. 2006. An optimized adaptive neural network for annual midterm energy forecasting. *IEEE Transactions on Power Systems* 21 (1): 385–391.

24. Jain, Rishee K, Kevin M Smith, Patricia J Culligan and John E Taylor. Forecasting energy consumption of multi-family residential buildings using support vector regression: Investigating the impact of temporal and spatial monitoring granularity on performance accuracy. *Applied Energy* 123: 168–178.
25. Ghelardoni, Luca, Alessandro Ghio, and Davide Anguita. 2013. Energy load forecasting using empirical mode decomposition and support vector regression. *IEEE Transactions on Smart Grid* 4 (1): 549–556.
26. Sánchez-Fernández, Matilde, Mario de Prado-Cumplido, Jerónimo Arenas-García, and Fernando Pérez-Cruz. 2004. Svm multiregression for nonlinear channel estimation in multiple-input multiple-output systems. *IEEE Transactions on Signal Processing* 52 (8): 2298–2307.
27. Zhou, Kaile, Fu. Chao, and Shanlin Yang. 2016. Big data driven smart energy management: From big data to big insights. *Renewable and Sustainable Energy Reviews* 56: 215–225.
28. Al-Saba, Tawfiq, and Ibrahim El-Amin. 1999. Artificial neural networks as applied to long-term demand forecasting. *Artificial Intelligence in Engineering* 13 (2): 189–197.
29. Engle, Robert F and Clive WJ Granger. 1987. Co-integration and error correction: Representation, estimation, and testing. *Econometrica: Journal of the Econometric Society* 251–276.
30. Lai, TM, Wai Ming To, WC Lo, YS Choy and KH Lam. 2011. The causal relationship between electricity consumption and economic growth in a gaming and tourism center: The case of macao sar, the People's Republic of China. *Energy* 36 (2): 1134–1142.
31. Apergis, Nicholas and James E Payne. 2011. Renewable and non-renewable electricity consumption–growth nexus: evidence from emerging market economies. *Applied Energy*, 88 (12): 5226–5230.
32. Findley, David F, Brian C Monsell, William R Bell, Mark C Otto and Bor-Chung Chen. 1998. New capabilities and methods of the x-12-arima seasonal-adjustment program. *Journal of Business & Economic Statistics* 16 (2): 127–152.
33. Guo, Hongye, Qixin Chen, Qing Xia and H Fan. 2016. Study on mid-term electricity load forecast considering time lag effects of economic factors. *Power System Technology* 40 (2): 514–520.
34. Antoniadis, Anestis, Xavier Brossat, Yannig Goude, Jean-Michel Poggi and Vincent Thouvenot. 2016. Automatic component selection in additive modeling of french national electricity load forecasting. In *Nonparametric Statistics*, 191–209. Springer.
35. Che, Jinxing, Youlong Yang, Li. Li, Xuying Bai, Shenghu Zhang, and Chengzhi Deng. 2017. Maximum relevance minimum common redundancy feature selection for nonlinear data. *Information Sciences* 409: 68–86.
36. Dickey, David A and Wayne A Fuller. 1981. Likelihood ratio statistics for autoregressive time series with a unit root. *Econometrica: Journal of the Econometric Society* 1057–1072.
37. Ma, Tao, Zhou Zhou, and Baher Abdulhai. 2015. Nonlinear multivariate time-space threshold vector error correction model for short term traffic state prediction. *Transportation Research Part B: Methodological* 76: 27–47.

Chapter 5
Probabilistic Load Forecasting

5.1 Introduction

The electrical load is affected by various factors such as weather conditions, distributed renewable integration, demand response implementation, energy policy, emergent events, etc. Traditional load forecasting only provides an expected value of future load. In modern power systems, characterizing the uncertainties of future load will benefit the reliability and economy of the whole system. More and more probabilistic load forecasting (PLF) methods have been proposed in recent years [1].

The whole process of load forecasting includes data preprocessing, feature engineering, model establishment, and optimization, result analysis and visualization. The quantile regression model is the most frequently used tool for PFL. An embedding-based quantile regression neural network (QRNN) is proposed in [2] for PLF where the discrete variables such as day types, days of the week, and hours of the day are modeled by the embedding matrix. Another improved QRNN was presented in [3]. The network involves Gaussian noise layers and dropout layers to overcome the overfitting issue. A "separate-forecasting-integrate" framework was proposed in [4] to forecast the net load with high penetration of behind-the-meter PV. The decomposed parts are forecasted by QRNN, quantile random forest, or quantile gradient boosting regression tree (GBRT); while the dependencies between load and PV uncertainties are modeled by copula function and describe dependent convolution (DDC) [5]. Quantile, density, and interval are the three main forms of probabilistic forecasts. Kernel density estimation can make a bridge between quantile forecast and density forecast. The triangle kernel function was used in [6] to transform the quantiles obtained by QRNN to density forecasts.

Since there have already existed various point load forecasting methods, it is a good idea to transform the point forecasts into probabilistic forecasts. A temperature scenario generation approach was proposed in [7], where various temperature scenarios are fed into the point forecasting model to generate various point forecasts and then the quantiles can be calculated based on the point forecasts. A simple quantile regression averaging model was established on multiple point forecasts generated by

Q. Chen et al., *Data Analytics in Power Markets*,
https://doi.org/10.1007/978-981-16-4975-2_5

sister forecasting models in [8]. A deep investigation of point forecasting residual was conducted in [9]. A residual simulation was implemented based on the normality assumption to produce probabilistic forecasts. Instead of focusing on specific PLF model, an ensemble approach was proposed in [10] to combine the quantile probabilistic forecasts of multiple models. The combining model is formulated as a linear programming (LP) problem to minimize the pinball loss of the final model.

Probabilistic load forecasting has also been applied for individual households in addition to system-level load [11]. Conditional kernel density estimation was used in [12] to model the conditional distribution of residential load on different time periods. In [13], the uncertainty within smart meter data was forecasted by boosting additive quantile regression models. Case studies show the superiority of the additive model compared with the normality assumption-based method. Gaussian process regression (GPR) was applied in [14] for residential load forecasting, where different forms of kernels were compared.

Most of the load forecasting literature focused on the establishment and optimization of the forecasting model; while, very few research focuses on the preprocessing and feature engineering phase, especially for PLF. Feature selection, as an important part of feature engineering, tries to select the most useful or relevant features from the original data to enhance the accuracy of load forecasts and increase the interpretability of the forecasting model. When the number of original features is too large, it can also be used to reduce the solving complexity of the model. Most of the research areas of feature selection are concentrated in the traditional point forecasting, such as the least absolute shrinkage and selection operator (LASSO) method [15]. However, feature selection in the field of probabilistic forecasting has rarely been studied. To the best of our knowledge, there is only one feature selection work for PLF introduced in [16] by varying the number of lag hours and lag days of the temperature. The proposed global feature selection method is quite similar to the exhaustive method. The method was tested in seven states of the United States. The main idea is to evaluate the performance of feature selection using pinball loss instead of root mean square error (RMSE).

In this chapter, we enrich the PLF feature selection literature by proposing a novel sparse penalized quantile regression method. To tackle the challenges of large computation burden and non-differentiable issue, Alternating Direction Method of Multipliers (ADMM) [17] is applied to decompose the optimization problem. The results show that the proposed feature selection method can improve the performance of the original probabilistic forecasting model compared with the original model and heuristic model.

The main contributions of this chapter can be summarized as follows:

1. Proposing a feature selection method for probabilistic load forecasting by introducing L_1-norm sparse penalty into quantile regression model;
2. Designing an ADMM algorithm to decouple the large scale optimization problem and boost the efficiency of the model training process;
3. Validating the superiority of the proposed sparse penalized quantile regression method by numerical simulation based on the cases from the open datasets.

The remainder of this chapter is organized as follows: Sect. 5.2 provides the used dataset and the regression models for load forecasting; Sect. 5.3 introduces a straightforward feature selection method which is used as a benchmark in the case studies; Sect. 5.4 introduces the proposed feature selection method and the ADMM based training method; Sect. 5.5 presents the implementation details of our proposed method; Sect. 5.6 conducts the case studies on the open datasets from Global Energy Forecasting Competition in the year of 2012 (GEFCom 2012).

5.2 Data and Model

5.2.1 Load Dataset Exploration

The load data used in this chapter is the open dataset used for GEFCom 2012 [18]. The dataset includes the 5 years of hourly load data of 20 regional power systems of North Carolina from 2004 to 2008. The corresponding hourly temperature data from 11 weather stations are also provided.

Figure 5.1 shows the four-year load data of the first zone and temperature data from the first weather station from 2004/01/01 to 2007/12/31. The load and weather show clear periodicity as well as large variations. We can also find an approximate quadratic relationship between the load and temperature data. To produce highly accurate forecasts, we need to carefully model how calendar dates and temperature influence the load.

5.2.2 Linear Regression Model Considering Recency-Effects

This section mainly introduces the forecasting model and the input features to be selected. The temperature has a complex impact on the electrical load. Both the current and past temperature have an influence on the load because of the inertia of air temperature change and consumers' perception. For example, when the temperature rises suddenly at a certain time, the consumers in this area do not often respond immediately, but take the cooling measures for several minutes or hours, such as opening the air conditioners. Thus, the load sequence may lag behind the temperature sequence. We call it the lag effect or recency-effect. In addition, the way how temperature affects load may also change during different days of the year and hours of the day. That is to say, there is a cross-effect between temperature and calendar date.

Multiple linear regression models, are widely used in the field of load forecasting by connecting the characteristic variables and dependent load. They can generate accurate load forecasts without consuming abundant computation resources. Since the load is highly dependent on temperature as well as calendar variables, such

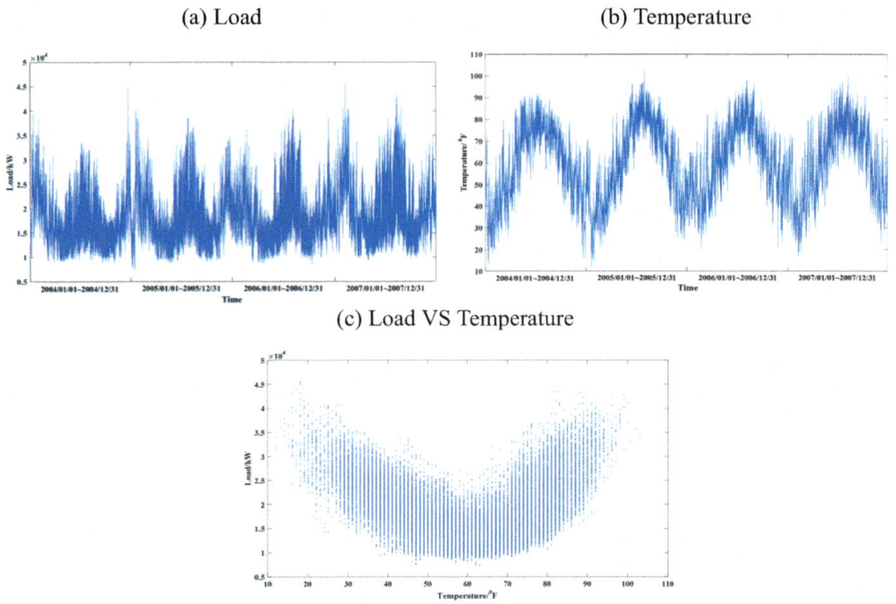

Fig. 5.1 Load and temperature data from 2004/01/01 to 2007/12/31

as hour, month, and week, Tao [19] proposed a naive vanilla benchmark model considering cross-effects of variables mentioned above corresponding to the time of load being forecasted. It can be formulated as:

$$
\begin{aligned}
\hat{y}_t &= \beta_0 + \beta_1 Trend_t + \beta_2 M_t + \beta_3 W_t \beta_4 H_t \\
&\quad + \beta_5 W_t H_t + f(T_t) \\
f(T_t) &= \beta_6 T_t + \beta_7 T_t^2 + \beta_8 T_t^3 + \beta_9 T_t M_t + \beta_{10} T_t^2 M_t \\
&\quad + \beta_{11} T_t^3 M_t + \beta_{12} T_t H_t + \beta_{13} T_t^2 H_t + \beta_{14} T_t^3 H_t
\end{aligned}
\tag{5.1}
$$

where $\hat{y}_t$ denotes the estimated load at time t; T_t denotes the temperature at time t, M_t,W_t and H_t stand for month-of-the-year, day-of-the-week, and hour-of-the-day classification variables corresponding to time t, respectively; $Trend_t$ denotes a series of integer numbers (i.e., 1, 2, 3, ...) to describe the long-term increase of the load; $f(\cdot)$ is a function of T_t that accounts for the polynomial relationship between temperature and load and the cross-effects among temperature, month-of-the-year, and hour-of-the-day.

An improved multiple linear regression model considering the recency-effect was proposed in [20], which takes the moving historical average temperature, the lagged temperature, and its interaction with the calendar variables into the regression model. The model with recency-effect can improve the performance of load forecast. It is expressed as follows:

$$\hat{y}_t = \beta_0 + \beta_1 Trend_t + \beta_2 M_t + \beta_3 W_t + \beta_4 H_t + \\ \beta_5 W_t H_t + f(T_t) + \underbrace{\sum_{d=1}^{N_D} f(\widetilde{T}_{t,d}) + \sum_{h=1}^{N_H} f(T_{t-h})}_{\text{recency effects}} \tag{5.2}$$

where, the last two terms represent the recency-effect; N_D and N_H denotes the number of days and hours of the lag temperature that will be considered as the recency-effect, respectively; d and h denote the index for the lag days and hours. The moving historical average temperature $\widetilde{T}_{t,d}$ is calculated as follows:

$$\widetilde{T}_{t,d} = \frac{1}{24} \sum_{h=24d-23}^{24d} T_{t-h}, \quad d = 1, 2, , ..., N_D \tag{5.3}$$

Thus, the model considering recency-effect in (5.2) can be neatly presented as follows:

$$\hat{y}_t = \boldsymbol{\beta}^\top \mathbf{X}_t \tag{5.4}$$

where, $\mathbf{X}_t$ denotes a collection of all the features; $\boldsymbol{\beta}$ denotes the vector of the coefficients to be trained.

The number of features N_F depends on the number of lag hours N_D and number of lag days N_H. The month-of-the-year M_t, day-of-the-week W_t, and hour-of-the-day H_t are all presented by the dummy variables. The dummy coding uses all-zero vector $\mathbf{0}$ to present one category of the classification variables. Thus, the dummy encoding method is one dimension smaller than that of one-hot encoding method. It can guarantee the final feature matrix is a full-rank matrix after adding a all-one column representing the intercept. The month-of-the-year, day-of-the-week, and hour-of-the-day need to be represented by 11, 6, and 23 classification variables. Thus, the total number of feature N_F is:

$$\begin{aligned} N_F &= 1 + 11 + 6 + 23 + 23 \times 6 \\ &+ (3 + 3 \times 11 + 3 \times 23)(1 + N_D + N_H) \\ &= 284 + 105(N_D + N_H) \end{aligned} \tag{5.5}$$

If we consider the temperature of lag 7 days and 12 h, the total number of features is 2279, which makes the regression model a very high dimensional problem and results in high computational burden. This is the main reason why to conduct feature selection. In the following two sections, two Lasso-based feature selection methods (Pre-Lasso and Quantile-Lasso) are introduced.

5.3 Pre-Lasso Based Feature Selection

This section first introduces a benchmark for feature selection named Pre-Lasso. The main idea of Pre-Lasso is to select the features based on the point forecasting model and then use the selected features for quantile model training.

A forecasting model is trained to minimize the total loss:

$$\hat{\boldsymbol{\beta}} = \arg\min_{\boldsymbol{\beta}} \sum_{t=1}^{N_T} l(r_t);\ r_t = y_t - \boldsymbol{\beta}^\top \mathbf{X}_t \tag{5.6}$$

where r is the fitting residual; $l(r)$ is the loss function. For traditional point forecasting, the loss function is square error ($l(r) = r^2$).

Lasso is an efficient and mature compression estimation method for feature selection and regularization [21]. It adds the L_1-norm sparse penalty to the original loss function of the regression model:

$$\hat{\boldsymbol{\beta}} = \arg\min_{\boldsymbol{\beta}} \sum_{t=1}^{N_T} l(r_t) + \lambda \|\boldsymbol{\beta}\|_1;\ r_t = y_t - \boldsymbol{\beta}^\top \mathbf{X}_t \tag{5.7}$$

where, N_T denotes the length of all time periods; $\|\boldsymbol{\beta}\|_1$ denotes the L_1-norm sparse penalty term; λ is the weight for the sparse penalty and can be determined by cross-validation. L_1-norm penalty can force the optimization process to change some regression coefficients to 0 or make the vector $\boldsymbol{\beta}$ sparse. The features with 0 value coefficients will be filtered out, thus can be regarded as feature selection.

Since feature selection has been widely studied for point load forecasting, a very intuitive way is to conduct feature selection using traditional Lasso and then use the selected features for quantile regression. We call this approach as Pre-Lasso, that is to say, the features are selected before training the probabilistic forecasting model.

The Pre-Lasso method can be divided into two stages. The first stage is to select features using traditional point forecasting based Lasso:

$$\hat{\boldsymbol{\beta}} = \arg\min_{\boldsymbol{\beta} \in \mathbb{R}^{N_f}} \sum_{t=1}^{N_T} (y_t - \boldsymbol{\beta}^\top \mathbf{X}_t)^2 + \lambda \|\boldsymbol{\beta}\|_1 \tag{5.8}$$

The problem in (5.8) is solved using Least Angle Regression (LARS) method [22].

The second stage is to conduct quantile regression based on the selected features:

$$\hat{\boldsymbol{\beta}}_{2,q} = \arg\min_{\boldsymbol{\beta}_{2,q} \in \mathbb{R}^{k_2}} \sum_{t=1}^{N_T} (\rho_q(y_t, \boldsymbol{\beta}_{2,q}^\top \mathbf{X'}_t)),\quad q = 1, 2, ..., Q \tag{5.9}$$

where, $\mathbf{X}'_t$ denotes the feature vector that have been selected in the first stage; $\boldsymbol{\beta}_{2,q}$ denotes the coefficient vector for the q-th quantile; Q denotes the number of quantiles to be forecasted. Note that the quantile regression model is trained individually for each quantile.

5.4 Sparse Penalized Quantile Regression (Quantile-Lasso)

The Pre-Lasso method introduced in the last section is straightforward and easily-implemented by directly applying traditional Lasso. This method has two drawbacks: (1) Pre-Lasso directly selects the input features according to the performance of the point load forecasting instead of the performance of the probabilistic forecasting. Different supervised metrics may lead to different selected features. (2) It is unreasonable to use the same selected features for all quantile regression models. Feature selection should be conducted individually for each quantile.

In this section, we propose a sparse penalized quantile regression method to select the features by directly modifying the objective function of the quantile regression model. To distinguish our method from Pre-Lasso, we name this method as Quantile-Lasso.

5.4.1 *Problem Formulation*

Traditional quantile regression model is to optimize the parameter $\boldsymbol{\beta}_q$ to minimize the total pinball loss:

$$\hat{\boldsymbol{\beta}}_q = \arg\min_{\boldsymbol{\beta}_q} \sum_{t=1}^{N_T} \rho_q(r_{q,t});\ r_{q,t} = y_t - \boldsymbol{\beta}_q^{\top}\mathbf{X}_t \tag{5.10}$$

where, $r_{q,t}$ denotes the q-th quantile error; for the q-th quantile load forecasting, the loss function is the pinball loss:

$$\rho_q(r_{q,t}) = \begin{cases} (1-q)r_{q,t} & r_{q,t} \leq 0 \\ q r_{q,t} & r_{q,t} > 0 \end{cases} \tag{5.11}$$

The Quantile-Lasso method can be easily formulated by adding an L_1-norm penalty into the objective function of the quantile regression:

$$\hat{\boldsymbol{\beta}}_q = \arg\min_{\boldsymbol{\beta}_q} \sum_{t=1}^{N_T} \rho_q(r_{q,t}) + \lambda_q \|\boldsymbol{\beta}_q\|_1;\ r_{q,t} = y_t - \boldsymbol{\beta}_q^{\top}\mathbf{X}_t \tag{5.12}$$

where, λ_q is the weight for the sparse penalty of the q-th quantile. For different quantiles, the best values of λ_q are different. Quantile-Lasso shares a similar strategy for feature selection with traditional Lasso. Quantile-Lasso model in (5.12) is a special form of (5.7) by substituting the loss function $l(r_t)$ with pinball loss $\rho_q(r_{q,t})$.

Since the pinball loss and L_1-norm are convex, it is easy to prove that the model in (5.12) is a convex optimization problem.

Even though the Quantile-Lasso model can be neatly represented like traditional Lasso, solving the optimization problem is not a trivial task. There are two main challenges:

1. Since the number of training samples and the number of features to be selected are very large, the feature selection process is cast to a big data problem and a large-scale convex optimization problem.
2. Both the pinball loss and L_1-norm are not differentiable everywhere. It is hard to use the traditional gradient descent-based optimization method to solve the problem.

5.4.2 ADMM Algorithm

We tackle the above two challenges by using ADMM to decompose each iteration of the large-scale convex optimization problem into two sub-optimization problems. The two sub-optimization problems can be solved using off-the-shelf methods.

ADMM can efficiently solve the optimization problem in the form of:

$$\begin{aligned} \min \quad & f(\mathbf{r}) + g(\boldsymbol{\beta}) \\ \text{s.t.} \quad & A\mathbf{r} + B\boldsymbol{\beta} = C \end{aligned} \tag{5.13}$$

where, $\mathbf{r}$ and $\boldsymbol{\beta}$ are decision variables; $f(\cdot)$ and $g(\cdot)$ are convex functions; A, B, and C are constant variables in the linear constraint. The Quantile-Lasso model for each quantile in (5.12) has the same form as (5.13).

The augmented Lagrangian function of (5.12) can be written as follows:

$$\begin{aligned} L_\gamma(\mathbf{r}, \boldsymbol{\beta}, \mathbf{u}) =& \rho_q(\mathbf{r}) + \lambda \|\boldsymbol{\beta}\|_1 + \mathbf{u}^\top(\mathbf{y} - \boldsymbol{\beta}^\top \mathbf{X} - \mathbf{r}) \\ &+ \frac{\gamma}{2}\|\mathbf{y} - \boldsymbol{\beta}^\top \mathbf{X} - \mathbf{r}\|_2^2 \end{aligned} \tag{5.14}$$

where, we omit subscript q and t in (5.12) for simplicity; $\mathbf{u}$ is the dual variable; γ is a defined positive constant to control the step of each iteration; $\mathbf{y}$ and $\mathbf{r}$ are vector form of y_t and $r_{q,t}$, respectively; $\rho_q(\mathbf{r})$ denotes the sum of all pinball loss, i.e., $\rho_q(\mathbf{r}) = \sum_{t=1}^{N_T} \rho_q(r_{q,t})$.

ADMM takes advantages of the decomposability of dual ascent and superior convergence properties of the multipliers [17]. The basic idea of ADMM is to minimize the values of two original decision variables $\mathbf{r}$ and $\boldsymbol{\beta}$, as well as update the dual variables. In this way, the augmented Lagrangian function $L_\gamma(\mathbf{r}, \mathbf{u}, \boldsymbol{\beta})$ decrease gradually. Thus, ADMM consists of two sub-optimization problems and one update in each iteration:

$$\begin{aligned}\boldsymbol{\beta}^{k+1} &:= \arg\min_{\boldsymbol{\beta}} L_\gamma(\boldsymbol{\beta}, \mathbf{r}^k, \mathbf{u}^k)\\ \mathbf{r}^{k+1} &:= \arg\min_{\mathbf{r}} L_\gamma(\boldsymbol{\beta}^{k+1}, \mathbf{r}, \mathbf{u}^k)\\ \mathbf{u}^{k+1} &:= \mathbf{u}^k + \gamma(\mathbf{y} - \mathbf{X}\boldsymbol{\beta}^{k+1} - \mathbf{r}^{k+1})\end{aligned} \tag{5.15}$$

If we define $\mathbf{s} = \mathbf{y} - \boldsymbol{\beta}^\top \mathbf{X} - \mathbf{r}$, then we have

$$\mathbf{u}^\top \mathbf{s} + \frac{\gamma}{2}\|\mathbf{s}\|_2^2 = \frac{\gamma}{2}\|\mathbf{s} + (1/\gamma)\mathbf{u}\|_2^2 - \frac{1}{2\gamma}\|\mathbf{u}\|_2^2 \tag{5.16}$$

Thus, the update of $\boldsymbol{\beta}$ can be rewritten as follows:

$$\begin{aligned}\boldsymbol{\beta}^{k+1} &:= \arg\min_{\boldsymbol{\beta}} \lambda\|\boldsymbol{\beta}\|_1 + \mathbf{u}^{(k)T}\mathbf{s} + \frac{\gamma}{2}\|\mathbf{s}^{(k)}\|_2^2\\ &\arg\min_{\boldsymbol{\beta}} \lambda\|\boldsymbol{\beta}\|_1 + \frac{\gamma}{2}\|\mathbf{s} + \frac{1}{\gamma}\mathbf{u}^k\|_2^2 - \frac{1}{2\gamma}\|\mathbf{u}^k\|_2^2\\ &\arg\min_{\boldsymbol{\beta}} \lambda\|\boldsymbol{\beta}\|_1 + \frac{\gamma}{2}\|\mathbf{y} - \mathbf{X}\boldsymbol{\beta} - \mathbf{r}^k + \frac{1}{\gamma}\mathbf{u}^k\|_2^2\end{aligned} \tag{5.17}$$

The sub-optimization problem in (5.17) has the same form as (5.8), which can also be solved using LARS method [22].

The update of $\mathbf{r}$ can be rewritten as follows:

$$\begin{aligned}\mathbf{r}^{k+1} &:= \arg\min_{\mathbf{r}} \rho_q(\mathbf{r}) + \mathbf{u}^{(k)T}\mathbf{s} + \frac{\gamma}{2}\|\mathbf{s}\|_2^2\\ &\arg\min_{\mathbf{r}} \rho_q(\mathbf{r}) + \frac{\gamma}{2}\|\mathbf{s} + (1/\gamma)\mathbf{u}^k\|_2^2 - \frac{1}{2\gamma}\|\mathbf{u}^k\|_2^2\\ &\arg\min_{\mathbf{r}} \rho_q(\mathbf{r}) + \frac{\gamma}{2}\|\mathbf{y} - \mathbf{X}\boldsymbol{\beta}^{k+1} - \mathbf{r} + \frac{1}{\gamma}\mathbf{u}^k\|_2^2\end{aligned} \tag{5.18}$$

The sub-optimization problem in (5.18) has the colse-form solution by using subdifferential calculus:

$$\mathbf{r}^{k+1} := S_{1/\gamma}(\mathbf{y} - \mathbf{X}\boldsymbol{\beta}^{k+1} + \mathbf{u}^k/\gamma - (2\mathbf{q} - \mathbf{1})/\gamma) \tag{5.19}$$

where, $\mathbf{q}$ and $\mathbf{1}$ are $N_T \times 1$ vectors with all the same element q and q, respectively; S denotes the soft thresholding operator, which is defined as:

$$S_a(b) = \begin{cases} b - a & b > a\\ 0 & |b| \le a\\ b + a & b < -a\end{cases} \tag{5.20}$$

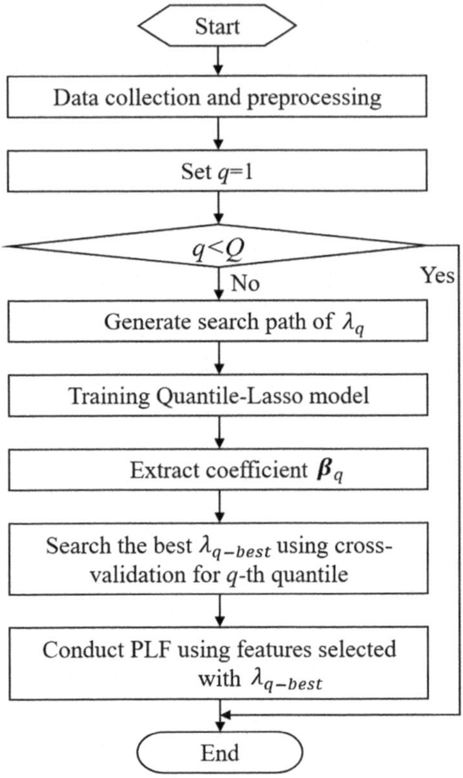

Fig. 5.2 Procedures for Quantile-Lasso implementation

To summarize, the large-scale Quantile-Lasso model is decomposed into two sub-optimization problems where one can be solved using the LARS method; the other one has a close-form solution. In this way, the Quantile-Lasso model can be solved efficiently and the global optimum can be obtained.

5.5 Implementation

This section will introduce the process to implement the proposed feature selection method for probabilistic forecasting. The whole procedures are shown in Fig. 5.2.

First, we collect the historical load data and its corresponding temperature data. Data preprocessing including data cleaning, normalization, and dataset split is also conducted. The load data are cleaned by exploring the relationship between load and temperature and detecting the sudden changes. Details are provided in our previous work [2]. All the features are normalized into [0, 1] using min-max scaling. The whole dataset is split into three parts for training, validation, and testing, respectively.

For the Quantile-Lasso method, feature selection is conducted individually for each quantile q. That is to say, the Quantile-Lasso method is implemented for Q times. Take the q-th quantile as an example, we generate the search path of λ_q, which includes a number of possible values of the adjustment parameter λ_q. Then, we conduct line search according to the search path of λ_q. For each possible value of λ_q, the ADMM algorithm proposed in Sect. 5.4.2 is employed to train the Quantile-Lasso model in (5.12). Different possible values of λ_q produce different coefficients $\boldsymbol{\beta}_q$, and thus result in different selected features. We evaluate the performance of the selected features and trained models using pinball loss on the validation dataset for each value λ_q. In this way, we can search the optimal adjustment parameter $\lambda_{q,best}$ according to their performances. Finally, we conduct PLF using the selected features with the best penalty coefficient $\lambda_{q,best}$ on the testing dataset and record their performance for comparison with other benchmarks such as the Pre-Lasso method.

The Pre-LASSO method has similar implementation procedures to the Quantile-Lasso method. The search path of λ is also generated and tested one by one using the validation dataset. However, the optimal adjustment parameter λ_{best} is searched according to RMSE instead of pinball loss on the validation dataset. After determining the optimal adjustment parameter λ_{best}, we retain the features with non-zero coefficients to train the quantile regression model on the validation dataset and test the performance on the testing dataset. For the Pre-Lasso method, the quantile regression model for different quantiles uses the same selected features.

After the two methods have been trained and validated, we compare the performances of the two methods on the same testing dataset in terms of pinball loss.

5.6 Case Study

5.6.1 Experiment Setups

The load and temperature data used in the case studies are from GEFCom 2012, of which the basic information is introduced in Sect. 5.2. We choose three-years load and temperature data from 2005 to 2007 as the training dataset, the first half-year data of 2008 as the validation dataset, and the second half-year data of 2008 as the test dataset. That is to say, the final performance of the load forecasting is evaluated on the second half-year data of 2008.

We use an average quantile score (AQS) to evaluate the performance of the proposed and competing method. AQS is defined as the average of all the quantiles' pinball loss:

$$AQS = \frac{1}{Q \times N_T} \sum_{q=1}^{Q} \sum_{t=1}^{N_T} \rho_q(\hat{y}_{q,t} - y_t) \tag{5.21}$$

where, ρ_q is the pinball loss defined in (5.11); $\hat{y}_{t,q}$ and y_t denote the forecasted q-th quantile and real value of the load, respectively. A total of 9 quantiles 0.1, 0.2, 0.3... 0.9, which are denoted as $q_1, q_2, q_3...q_9$, are used to form the probabilistic forecasts in this chapter.

Since the proposed feature selection is designed for linear regression models, two base competing methods are original linear quantile regression and the linear quantile regression based on Pre-Lasso. The base forecasting model is illustrated in Sect. 5.2 to consider the recency-effects of temperature on loads. There are two variables to be determined: the number of lag days N_D and the number of lag hours N_H. We choose two variable pairs (N_D, N_H) as (3, 4) and (7, 12) respectively, in our case studies, which are denoted as D3-H4 model and D7-H12 model. The search path of λ_q in the Quantile-Lasso model and λ in the Pre-Lasso model for the L_1-norm penalty is the same for both the D3-H4 model and D7-H12 model. If the recency-effect model is the D3-H4 model or D7-H12 model, we have 1019 and 2279 features to be selected, respectively, without consideration of intercept according to (5.5). Neural networks and GBRT are commonly used and powerful regression models for load forecasting. They have been widely used in GEFCom 2012 and 2014 [1, 18]. To show the superiority of the MLR model for load forecasting, these two nonlinear quantile regression models, quantile regression neural network (QRNN) [23] and quantile gradient boosting regression tree (QGBRT) [24] with default parameters are also tested for comparisons.

5.6.2 *Results*

Figures 5.3 and 5.4 present the changes of AQS on testing dataset with variations of λ for two randomly selected areas (Area #6, and Area #9) using Pre-Lasso. The search path of λ varies from 0.0001 to 100, i.e., $-\log(\lambda)$ varies from -2 to 4. The blue lines represent the number of features that have been selected (NoS). With the decrease of λ, i.e., the increase of $-\log(\lambda)$, NoS shows clear increase trends for all cases. When $-\log(\lambda) > 3$, NoS is close to the number of original features.

The red lines represent the performances of the quantile regression model with selected features on the testing dataset. A lower AQS means a better performance of the model. The four red lines of two models in two areas have similar rough trends: decrease first, then oscillate before getting a stable value. The final stable value of the red lines corresponding to the case of $-\log(\lambda) = 4$ can be viewed as the performance of the model without feature selection. It means that we can find a value of λ that corresponds to the minimum of AQS before the AQS gets stable according to the cross validation. The black dot line marks the optimal value of $-\log(\lambda)$ that produces the minimum of AQS. The optimal values of adjustment parameter $-\log(\lambda)$ are different for different models and in different areas.

For the D3-H4 model in Area #6 and Area #9, the numbers of selected features with the optimal adjustment parameters of Pre-Lasso are about 550 and 300, respectively, which is only 55% and 30% of the number of original features. We can see only

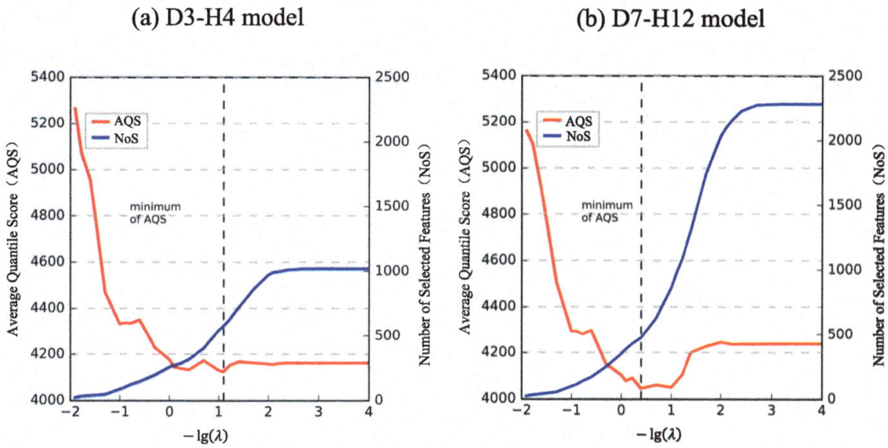

Fig. 5.3 Changes of AQS on testing dataset with variation of λ for Area #6

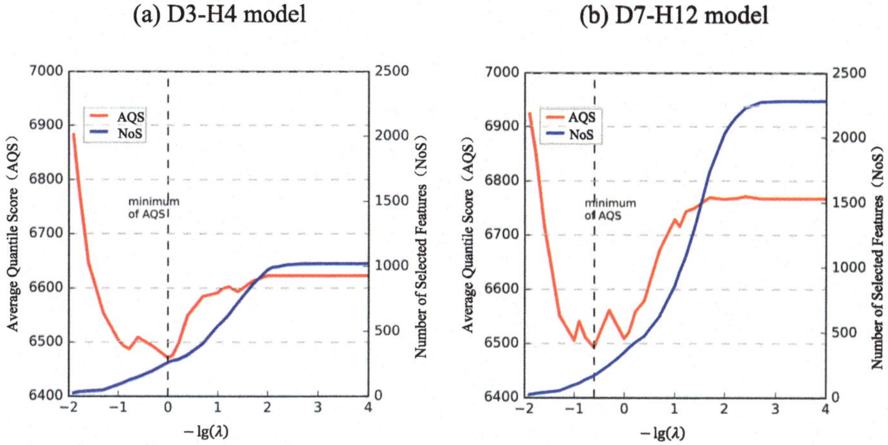

Fig. 5.4 Changes of AQS on testing dataset with variation of λ for Area #9

0.9% and 3.2% improvements in terms of AQS. However, for the D7-H12 model, only 20% and 10% of the original features have been selected for Area #6 and Area #9, and gain 4.5% and 4.1% improvements, respectively.

D3-H4 model-based feature selection in Area #6 fails to bring a significant improvement on the AQS performance. D7-H12 model-based feature selection gains greater improvement compared with the D3-H4 model. A possible reason is that the D7-H12 model contains more recency-effects of temperature and the effective features are selected. The Pre-Lasso method selects different features and has different improvements for different areas and different models, because the loads in different areas have different responsiveness on the current and lag temperatures. For example, if the load of an area is less affected by the temperature, the elimination of the

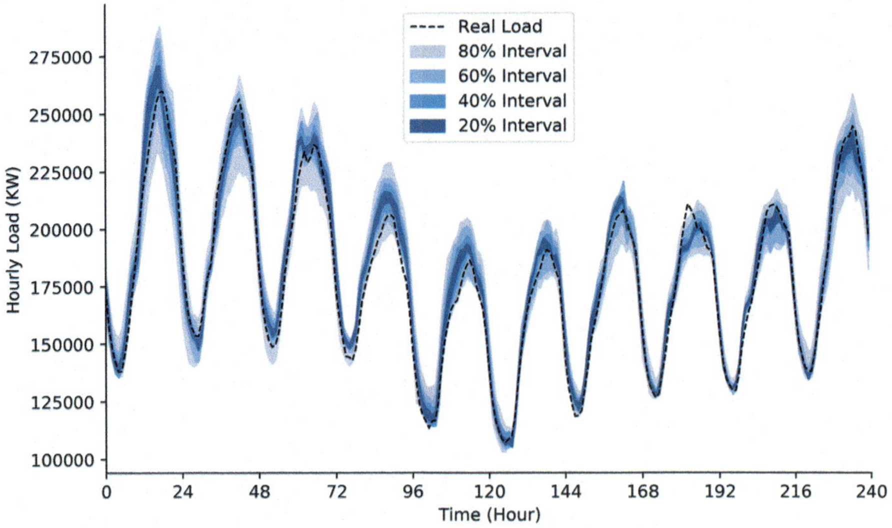

Fig. 5.5 Probabilistic forecasts of Area #2 from 17 July, 2007 to 26 July, 2007

Table 5.1 Number of features before and after L_1-norm panelized feature selection (D7-H12)

Area	Number of original features	Quantile-Lasso for each quantile									Pre-Lasso
		q_1	q_2	q_3	q_4	q_5	q_6	q_7	q_8	q_9	
1	2279	1067	1781	1796	1841	1820	1784	1259	1171	1192	1459
2		223	591	1227	1240	1251	1507	1800	1608	1573	1372
3		224	591	1225	1240	1248	1505	1800	1607	1677	1373
4		55	262	541	621	622	699	670	729	1454	486
5		62	67	114	126	227	225	302	307	374	362
6		234	596	839	1240	1238	1505	1803	1768	1693	723
7		224	591	1225	1240	1248	1505	1800	1607	1677	1373
8		1315	1224	1358	1794	1667	1501	1470	1593	1211	2193
9		372	539	650	631	1192	1192	911	1089	256	310
10		1057	1092	1124	1189	969	984	1181	1091	838	925

temperature-related features will less change the performance of the PLF model in this area. However, for the area whose load is very sensitive to temperature, feature selection may show higher gains on the performance. Figure 5.5 presents the forecasted quantiles and the real load of Area #2 over ten days from 17 July, 2007 to 26 July, 2007. The forecasts are obtained using the D7-H12 model with Quantile-Lasso.

Table 5.1 gives the number of features that are selected by Quantile-Lasso method with the optimal adjustment parameters $\lambda_{q,best}$ for different areas from q_1 to q_9. The Quantile-Lasso method is implemented based on the D7-H12 model. Interestingly, the numbers of selected features show increasing trends in the first four quantiles

Table 5.2 Performance of different feature selection methods

Area	D3-H4					D7-H12				
	Pre-Lasso	Quantile-Lasso	MLR	QRNN	QGBRT	Pre-Lasso	Quantile-Lasso	MLR	QRNN	QGBRT
1	452.5	**437.1**	456.4	798	651	442.4	**428.9**	457.8	1274	592
2	3849	3751	3918.8	8369	**3571**	3825.1	**2685.3**	3988.6	8579	3935
3	4153	4047.3	4228.5	9363	**3734**	4127.6	**3976.4**	4303.6	10832	4017
4	16.8	**16.7**	17.9	38	25	16.7	**16.5**	17.3	39	25
5	341.8	**309.4**	374.8	884	623	344.8	**308.6**	397.3	922	668
6	4122.3	**3977.4**	4160	9443	4306	4042.7	**3886.4**	4237.4	9148	4516
7	4153	4047.3	4228.5	9796	**3715**	4127.6	**3976.4**	4303.6	10536	4336
8	92.8	**92.7**	95.4	159	139	95.2	**92.1**	97.2	172	130
9	6469.9	**6400**	6621.9	14291	10630	6491.8	**6452.1**	6767	14573	12250
10	599.4	**594.8**	601.2	1427	717	619.6	**613.3**	647.1	1681	809

(from q_1 to q_4). However, there is no clear trend for larger quantiles. The lower quantile may correspond to the baseload and is less influenced by complex recency-effects. Thus the number of selected features is much smaller.

The number of features selected by Pre-Lasso does not change for different quantiles. Compared with Pre-Lasso, the Quantile-Lasso method can dynamically select features by adjusting the sparse penalty (adjustment parameter λ) according to the performance of the model on the validation dataset. In this way, Quantile-Lasso can produce better probabilistic forecasts after feature selection compared with Pre-Lasso.

Table 5.2 presents the AQS of Quantile-Lasso method, Pre-Lasso method, original model without feature selection (MLR), QRNN, and QGBRT, where the inputs of QRNN and QGBRT are the same as MLR model without feature selection. Both Pre-Lasso and Quantile-Lasso with D3-H4 method and D7-H12 model have lower AQS than the original model without feature selection. In addition, the results again verify that Quantile-Lasso has a better performance compared with Pre-Lasso and the original method for all ten areas by selecting features individually for each quantile. QRNN has the worst performance because of a large number of input features without selection. For the D3-H4 model, QGBRT, a powerful regression technique, has a better performance compared with Quantile-Lasso in three areas. However, for the D7-H12 model, Quantile-Lasso has the best performance in all areas. QGBRT may have worse performance with a larger number of input features which also indicates the importance of feature selection for probabilistic forecasting.

5.7 Concluding Remarks

This chapter provides a Quantile-Lasso method for feature selection in the PLF model by adding L_1-norm penalty into the loss function. ADMM algorithm is proposed to solve the large-scale optimization problem. The method is compared with the Pre-LASSO method and original model without feature selection. Even though Pre-Lasso is easy to implement using an off-the-shelf algorithm but has limited improvement. Quantile-Lasso has a higher degree of freedom to adaptively select the features for different quantiles. The averaged relative improvements of Quantile-Lasso are 10.46% and 6.06% compared with traditional quantile regression without feature selection and Pre-Lasso in terms of pinball loss, respectively. Future works will focus on parallel implementation of the proposed ADMM algorithm on larger datasets and the applications to probabilistic wind power or PV forecasting.

References

1. Hong, Tao, and Shu Fan. 2016. Probabilistic electric load forecasting: A tutorial review. *International Journal of Forecasting* 32 (3): 914–938.
2. Gan, Dahua, Yi. Wang, Shuo Yang, and Chongqing Kang. 2018. Embedding based quantile regression neural network for probabilistic load forecasting. *Journal of Modern Power Systems & Clean Energy* 6 (2): 1–11.
3. Zhang, Wenjie, Hao Quan, and Dipti Srinivasan. 2019. An improved quantile regression neural network for probabilistic load forecasting. *IEEE Transactions on Smart Grid* 10 (4): 4425–4434.
4. Wang, Yi, Ning Zhang, Qixin Chen, Daniel S Kirschen, Pan Li, and Qing Xia. 2018. Data-driven probabilistic net load forecasting with high penetration of behind-the-meter PV. *IEEE Transactions on Power Systems* 33 (3): 3255–3264.
5. Wang, Yi., Ning Zhang, Chongqing Kang, Miao Miao, Rui Shi, and Qing Xia. 2018. An efficient approach to power system uncertainty analysis with high-dimensional dependencies. *IEEE Transactions on Power Systems* 33 (3): 2984–2994.
6. He, Yaoyao, Xu. Qifa, Jinhong Wan, and Shanlin Yang. 2016. Short-term power load probability density forecasting based on quantile regression neural network and triangle kernel function. *Energy* 114: 498–512.
7. Xie, Jingrui, and Tao Hong. 2018. Temperature scenario generation for probabilistic load forecasting. *IEEE Transactions on Smart Grid* 9 (3): 1680–1687.
8. Liu, Bidong, Jakub Nowotarski, Tao Hong, and Rafał Weron. 2017. Probabilistic load forecasting via quantile regression averaging on sister forecasts. *IEEE Transactions on Smart Grid* 8 (2): 730–737.
9. Xie, Jingrui, Tao Hong, Thomas Laing, and Chongqing Kang. 2017. On normality assumption in residual simulation for probabilistic load forecasting. *IEEE Transactions on Smart Grid* 8 (3): 1046–1053.
10. Wang, Yi, Ning Zhang, Yushi Tan, Tao Hong, Daniel S Kirschen, and Chongqing Kang. 2018. Combining probabilistic load forecasts. *IEEE Transactions on Smart Grid* 10 (4): 3664–3674.
11. Wang, Yi, Qixin Chen, Tao Hong, and Chongqing Kang. 2018. Review of smart meter data analytics: Applications, methodologies, and challenges. *IEEE Transactions on Smart Grid* 10 (3): 3125–3148.
12. Arora, Siddharth, and James W. Taylor. 2016. Forecasting electricity smart meter data using conditional kernel density estimation. *Omega* 59: 47–59.

13. Taieb, Souhaib Ben, Raphaël Huser, Rob J. Hyndman, and Marc G. Genton. 2016. Forecasting uncertainty in electricity smart meter data by boosting additive quantile regression. *IEEE Transactions on Smart Grid* 7 (5): 2448–2455.
14. Shepero, Mahmoud, Dennis Van Der Meer, Joakim Munkhammar, and Joakim Widèn. 2018. Residential probabilistic load forecasting: A method using gaussian process designed for electric load data. *Applied Energy* 218: 159–172.
15. Tibshirani, Robert. 2011. Regression shrinkage and selection via the lasso. *Journal of the Royal Statistical Society* 73 (3): 267–288.
16. Xie, Jingrui, and Tao Hong. 2018. Variable selection methods for probabilistic load forecasting: Empirical evidence from seven states of the united states. *IEEE Transactions on Smart Grid* 9 (6): 6039–6046.
17. Boyd, Stephen, Neal Parikh, Eric Chu, and Borja Peleato. 2010. Distributed optimization and statistical learning via the alternating direction method of multipliers. *Foundations & Trends in Machine Learning* 3 (1): 1–122.
18. Hong, Tao, Pierre Pinson, and Shu Fan. 2014. Global energy forecasting competition 2012. *International Journal of Forecasting* 30 (2): 357–363.
19. Hong, Tao. 2010. Short Term Electric Load Forecasting. PhD thesis, North Carolina State University. https://repository.lib.ncsu.edu/handle/1840.16/6457.
20. Wang, Pu., Bidong Liu, and Tao Hong. 2016. Electric load forecasting with recency effect: A big data approach. *International Journal of Forecasting* 32 (3): 585–597.
21. Tibshirani, Robert. 1996. Regression shrinkage and selection via the lasso. *Journal of the Royal Statistical Society. Series B (Methodological)* 58 (1): 267–288.
22. Efron, Bradley, Trevor Hastie, Iain Johnstone, and Robert Tibshirani. 2004. Least angle regression. *Annals of Statistics* 32 (2): 407–451.
23. Cannon, Alex J. 2011. Quantile regression neural networks: Implementation in R and application to precipitation downscaling. *Computers & Geosciences* 37 (9): 1277–1284.
24. Ridgeway, Greg. 2007. Generalized boosted models: A guide to the gbm package. https://pbil.univ-lyon1.fr/CRAN/web/packages/gbm/vignettes/gbm.pdf.

Part II
Electricity Price Modeling and Forecasting

Chapter 6
Subspace Characteristics of LMP Data

6.1 Introduction

The locational marginal prices (LMPs) are the foundation of modern electricity markets. They reflect the marginal operation and generation cost of the whole power system at each node [1, 2]. The data of LMPs are calculated and publicized by an independent system operator (ISO) or regional transmission organization (RTO) according to the price bid & offer collected in the day-ahead and real-time electricity market. The problem of security-constrained economic dispatch (SCED) is formulated and solved using linear programming or convex optimization tools. The system information, including network topology, transmission capacity, and line impedance, is included in the SCED and partially determines the distribution of LMP data.

Generally, detailed market structure information (e.g., topology and line parameters) is private due to security concerns. Thus, the market participants may not have access to such information. Researchers aim to study the publicized LMP data with additional system information in data-driven approaches. In [3], Geng et al. proposed a multi-class support vector machine (SVM)-based approach to fit the LMP-load coupling. The LMP in a certain power system was modeled as a function of loads at all nodes. The distribution of LMPs in the load space was analyzed and fitted precisely using an SVM. Kekatos et al. aimed to recover network topology from LMP data in [4]. The low-rank feature of the LMP data and sparsity of the topology adjacency matrix were utilized to formulate a convex optimization problem. It has been proved that as long as enough LMP data are given, the connectivity of certain lines could be approximately recovered and tracked. In [5], linear regression was used to recover the potential market structure based on the released LMP data and Lagrange multipliers of the inequality constraints in the SCED problem. However, in some electricity markets, the exact values of the Lagrange multipliers are not publicized.

Recent research on LMP [3, 6] has revealed that if the network parameters and generation offers remain the same, the congestion component of the LMP data lies in a discrete space w.r.t. nodal loads. In other words, for the nodal loads that belong to the same system pattern region (SPR), the network congestion status and the congestion LMP vector remain the same when the generation offers are constant. A simple proof using multiparametric linear programming (MLP) was provided in [3], where the SPR was defined as the critical region (CR) in the theory of MLP. A

Q. Chen et al., *Data Analytics in Power Markets*,
https://doi.org/10.1007/978-981-16-4975-2_6

much more rigorous proof and discussion on the SPRs of LMP under both linear and quadratic cases were given in [6].

Most previous works focus on learning the coupling between additional system information and LMP data. We want to utilize the aforementioned characteristics of the congestion LMP vectors and extract hidden messages from them without any additional system information (e.g., nodal load data, or topology data, which are usually unavailable). The core work here is to identify the network congestion status/SPRs using only the publicized LMP data. If the SPR of the LMP is identified, the distribution of the congestion LMP vector in the high-dimensional Euclidean space is limited to certain subspaces, which could help in LMP forecasting [7, 8]. Additionally, through congestion status identification, we can know the borderline conditions of the system parameters as well as load and generator quotations. These results can help market participants enhance their understanding of the market structure and thus help to determine their bidding strategy.

Some studies show the benefit of SPR/CR information in LMP forecasting and other applications. Zhou et al. exploited the structural aspects underlying the power market operations using SPR to reduce the congestion forecast error [8]. The convex hull of the zonal load was fitted in high-dimensional load space with the QuickHull algorithm. In [9], the short-term probabilistic LMP forecasting problem was considered with load uncertainty from a system operator perspective. CRs were used to map the uncertainty from the load space to the LMP space.

Congestion status is also important for bidding in energy markets and financial transmission right (FTR) markets. For example, in [10] the impact of local transmission congestion on energy storage arbitrage opportunities was discussed. A bilevel imperfect competition model was proposed and the results indicated that the network congestion had a large impact on the energy storage profits. In [11], a strategic bidding method considering FTR and demand response was proposed for load serving entities. In [12], Lo Prete et al. studied the usage of unprofitable virtual transactions aiming at altering the network congestion status and improving the FTR revenue based on an equilibrium model. The information of congestion scenarios is fundamental in these models. In [13], an inverse optimization-based approach for estimating rivals' production cost functions in electricity markets was proposed, and a discussion on its extension was given to take into account the network congestion status.

A static generation offering behavior is usually assumed in the application of SPR/CR. Actually, the congestion LMP vector changes within the same subspace when the marginal generation offer fluctuates [14]. Thus, the discrete space that the LMP vector lies in is turned into a continuous space, and we cannot easily identify the congestion status from the LMP. In this chapter, we propose an unsupervised method to cluster the LMP data into different subspaces in hierarchical ways with some efficient machine learning techniques. The efficacy of the method is proved by numerical experiments on the IEEE 30-bus system, IEEE 118-bus system, Illinois 200-bus system [15], and Southwest Power Pool (SPP) [16] with real market data. The contribution of this chapter is threefold.

1. Problem modeling: We study the problem of identifying the congestion status of the LMP in a data-driven approach. The problem is modeled as recovering basis vectors for the congestion LMP data in high-dimensional Euclidean space.
2. Highly efficient solution framework: By applying efficient machine learning techniques, including principal component analysis, spectral clustering, and dual principal component pursuit in bottom-up and top-down hierarchical searching, the hidden subspaces in LMP data are identified one by one. The bottom-up algorithm can directly find the basis vectors that span the subspaces when there is no basis gap. The top-down algorithm can find the potential subspaces by detecting the norm vectors and addressing the basis gap.
3. Comprehensive experiments: Four numerical cases, including one with real market data, are presented to give a visualized procedure for the hierarchical searching methods. The cases verify the high accuracy and low time consumption of the proposed method.

This chapter is structured as follows. Section 6.2 describes the widely adopted LMP models and analyzes the distribution of congestion LMP vectors. Section 6.3 gives the proposed methodology and detailed algorithms. Three numerical cases are conducted in Sect. 6.4. Finally, Sect. 6.5 discusses the application of the proposed method and draws the conclusion.

6.2 Model and Distribution of LMP

The DC optimal power flow (DC-OPF) model has been widely adopted in SCED. The lossless DC-OPF model is formulated as follows:

$$\min_{\mathbf{P}_G} \sum_{i \in \mathcal{N}} c_i(P_{Gi}) \tag{6.1a}$$

$$\text{s.t. } \mathbf{1}^\top \mathbf{P}_G = \mathbf{1}^\top \mathbf{P}_D \qquad :\lambda \tag{6.1b}$$

$$|\mathbf{T}(\mathbf{P}_G - \mathbf{P}_D)| \preceq \mathbf{f}^{\max} \qquad :\boldsymbol{\mu}^+, \boldsymbol{\mu}^- \tag{6.1c}$$

$$\mathbf{P}^{\min} \preceq \mathbf{P}_G \preceq \mathbf{P}^{\max} \qquad :\gamma_{\min}, \gamma_{\max}. \tag{6.1d}$$

where $\mathcal{N}$ is the set of all nodes in the electricity market, c_i is the cost function offered at node i, $\mathbf{P}_G$ is the generated active power vectors, $\mathbf{T}$ is the power transfer distribution factor (PTDF) matrix, $\mathbf{f}^{\max}$ is the transmission capacity vector, $\mathbf{P}^{\min}$ and $\mathbf{P}^{\max}$ are the generation output limit, and $\mathbf{1}$ is the all-ones vector. λ is the shadow price for the balance constraint. $\boldsymbol{\mu}^+$ and $\boldsymbol{\mu}^-$ are the shadow prices for transmission constraints. $\gamma_{\min}$ and $\gamma_{\max}$ are the shadow prices for generation constraints.

The LMP vector $\boldsymbol{\pi}$ is the derivation of the Lagrangian function of (6.1) w.r.t. $\mathbf{P}_D$:

$$\boldsymbol{\pi} = \lambda \cdot \mathbf{1} + \mathbf{T}^\top \boldsymbol{\mu}. \tag{6.2}$$

where $\boldsymbol{\mu} = \boldsymbol{\mu}^- - \boldsymbol{\mu}^+$.

The LMP vector in (6.2) can be decomposed into two parts, the energy component $\pi^{\mathcal{E}}$ and the congestion component π^{C}:

$$\pi^{\mathcal{E}} = \lambda \mathbf{1}; \tag{6.3a}$$
$$\pi^{C} = \mathbf{T}^{\top} \boldsymbol{\mu}. \tag{6.3b}$$

The lossy DC-OPF model has been a hot topic for years, and different ISOs have slightly different ways of handling network loss in LMP assignment [17]. Here, we use the widely adopted model from Litvinov et al. [18]:

$$\min_{\mathbf{P}_G} \sum_{i \in N} c_i(P_{Gi}) \tag{6.4a}$$
$$\text{s.t. } \mathbf{1}^{\top}\mathbf{P}_G = \mathbf{1}^{\top}\mathbf{P}_D + l \qquad :\lambda \tag{6.4b}$$
$$l = l_0 + LF^{\top}(\mathbf{P}_G - \mathbf{P}_D) \qquad :\sigma \tag{6.4c}$$
$$|\mathbf{T}(\mathbf{P}_G - \mathbf{P}_D - l \cdot \mathbf{d})| \leq \mathbf{f}_{\max} \qquad :\boldsymbol{\mu}^{+}, \boldsymbol{\mu}^{-} \tag{6.4d}$$
$$\mathbf{P}^{\min} \leq \mathbf{P}_G \leq \mathbf{P}^{\max} \qquad :\gamma_{\min}, \gamma_{\max}. \tag{6.4e}$$

where l is the total network loss, l_0 is the intercept for the linearized loss constraint, LF is the linearized factors for network loss, $\mathbf{d}$ is the loss distribution factor, and σ is the shadow price for the linearized loss constraint. According to [18], the lossy LMP is derived as:

$$\pi = \sigma \cdot \mathbf{1} - \sigma \cdot LF + \mathbf{T}^{\top}\boldsymbol{\mu} - \mathbf{1} \cdot \mathbf{d}^{\top}\mathbf{T}^{\top}\boldsymbol{\mu}, \tag{6.5}$$

and an additional loss component $\pi^{\mathcal{L}}$ can be decomposed:

$$\pi^{\mathcal{E}} = \sigma \cdot \mathbf{1}; \tag{6.6a}$$
$$\pi^{C} = \underbrace{\mathbf{T}^{\top}\boldsymbol{\mu}}_{\pi_1} - \underbrace{\mathbf{1} \cdot \mathbf{d}^{\top}\mathbf{T}^{\top}\boldsymbol{\mu}}_{\pi_2}; \tag{6.6b}$$
$$\pi^{\mathcal{L}} = -\sigma \cdot LF. \tag{6.6c}$$

Generally, ISOs release data of all three LMP components on their websites. In fact, only a very small proportion of lines in a transmission system can be congested during a short market period [7]. In the California ISO, only two lines are typically congested [19]. Thus, the vector of π^{C} always lies in the subspaces spanned by a few column vectors of matrix $\mathbf{T}^{\top}$ (lossless) or $(\mathbf{T}^{\top} - \mathbf{1} \cdot \mathbf{d}^{\top}\mathbf{T}^{\top})$ (lossy). In lossy markets where the loss distribution vector $\mathbf{d}$ is updated in a timely manner, the subspace feature of π^{C} might be unclear. In (6.6b) the price consists of two parts, π_1 spanned by constant vectors and π_2 spanned by inconstant vectors. Note that $\mathbf{1}$ is an all-ones vector and $\mathbf{d}^{\top}\mathbf{T}^{\top}\boldsymbol{\mu}$ is a scalar, indicating that π_2 is identical at every node. Thus, the effect of π_2 can be eliminated by subtracting the value of π^{C} at a certain node from the π^{C} at every node.

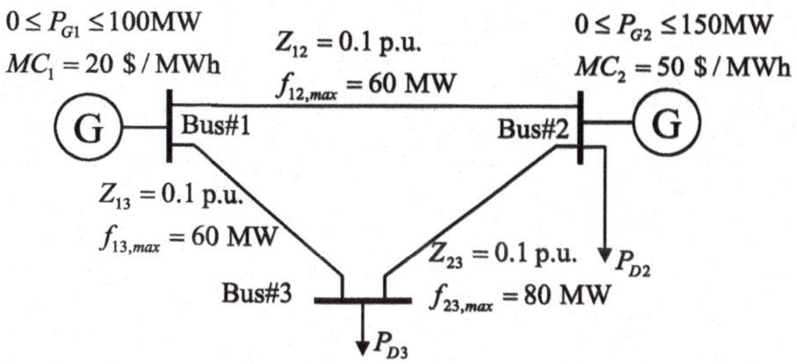

Fig. 6.1 The 3-bus system in [3]

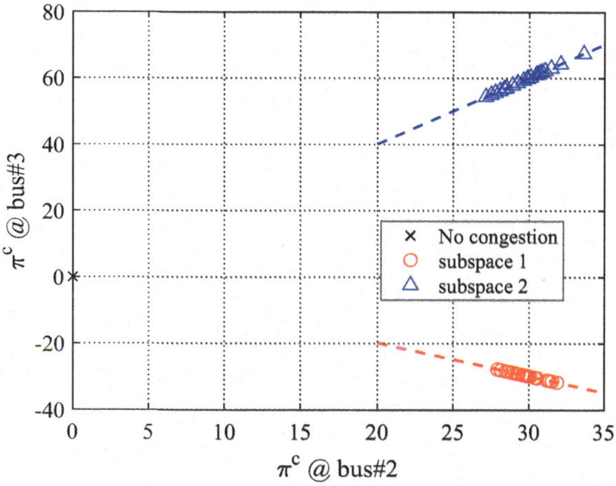

Fig. 6.2 Two LMP patterns of the 3-bus system

Currently, some electricity markets are considering adopting multi-interval SCED or look-ahead SCED [20, 21] in market operation. Ramping constraints and other complicated constraints would be included in the constraints of these SCED models. These constraints will impact the solution of dual variables μ and σ. However, the formulation of (6.5) and (6.6) will not be changed explicitly, which can be proved by applying Kuhn-Tucker conditions and similar steps in [18] for these SCED models. Our analysis and the proposed method in this chapter are still applicable for look-ahead SCED models.

We use the 3-bus system in [3] as an example (Fig. 6.1) and assume that bus#1 is the reference bus. A small Gaussian fluctuation is added to the marginal cost of the two generators. The loads at bus#2 and bus#3 are chosen from within the range of -150 to 300 MWh to generate different LMP vectors with the lossless model in (6.1). Figure 6.2 shows the distribution of the LMP vectors in Euclidean space. Two congestion statuses can be easily noticed when all the vectors with congestion fall in two linear subspaces.

The LMP subspaces are more complex in a network with more nodes and transmission lines. It is hard to visualize these subspaces when the number of nodes is more than 4. However, we can refer to Fig. 6.3 as a simplified illustration. The vector of π^C falls in subspace S_4 when both lines of S_1 and S_2 are congested together.

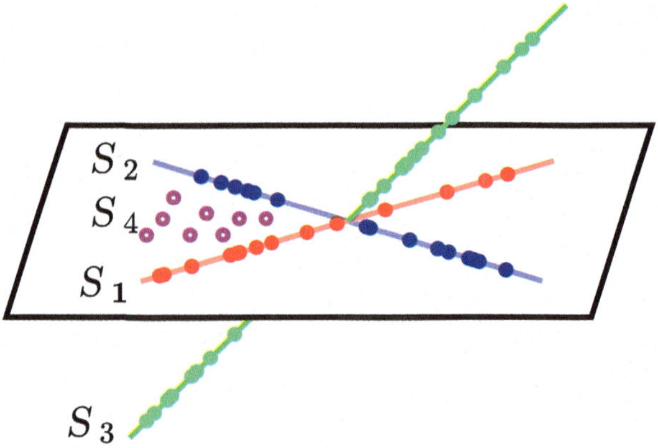

Fig. 6.3 The overlapping subspaces: S_1, S_2, and S_3 have one congested line, while S_4 has two congested lines, corresponding to those of S_1 and S_2

It is important to note that, when the system has only a few nodes, the LMP patterns (similar to those in Fig. 6.2) can be easily identified using subspace clustering methods such as [22, 23]. However, in real applications, the LMP patterns (similar to those in Fig. 6.3 but with more complexity) are distributed in overlapping subspaces with outliers. The subspace S_4 in Fig. 6.3 entirely covers S_1 and S_2, while most of the subspace clustering methods can handle only independent or disjoint subspaces [24, 25] and separate S_3 from S_4. Therefore, a specialized method for identifying overlapping LMP subspaces with outliers should be proposed.

6.3 Methodology

The problem formulation, the basic framework of the proposed method, and detailed algorithms used are introduced in this section.

6.3.1 Problem Formulation

Assume that we have collected the time series of $\boldsymbol{\pi}^C$ at all nodes and denote it as $\mathbf{X}$:

$$\mathbf{X} = [\pi_t^C]_{t\in\mathcal{T}}, \quad |\mathcal{T}| = M. \tag{6.7}$$

where $\mathcal{T}$ is the set of time intervals, M is the number of time intervals.

Let n_b and n_l denote the number of buses and lines, respectively, and assume that k is the number of ever-congested lines. Let $j_1, j_2, \ldots, j_k$ ($k << n_l$) denote the indexes of ever-congested lines during the time period $\mathcal{T}$, and let $T_1, T_2, \ldots, T_{n_b}$

Table 6.1 The appeared combinations of congested lines and values of $\mathfrak{X}_t$ in the numerical test of IEEE case30

Number of congested lines	Line No.	Value of $\mathfrak{X}_t$
4	#(10,29,30,35)	$(1, 1, 1, 1)^\top$
3	#(10,30,35),#(29,30,35)	$(1, 0, 1, 1)^\top$, $(0, 1, 1, 1)^\top$
2	#(10,30),#(10,35) #(29,35),#(30,35)	$(1, 0, 1, 0)^\top$, $(1, 0, 0, 1)^\top$ $(0, 1, 0, 1)^\top$, $(0, 0, 1, 1)^\top$
1	#30, #35	$(0, 0, 1, 0)^\top$, $(0, 0, 0, 1)^\top$

denote all the column vectors of matrix $\mathbf{T}^\top$. As we mentioned in Sect. 6.2, $\mathbf{X}$ can be expressed by the linear combination of $T_{j_1}, T_{j_2}, \ldots, T_{j_k}$:

$$\mathbf{X} = \left[\mathbf{x}_1, \mathbf{x}_2, \ldots, \mathbf{x}_M\right] = \left[T_{j_1}, T_{j_2}, \ldots, T_{j_k}\right] \cdot \left[\chi_t\right]_{t\in\mathcal{T}}, \tag{6.8}$$

where $\chi_t \in \mathbb{R}^k$ is the linear coefficient determined by the dual variable of line congestion μ at time t. Define $\mathfrak{X}_t$ to show the nonzero entries of χ_t:

$$\mathfrak{X}_t = \mathbb{1}(\chi_t \neq 0) \approx \mathbb{1}(|\chi_t| > \epsilon), \tag{6.9}$$

where ϵ is a small number. The value of $\mathfrak{X}_t$ determines which congestion status the vector of π_t^C belongs to. If $\mathfrak{X}_t$ is recovered from $\mathbf{X}$, the congestion information of the electricity market at time t can be revealed. Theoretically, there are 2^k possible values of $\mathfrak{X}_t$, but usually the number of congestion statuses is much smaller in reality. In the numerical experiments of the IEEE 30-bus system in Sect. 6.4, there are only 9 appearing values of $\mathfrak{X}_t$ with 4 possible congested lines, as shown in Table 6.1.

6.3.2 Basic Framework

The basic target is to recover $\mathfrak{X}_t$ from $\mathbf{X}$ without any prior knowledge on $\mathbf{T}$. Since all the M column vectors of $\mathbf{X}$ lie in the subspace spanned by a common vector set $\{T_{l_1}, T_{l_2}, \ldots, T_{l_k}\}$, we need to identify the common basis vectors $\mathbf{B}$ for $\mathbf{X}$. First, the dimension of $\mathbf{X}$ is reduced from n_b to k, which reduces the computational burden for the following steps. Then a bottom-up approach is adopted to detect explicit basis vectors. If there are implicit basis vectors that result in a basis gap (introduced in the next subsection), a top-down approach is used to cover this situation. Once $\mathbf{B}$ is recovered, $\mathfrak{X}_t$ can be encoded as $\mathbb{1}(\mathbf{B}^\dagger \mathbf{x}_t \neq 0)$, where $\mathbb{1}(\cdot)$ is the boolean function for a given condition. Figure 6.4 shows the basic framework of our method.

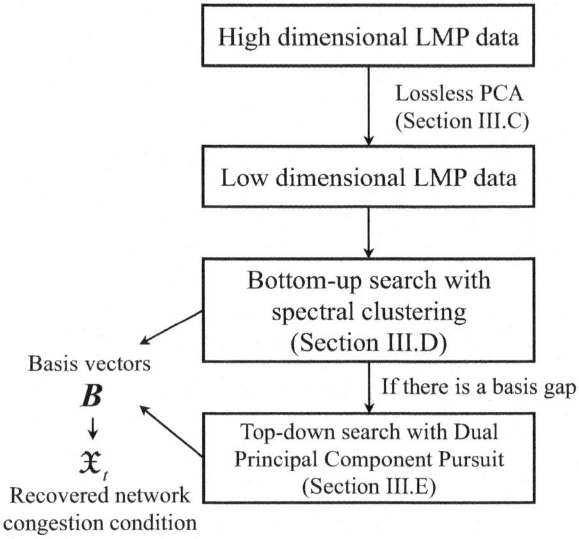

Fig. 6.4 A simple framework of the proposed method

6.3.3 Principal Component Analysis

In a large-scale power system with a long-term price record, $\mathbf{X}$ can be a high-dimensional matrix. Because the number of ever-congested lines k is usually small, $\mathbf{X}$ is low-rank, and its dimension can be reduced using principal component analysis (PCA). After PCA, k components can be used to represent the original data $\mathbf{X}$ without any loss. By abuse of notation, we continue to use $\mathbf{X}$ to denote the dimensionally reduced data matrix. The time complexity for PCA is $O(n_b^2 M + n_b^3)$.

6.3.4 Recursive Basis Search (Bottom-Up)

It is difficult to find all the overlapping subspaces that have different dimensions in $\mathbb{R}^k$ space with a large number of data points. However, it is much easier to identify 1-dimensional subspaces (i.e., straight lines passing through the origin) among these data points. If two vectors $\mathbf{v}_1, \mathbf{v}_2$ belong to the same 1-dimensional subspace, their cosine similarity should be 1:

$$\cos \angle(\mathbf{v}_1, \mathbf{v}_2) = \frac{\mathbf{v}_1^\top \mathbf{v}_2}{|\mathbf{v}_1||\mathbf{v}_2|} = 1. \tag{6.10}$$

Thus, an affinity matrix $\mathbf{A}$ can be constructed using cosine similarity. To filter out the similarity for those vectors belonging to different 1-dimensional subspaces, an elementwise cutoff kernel function

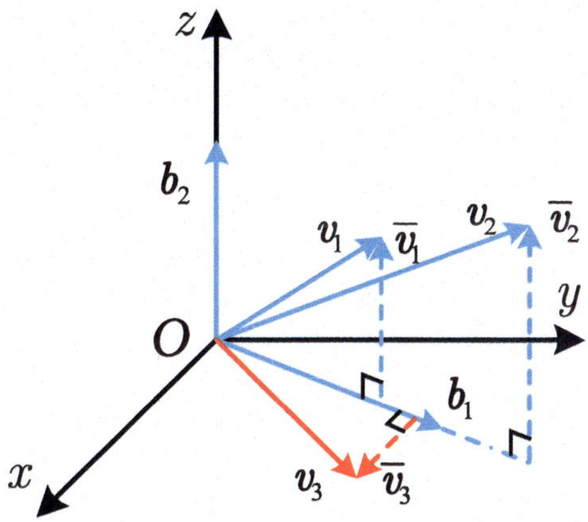

Fig. 6.5 Bottom-up basis search using orthogonalization. $\mathbf{b}_1$ is the basis of the bottom subspace, and $\mathbf{v}_1, \mathbf{v}_2 \in span(\mathbf{b}_1, \mathbf{b}_2)$ are in the upper subspace. $\mathbf{v}_3$ belongs to other subspaces. $\bar{\mathbf{v}}_i (i = 1, 2, 3)$ are the orthogonalized vectors w.r.t. $\mathbf{b}_1$

$$\kappa(x) = \begin{cases} 1, & (|x| > 1 - \varepsilon) \\ 0, & \text{otherwise} \end{cases} \tag{6.11}$$

is used to obtain $\mathbf{A}' = \kappa(\mathbf{A})$, where ε is a small number. $\mathbf{A}'$ is the sparse affinity matrix that contains only the affinity of vectors of the same 1-dimensional subspaces. Finding the subspaces in $\mathbf{A}'$ is equivalent to finding the clusters according to $\mathbf{A}'$. Then, a spectral clustering technique that maximizes the relative eigengap [26] can be used to identify the 1-dimensional subspaces automatically. Let $\gamma_1, \ldots, \gamma_M$ denote the eigenvalues of $\mathbf{A}'$ in ascending order. The relative eigengap is defined as:

$$\xi_i = \frac{\gamma_{i+1} - \gamma_i}{\gamma_i}, \quad i \geq 2. \tag{6.12}$$

A high value of ξ_i indicates that $\mathbf{A}'$ admits a good decomposition into at least i clusters [26]. Thus, k-means with i as the number of partition sets is applied with the corresponding eigenvectors. The vector bases of these 1-dimensional subspaces can be calculated once the clusters are detected. Due to space limitation, the detail of the algorithm is not presented here.

After we have found the data points that lie in the 1-dimensional subspaces, the next step is to search for the 2-dimensional subspaces. Take the vectors in Fig. 6.5 as an example. Assume that $\mathbf{v}_1, \mathbf{v}_2$ belongs to the 2-dimensional subspace spanned by $\mathbf{b}_1$ and $\mathbf{b}_2$, while $\mathbf{v}_3$ lies in another subspace. If we have found vector $\mathbf{b}_1$ as a basis using spectral clustering, $\mathbf{v}_1, \mathbf{v}_2, \mathbf{v}_3$ can be orthogonalized into $\bar{\mathbf{v}}_1, \bar{\mathbf{v}}_2, \bar{\mathbf{v}}_3$ w.r.t. $\mathbf{b}_1$ (i.e., projection onto $span(\mathbf{b}_1)^{\perp}$). After orthogonalization, $\bar{\mathbf{v}}_1$ and $\bar{\mathbf{v}}_2$ lie in a 1-dimensional subspace, and the basis $\mathbf{b}_2$ can be easily calculated. This procedure can be repeated until all the subspaces and their bases are found. The pseudo code for the whole bottom-up searching procedure is shown in Algorithm 6.1.

The time complexity for calculating the affinity matrix $\mathbf{A}$ is $O(k^2M)$. For spectral clustering, full singular vector decomposition (SVD) is applied with time complexity of $O(M^3)$. Then k-means clustering is used to group the vectors in the latent space.

Recall the structure in Table 6.1. Our recursive basis search process is to determine the overlapping subspaces from the lowest one-dimensional space to the highest dimensional space (bottom-up). This method is powerful when dealing with structures with continuously increasing bases and will have good performance in situations where only a few lines are possibly congested. However, a problem arises when there is a **basis gap** after the orthogonalization, i.e., all the data points belong to subspaces whose dimension is higher than 1, as shown in Fig. 6.6. In this case, no obvious lines can be detected, and a top-down search method can be used to bridge the basis gap.

Algorithm 6.1 Bottom-up subspace search

procedure BOTTOM- UP($\mathbf{X}, \varepsilon$) ▷ $\mathbf{X}$ has been dimensionally-reduced with PCA.
Initialization : $\mathcal{B} \leftarrow \emptyset$;
  repeat
    normalize $\mathbf{X}$;
    $\mathbf{A} \leftarrow |\mathbf{X}^\top \mathbf{X}|$; $\mathbf{A}' \leftarrow \kappa(\mathbf{A}, \varepsilon)$;
    $\mathbf{X}_1, \ldots, \mathbf{X}_K$, and $K \leftarrow$ SPECTRALCLUSTERING($\mathbf{X}, \mathbf{A}'$);
    for $i \leftarrow 1 : K$ **do**
      if $Rank(\mathbf{X}_i) == 1$ **then**
        $\mathcal{B} \leftarrow \mathcal{B} \cup \hat{\mathbf{X}}_i$
      end if
    end for
    $\mathbf{X} \leftarrow$ Projection($\mathbf{X}, \mathcal{B}^\perp$) ▷ Dimension Reduction
  until $K = 1$
  return $\mathcal{B}$
end procedure

6.3.5 *Hyperplane Detection (Top-down)*

Assume that after the recursive basis search method in Sect. 6.3.4, the distribution of data points is as in Fig. 6.6. The task of the top-down search is to find the basis vectors or the norm vector of H_1 among outliers with full rank. The problem of top-down searching can be formulated as follows:

$$\min_{\mathbf{n}} ||\mathbf{X}^\top \mathbf{n}||_0 \quad s.t. \quad ||\mathbf{n}||_2 = 1, \tag{6.13}$$

which aims to find the norm vector $\mathbf{n}$ having as many points as possible contained in H_1. The ℓ_0-norm in the objective function and the quadratic constraint in (6.13) make the problem nonconvex. Using any algorithm, there are three cases of possible solutions $\mathbf{n}$ and H_1 for (6.13):

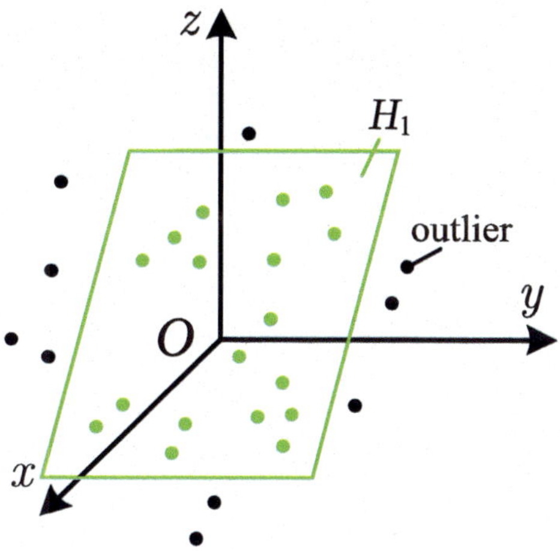

Fig. 6.6 An example of the basis gap in $\mathbb{R}^3$. H_1 is a 2-dimensional subspace where inliers are distributed. The outliers are in 3-dimensional space

1. Global optimum. H_1 contains as many as possible data points.
2. Non-trivial local optimum. H_1 contains some data points exceeding a certain threshold.
3. Trivial solution. H_1 contains only a few data points.

For a global optimum or non-trivial local optimum, the result is acceptable because we have found a hyperplane that contains some data points and forms a congestion status. Thus, we do not have to obtain the global optimum for (6.13) since a local optimal vector $\tilde{\mathbf{n}}$ could also reveal a satisfactory subspace but containing a bit fewer points in it. The problem of (6.13) in high-dimensional space can be suitably addressed by a recently developed method in machine learning named dual principal component pursuit (DPCP) [27]. The brief procedure of DPCP is introduced here.

First, the ℓ_0-norm of the objective function can be relaxed as the ℓ_1-norm. However, the constraint of (6.13) is still nonconvex:

$$\min_{\mathbf{b}} ||\mathbf{X}^\top \mathbf{n}||_1 \quad s.t. \quad ||\mathbf{n}||_2 = 1. \tag{6.14}$$

To solve (6.14), the constraint is further relaxed by a sequence of linear programs:

$$\mathbf{n}_{i+1} := \underset{\mathbf{m}^\top \hat{\mathbf{n}}_i = 1}{argmin} ||\mathbf{X}^\top \mathbf{m}||_1, \quad i \in \mathbb{N}, \tag{6.15}$$

where $\hat{\cdot}$ indicates normalization to unit ℓ_2-norm, and $\mathbf{n}_0$ is initialized to minimize $||\mathbf{X}^\top \mathbf{n}||_2$ using SVD. The outliers and inliers can be separated according to their inner product with the converged normal vector $\mathbf{n}$. In cases where the proportion of outliers is low, $\{\mathbf{n}_i\}$ can converge to the global optimum of (6.14).

Please note that there are several methods aiming at solving (6.14) besides DPCP [27], for example, Exact Recovery of Sparsely-Used Dictionaries (EP-SpUD) [28] and the algorithm based on Alternating Direction Method (ADM) [29].

However, no algorithm is able to perfectly solve (6.14) for any given $\mathbf{X}$. DPCP is adopted because it performs well even in the challenging regime of large outlier ratios and high subspace relative dimensions [27, 30].

Assume that the dimension of the full-rank space is k and that the rank of H_1 is $k-1$. Let q denote the proportion of inliers in the total M points. It has been proved in [30] that DPCP can tolerate approximately $O(\frac{1}{(k-1)k\log k^2}q^2M^2)$ outliers. However, if the number of inliers is relatively small, the algorithm cannot guarantee a positive result. When there are no positive results from DPCP, an alternative algorithm based on random sampling (RS) can be used to reveal the potential hyperplane. Randomly choose $k-1$ column vectors $\mathbf{x}_{r_1}, \ldots, \mathbf{x}_{r_{k-1}}$ from $\mathbf{X}$ and the norm vector $\tilde{\mathbf{n}}$ for $span(\mathbf{x}_{r_1}, \ldots, \mathbf{x}_{r_{k-1}})$ can be constructed using Gaussian elimination or other algebraic methods. Then, check the following condition:

$$\Theta(\mathbf{X}^\top \tilde{\mathbf{n}}) > pM, \tag{6.16}$$

where p is a predefined threshold, and $\Theta(\cdot)$ is the function that counts the number of zero entries of a given vector. If (6.16) is satisfied, the hyperplane defined by $\tilde{\mathbf{n}}$ contains at least pM points. Otherwise, the random sampling procedure is repeated until the maximum number of sampling N is reached.

The probability of finding the expected norm vector in a single procedure is:

$$\frac{\binom{qM}{h-1}}{\binom{M}{k-1}} \approx q^{k-1}, \quad k-1 \ll qM. \tag{6.17}$$

The probability that the norm vector is not found within N procedures is

$$Prob = \left(1 - \frac{\binom{qM}{k-1}}{\binom{M}{k-1}}\right)^N \approx (1 - q^{k-1})^N. \tag{6.18}$$

If N is large enough, an upper bound for $Prob$ can be guaranteed as $Prob$ decreases exponentially w.r.t. N. The pseudo code for one iteration of the top-down subspace identification method is given in Algorithm 6.2.

Since (6.15) is a convex problem, it can be solved by efficient commercial solvers. The time complexity for RS is $O(Nk^3) = O(\frac{\log \varepsilon}{\log(1-p^{k-1})}k^3) \approx O(-\log \varepsilon \cdot (\frac{1}{p})^{k-1}k^3)$ for keeping the failing probability $Prob < \varepsilon$.

Theoretically, the top-down subspace search is applicable regardless of the basis gap. However, using the bottom-up algorithm is better in the first place. As illustrated in Fig. 6.10 later, the top-down searching considers the problem in a tree structure that would have 2^k nodes in worst conditions. Thus, Algorithm 6.2 is inefficient when k is large. If Algorithm 6.1 is used first, part of the basis vectors might be obtained, and the dimension of data can be reduced to improve the overall efficiency.

Algorithm 6.2 Top-down subspace search

```
procedure TOP- DOWN(X,N)
Initialization : ñ ← ∅;
  ñ ←DPCP(X);
  if (6.16) holds then
    return ñ
  else
    for i ← 1 : N do
      ñ ←RS(X);
      if (6.16) holds then
        return ñ
      end if
    end for
  end if
end procedure
```

6.3.6 *Short Summary*

In summary, the whole process of our unsupervised congestion status identification method has the following steps:

1. Data preprocessing: remove the non-congestion vectors from matrix **X**. Use PCA for dimension reduction.
2. Run procedure BOTTOM-UP(**X**, ε) in Algorithm 6.1, and obtain the basis set $\mathcal{B}$.
3. Project **X** onto $\mathcal{B}^{\perp}$, and remove the all-zero columns from **X**. Let $\mathbf{X}^{(2)}$ denote the result. Usually the process terminates here, because $\mathbf{X}^{(2)}$ is empty.
4. If $\mathbf{X}^{(2)}$ is not empty, procedure TOP- DOWN($\mathbf{X}^{(2)}$, N) from Algorithm 6.2 can be used iteratively to find the subspaces and their normal vectors.
5. Construct the basis matrix **B** from the basis set $\mathcal{B}$ of step 2 and the normal vectors of step 4. If $Rank(\mathbf{B})$ equals $Rank(X)$, then we have all the basis to represent **X**. The dual variables χ_t can be recovered by:

$$\hat{\chi}_t = \mathbf{B}^{\dagger}\mathbf{X} \tag{6.19}$$

where † means the Moore-Penrose pseudoinverse. And $\mathfrak{X}_t$ can be recovered using (6.9).

6.4 Case Study

In this section we show four cases based on the IEEE 30-bus system, the IEEE 118-bus system, the Illinois 200-bus system, and the SPP market, to demonstrate the effectiveness of the proposed method. All numerical experiments are conducted on an Intel Core i7-7500U@2.90 GHz laptop with MATLAB R2016a. The market is cleared every 5 min. In each simulated case, 576 vectors of π_t^C with congestion are collected.

To evaluate the identification accuracy, assume that $\mathbf{B}$ is the recovered basis matrix. We can recover $\mathfrak{X}_t$ from $\mathbf{B}^{\dagger}\mathbf{X}$. Let $\mathfrak{C}_t$ denote the recovered code $\mathbb{1}(\mathbf{B}^{\dagger}\mathbf{x}_t)$, and the following miscode rate is used to evaluate the result:

$$misrate = 1 - \sum_{t \in \mathcal{T}} \frac{\Theta(\mathfrak{X}_t - \mathfrak{C}_t)}{k \times M}. \tag{6.20}$$

For each row of $\mathfrak{X}_t$, which corresponds to the congestion status of a single transmission line, the *misrate* can be calculated to show the identification accuracy.

6.4.1 Case 1: IEEE 30-Bus System

This case was conducted on the IEEE 30-bus system with 6 generations at buses #1, #2, #13, #22, #23, and #27. The block offers of the generations are set the same as in [4]. The load data at each bus are generated from a Kaggle open source dataset [31] in January 2008. All the block offers and loads are added with Gaussian fluctuations of 3% relative standard deviation. The corresponding values of $\mathfrak{X}_t$ are shown in Fig. 6.7.

This case can be solved by bottom-up searching since there is no basis gap. After data preprocessing with PCA, the dimension of the data matrix $\mathbf{X}$ is reduced to (4, 576). Following the steps in Algorithm 6.1, we construct the $\mathbf{A}$ matrix and apply the cutoff kernel κ with $\varepsilon = 0.005$. Figure 6.8 shows the clustering results in the first iteration. Four clusters are detected, and clusters 1 and 3 are 1-dimensional subspaces. Thus, two basis vectors $\mathbf{B} = [\mathbf{b}_1, \mathbf{b}_2]$ can be constructed from clusters 1 and 3. Then, $\mathbf{X}$ is projected on the subspace $\mathcal{B}^{\perp} = span(\mathbf{B})^{\perp}$, and the all-zero columns are removed to obtain $\mathbf{X}^{(2)}$.

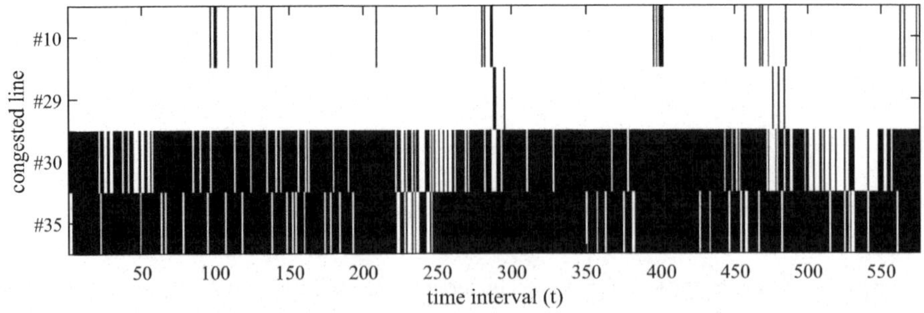

Fig. 6.7 The 576 values of $\mathfrak{X}_t$ in the numerical test of IEEE case30. Dark blocks indicate congestion (or 1-entries in $\mathfrak{X}_t$), while light blocks indicate non-congestion (or 0-entries in $\mathfrak{X}_t$)

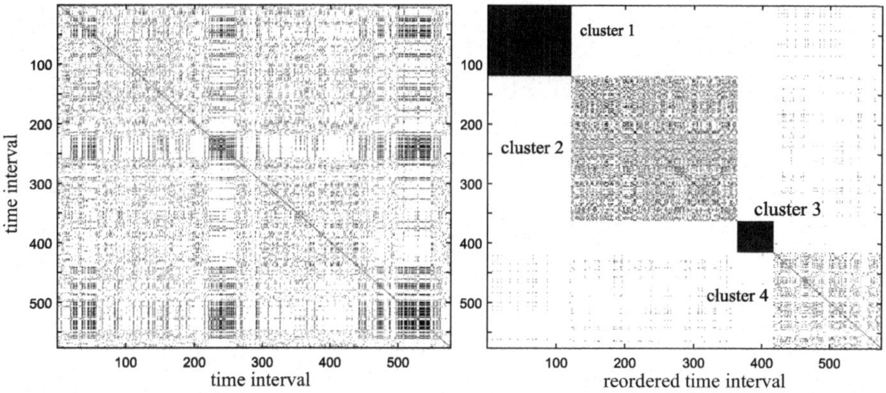

Fig. 6.8 Bottom-up searching with spectral clustering. Left: the **A** matrix with the cutoff kernel. Right: the spectral clustering result. Dark blocks indicate 1

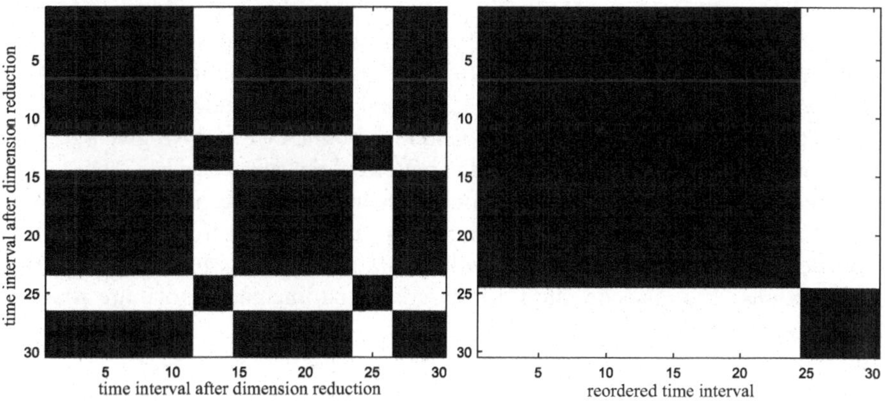

Fig. 6.9 Bottom-up searching result in Round 2. Left: **A** matrix updated from $\mathbf{X}^{(2)}$. Right: **A** matrix after spectral clustering and index reordering

In this case, $\mathbf{X}^{(2)}$ still contains 30 points after one iteration. Thus, the bottom-up searching continues, and **A** is updated in Round 2. Figure 6.9 shows the clustering result; two clusters are detected in this round. Two additional basis vectors can be added to **B**. Now that $Rank(\mathbf{B}) = 4$, we can recover $\mathfrak{X}_t$ from $\mathbf{B}^{\dagger}\mathbf{X}$. As we can see from Table 6.3, the miscode rate is very low for all the lines in this case, indicating a precise identification result.

6.4.2 Case 2: IEEE 118-Bus System

This case is intended to test the efficiency of the top-down searching algorithm. The regional load and load participation factors at different buses were set using the NREL extended case data of December 2012 provided in [32]. To simplify the combinations of congested lines, we set transmission capacity for only 7 interregional lines. The block offers of the generators were set so that Region 1 had a higher generation cost than the other two regions, and 3% Gaussian fluctuations were added to the load and block offers in the real-time market.

Table 6.2 shows the congestion combinations that appeared in this case. At least two lines are congested at any time, and the bottom-up searching algorithm cannot detect any rank-1 clusters. To address the basis gap in this case, the top-down searching algorithm is used. Figure 6.10 shows the detailed searching process. First, Algorithm 6.2 is applied to find the norm vector $\tilde{\mathbf{n}}_1$ for a hyperplane among the 576 points. The result of DPCP is positive, and the points are classified according to their orthogonality with $\tilde{\mathbf{n}}_1$. Here, 529 points lie in the detected hyperplane, and the searching continues with them. DPCP has a negative result in Round 2, and RS is used alternatively. $\tilde{\mathbf{n}}_2$ is detected by RS, and 230 points are orthogonal to $\tilde{\mathbf{n}}_2$. Additionally, RS is applied to the 47 outliers in Round 1 to reveal hidden subspaces. $\tilde{\mathbf{n}}_3$ is detected with 19 orthogonal points. In Round 3, both DPCP and RS give a negative result, which means we have reached the bottom of the subspaces.

To construct the basis vectors $\mathbf{B}$ from the hierarchical structure of Fig. 6.10, we start from the 230 points in the bottom. $\mathbf{b}_1, \mathbf{b}_2$ are extracted from them, and $\mathbf{b}_3$ is extracted from the other 299 points with $Rank = 3$. Finally, $\mathbf{b}_4$ is extracted from the 19 points. The miscode rate for each congested line and in total are shown in Table 6.3.

6.4.3 Case 3: Illinois 200-Bus System

To test the proposed method in a large power system, simulation of the electricity market is conducted on the Illinois 200-bus system [15]. The load data are randomly generated based on the values given in the standard case in MATPOWER [33]. The

Table 6.2 The appearing combinations of congested lines, values of $\mathfrak{X}_t$, and their frequency in case 2

Number of congested lines	Line No.	Value of $\mathfrak{X}_t$ and frequency
4	#(44,45,54,128)	$(1, 1, 1, 1)^\top$, 4.86%
3	#(44,45,54) #(44,54,128)	$(1, 1, 1, 0)^\top$, 3.30%
		$(1, 0, 1, 1)^\top$, 51.9%
2	#(44,54)	$(1, 0, 1, 0)^\top$, 39.9%

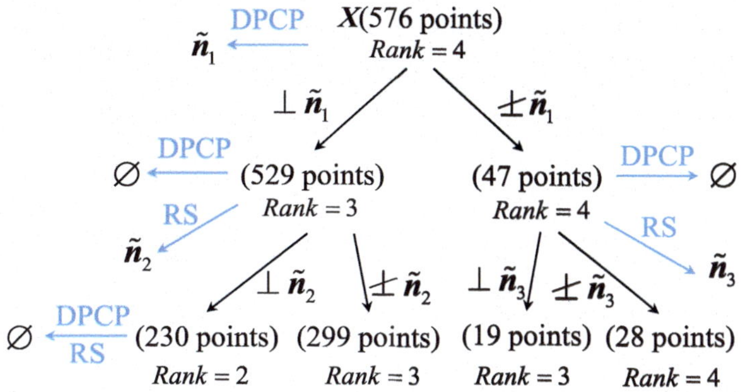

Fig. 6.10 Iterative top-down searching process of case 2

Table 6.3 Miscode rates in the three numerical cases

Misrate	IEEE 30-bus system	IEEE 118-bus system	Illinois 200-bus system	SPP market
Total	0.39%	0	0	0.54%
1st Line	0	0	0	0
2nd Line	0	0	0	0.80%
3rd Line	1.22%	0	0	1.61%
4th Line	0.35%	0	0	0
5th Line	–	–	0	0
6th Line	–	–	0	0.80%

block offers of generators are set the same as two blocks of 65% generator capacity with a price of 30 and 35 \$/MWh. Six of the 245 lines are ever-congested during the test time period.

In this case, the basis gap also exists, and the top-down algorithm is used to identify all the subspaces. Similar to Case 2, DPCP and RS are used iteratively to conduct a top-down hierarchical classification of the whole dataset. After 5 rounds, 6 basis vectors are detected to form the basis matrix **B**. The miscode rates are shown in Table 6.3.

6.4.4 Case 4: Southwest Power Pool (SPP)

We collected data from the real-time market of the SPP, which is an RTO serving 14 states in the USA. LMP data from 7 408 electric nodes on December 7, 2018 are used. We merged the nodes that differed only in name and had identical LMPs all the time, leaving 3 300 electric nodes. The SPP publishes the real-time π^C along with the

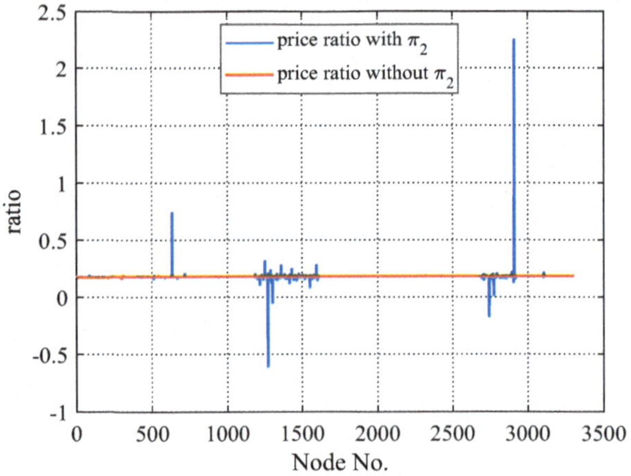

Fig. 6.11 The nodal congestion price ratio of the 5th and 3rd operation interval in SPP on December 7, 2018

binding constraints and their shadow prices μ on its website. Among the 288 time intervals, 249 intervals have one or more binding constraints. Six lines have been congested once or more during the day. Since the SPP uses the lossy OPF model, the raw π^C data need to be preprocessed using the method mentioned in Sect. 6.2 to eliminate the impact of π_2. Figure 6.11 shows an example on the results of the elimination. The 3rd and 5th operation interval on December 7, 2018 belong to the same congestion status with only one line congested. The congestion price ratio is nearly the same for all the nodes after removing π_2.

In this case, bottom-up searching is used. The clustering results in Round 1 are shown in Fig. 6.12, and 2 rank-1 clusters are detected. The size of cluster 1 and cluster 2 is 147 and 17, respectively. Thus, we obtain 2 basis vectors after Round 1. There are still 85 time intervals after dimension reduction, and a new round of search starts for the remaining data. In Round 2, three basis vectors are detected, as shown in Fig. 6.13. The sizes of the three clusters are 39, 20, and 10, respectively. Finally, the last basis vector is found in Round 3. The detailed miscode rates are shown in Table 6.3.

6.4.5 *Time Consumption*

To test the time consumption of the proposed method in systems of different scales, we repeated the identification method in the three cases for 50 times. The average time consumption values are listed in Table 6.4. The entire procedure of the method costs only a few seconds or even less in some cases.

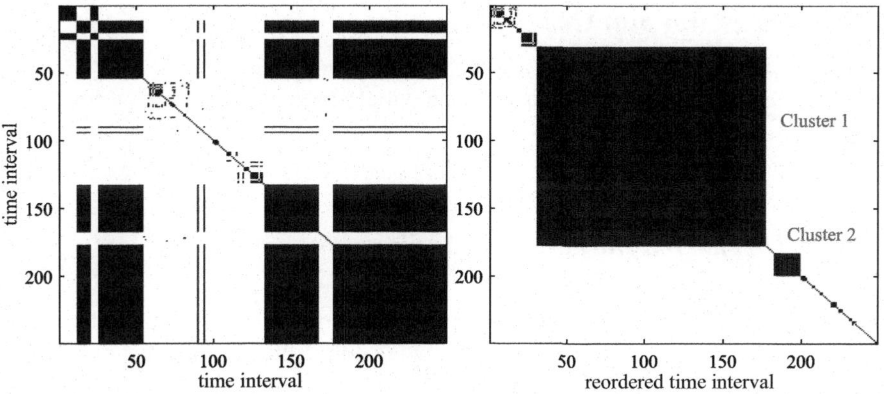

Fig. 6.12 Bottom-up searching with SPP data in Round 1. Left: the original **A** matrix. Right: **A** matrix after spectral clustering and index reordering

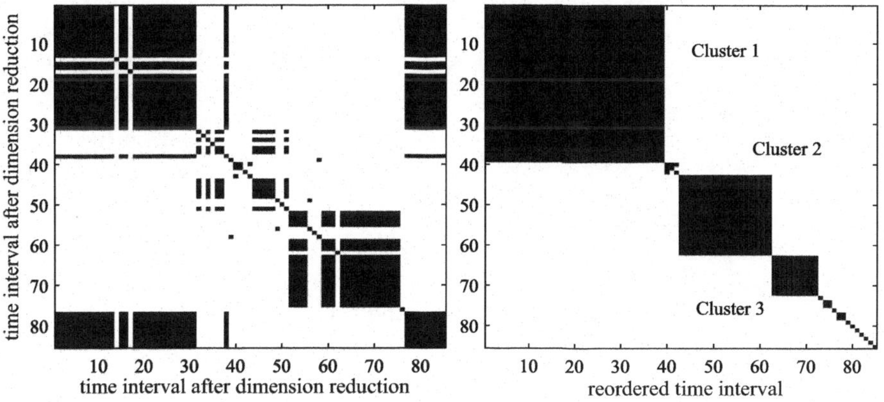

Fig. 6.13 Bottom-up searching with SPP data in Round 2. Left: **A** matrix updated from $\mathbf{X}^{(2)}$. Right: **A** matrix after spectral clustering and index reordering

Table 6.4 Average time consumptions (Second) of the four cases

Step	IEEE 30-bus system	IEEE 118-bus system	Illinois 200-bus system	SPP market
PCA	0.002	0.006	0.013	0.08
Rd.1	0.0230	0.480 (DPCP)	1.220 (DPCP)	0.225
Rd.2	0.0150	1.105 (DPCP/RS)	0.370 (DPCP)	0.071
Rd.3	–	–	0.546 (DPCP)	1E-4
Rd.4	–	–	0.366 (DPCP)	–
Rd.5	–	–	0.554 (DPCP/RS)	–

6.5 Discussion and Conclusion

This section presents a discussion on the potential application of the proposed method and concludes the chapter.

6.5.1 *Discussion on Potential Applications*

Essentially, the proposed method determines the network congestion status from public LMP data. The method has the following applications for market participants.

- Revealing the information of the congested status for each time period, especially for emerging markets where such information is not publicized, such as markets in China and Mexico.
- LMP forecast calibration or reconciliation. The basis vectors obtained from the method are useful for LMP forecasting. The forecasted LMP data achieved from traditional LMP forecasting models (e.g., neural networks, SVM, etc.) do not naturally follow the distribution mentioned in Sect. 6.2. Thus, it would be necessary to reconcile the forecasting results according to the revealed basis vectors with quadratic programming [34] or other related methods and get structurized forecasting results.
- Integrating with exogenous market variables. The congestion status is a function of nodal loads, generation bids, and transmission capacity. Such function can be fitted with exogenous market variables such as nodal loads [3, 8] and generation mix [7] and be further used for congestion forecasting, FTR valuation, and quantifying the market power of certain generators.

6.5.2 *Conclusion*

This chapter proposes a congestion status identification method based on unsupervised machine learning techniques with LMP data. The low-rank property of the congestion component of LMP is analyzed theoretically. Then the original problem is formulated as a subspace detection problem, which is solved by constructing the basis vector set and determining the coefficient of the basis vectors for each LMP data point. Two hierarchical basis vector searching algorithms are proposed to address the subspace detection problem. The bottom-up search can efficiently detect the basis vectors when there is no basis gap, and recover part of the power transfer distribution factor matrix of the power system. The top-down search can find the subspaces in cases with a basis gap. The results from four numerical cases show that the proposed method achieves nearly 100% accuracy within only a few seconds.

Future work includes integrating the identification method with LMP forecasting and market bidding strategy making.

References

1. Conejo, Antonio J, Enrique Castillo, Roberto Mínguez, and Federico Milano. 2005. Locational marginal price sensitivities. *IEEE Transactions on Power Systems* 20 (4): 2026–2033.
2. Yang, Zhifang, Anjan Bose, Haiwang Zhong, Ning Zhang, Jeremy Lin, Qing Xia, and Chongqing Kang. 2017. LMP revisited: A linear model for the loss-embedded LMP. *IEEE Transactions on Power Systems* 32 (5): 4080–4090.
3. Geng, Xinbo, and Le Xie. 2017. Learning the LMP-load coupling from data: A support vector machine based approach. *IEEE Transactions on Power Systems* 32 (2): 1127–1138.
4. Kekatos, Vassilis, Georgios B Giannakis, and Ross Baldick. 2016. Online energy price matrix factorization for power grid topology tracking. *IEEE Transactions on Smart Grid* 7 (3): 1239–1248.
5. Birge, John R, Ali Hortaçsu, and J Michael Pavlin. 2017. Inverse optimization for the recovery of market structure from market outcomes: An application to the MISO electricity market. *Operations Research* 65 (4): 837–855.
6. Mather, Jonathan, Enrique Baeyens, Dileep Kalathil, and Kameshwar Poolla. 2020. The geometry of locational marginal prices. Preprint. http://people.tamu.edu/~dileep.kalathil/Download/GLMP-TPS.pdf.
7. Radovanovic, Ana, Tommaso Nesti, and Bokan Chen. 2019. A holistic approach to forecasting wholesale energy market prices. *IEEE Transactions on Power Systems* 34 (6): 4317–4328.
8. Zhou, Qun, Leigh Tesfatsion, and Chen-Ching Liu. 2011. Short-term congestion forecasting in wholesale power markets. *IEEE Transactions on Power Systems* 26 (4): 2185–2196.
9. Ji, Yuting, Robert J Thomas, and Lang Tong. 2016. Probabilistic forecasting of real-time LMP and network congestion. *IEEE Transactions on Power Systems* 32 (2): 831–841.
10. Wang, Yishen, Yury Dvorkin, Ricardo Fernández-Blanco, Bolun Xu, and Daniel S Kirschen. 2017. Impact of local transmission congestion on energy storage arbitrage opportunities. In *2017 IEEE Power Energy Society General Meeting*, 1–5.
11. Fang, Xin, Fangxing Li, and Qinran Hu, and Yanli Wei. 2016. Strategic CBDR bidding considering FTR and wind power. *IET Generation, Transmission Distribution* 10 (10): 2464–2474.
12. Prete, Chiara Lo, Nongchao Guo, and Uday V Shanbhag. 2019. Virtual bidding and financial transmission rights: An equilibrium model for cross-product manipulation in electricity markets. *IEEE Transactions on Power Systems* 34 (2): 953–967.
13. Chen, Ruidi, Ioannis Ch Paschalidis, Michael C Caramanis, and Panagiotis Andrianesis. 2019. Learning from past bids to participate strategically in day-ahead electricity markets. *IEEE Transactions on Smart Grid* 10 (5): 5794–5806.
14. Cheverez-Gonzalez, Daniel, and Christopher L DeMarco. 2009. Admissible locational marginal prices via laplacian structure in network constraints. *IEEE Transactions on Power Systems* 24 (1): 125–133.
15. Birchfield, Adam B, Ti Xu, Kathleen M Gegner, Komal S Shetye, and Thomas J Overbye. 2017. Grid structural characteristics as validation criteria for synthetic networks. *IEEE Transactions on Power Systems* 32 (4): 3258–3265.
16. Southwest Power Pool, Inc. Real-time Balancing Market. 2019. https://marketplace.spp.org/groups/real-time-balancing-market.
17. Eldridge, Brent, Richard P O'Neill, and Anya Castillo. 2017. Marginal loss calculations for the DCOPF. *Federal Energy Regulatory Commission*, Technical Report. https://www.ferc.gov/sites/default/files/2020-05/marginallosscalculations.pdf.
18. Litvinov, Eugene, Tongxin Zheng, Gary Rosenwald, and Payman Shamsollahi. 2004. Marginal loss modeling in LMP calculation. *IEEE Transactions on Power Systems* 19 (2): 880–888.
19. Price, James E, and John Goodin. 2011. Reduced network modeling of WECC as a market design prototype. In *2011 IEEE Power and Energy Society General Meeting*, 1–6, July 2011.
20. Thatte, Anupam A, Dae-Hyun Choi, and Le Xie. 2014. Analysis of locational marginal prices in look-ahead economic dispatch. In *2014 Power Systems Computation Conference*, 1–7, Aug 2014.
21. Hua, Bowen, Dane A Schiro, Tongxin Zheng, Ross Baldick, and Eugene Litvinov. 2019. Pricing in multi-interval real-time markets. *IEEE Transactions on Power Systems* 34 (4): 2696–2705.

22. Vidal, Rene, Yi Ma, and Shankar Sastry. 2005. Generalized principal component analysis (GPCA). *IEEE Transactions on Pattern Analysis and Machine Intelligence* 27 (12): 1945–1959.
23. Elhamifar, Ehsan, and Rene Vidal. 2013. Sparse subspace clustering: Algorithm, theory, and applications. *IEEE Transactions on Pattern Analysis and Machine Intelligence* 35 (11): 2765–2781.
24. Tierney, Stephen, Yi Guo, and Junbin Gao. 2015. Segmentation of subspaces in sequential data. arXiv:1504.04090.
25. Vidal, René. 2011. Subspace clustering. *IEEE Signal Processing Magazine* 28 (2): 52–68.
26. Sánchez-García, Rubén J, Max Fennelly, Seán Norris, Nick Wright, Graham Niblo, Jacek Brodzki, and Janusz W Bialek. 2014. Hierarchical spectral clustering of power grids. *IEEE Transactions on Power Systems* 29 (5): 2229–2237.
27. Tsakiris, Manolis C, and René Vidal. 2018. Dual principal component pursuit. *The Journal of Machine Learning Research* 19 (1): 684–732.
28. Spielman, Daniel A, Huan Wang, and John Wright. 2012. Exact recovery of sparsely-used dictionaries. In *Proceedings of the 25th Annual Conference on Learning Theory, PMLR* 23: 37.1–37.18.
29. Qu, Qing, Ju Sun, and John Wright. 2014. Finding a sparse vector in a subspace: Linear sparsity using alternating directions. In *Advances in Neural Information Processing Systems*, 3401–3409.
30. Zhu, Zhihui, Yifan Wang, Daniel Robinson, Daniel Naiman, Rene Vidal, and Manolis Tsakiris. 2018. Dual principal component pursuit: Improved analysis and efficient algorithms. In *Advances in Neural Information Processing Systems*, 2172–2182.
31. Kaggle. 2012. Global energy forecasting competition 2012-load forecasting: A hierarchical load forecasting problem: Backcasting and forecasting hourly loads (in kW) for a US utility with 20 zones. https://www.kaggle.com/c/global-energy-forecasting-competition-2012-load-forecasting/data.
32. Pena, Ivonne, Carlo Brancucci Martinez-Anido, and Bri-Mathias Hodge. An extended IEEE 118-bus test system with high renewable penetration. *IEEE Transactions on Power Systems* 33(1): 281–289.
33. Zimmerman, Ray Daniel, Carlos Edmundo Murillo-Sánchez, Robert John Thomas, et al. 2011. Matpower: Steady-state operations, planning, and analysis tools for power systems research and education. *IEEE Transactions on Power Systems* 26 (1): 12–19.
34. Hyndman, Rob J, Alan J Lee, and Earo Wang. 2016. Fast computation of reconciled forecasts for hierarchical and grouped time series. *Computational Statistics & Data Analysis* 97: 16–32.

Chapter 7
Day-Ahead Electricity Price Forecasting

7.1 Introduction

Accurate forecasting of locational marginal prices (LMPs) is of vital importance to market participants, including electricity retailers, large customers, virtual power plants (VPPs), and financial institutions in virtual trading [1–3]. They usually optimize their scheduled operation or bidding strategy in day-ahead and real-time markets according to the expected value of the corresponding LMPs in the short term. Although LMP forecasting is fundamental, it is very difficult to achieve a high accuracy due to the complexity of the factors and mechanisms that lie behind LMP. In most of the electricity spot markets in the U.S., the LMPs are derived from the co-optimization of energy and ancillary reserves [4], which can be influenced by network constraints, energy, and ancillary requirements, and bids from generators.

The methods commonly employed for electricity price forecasting in the literature can be roughly classified into two categories: market-simulation-based and historical-data-based approaches [5]. The former approaches mimic the market clearance procedure with a security-constrained unit commitment and economic dispatch (SCUC and SCED) from the perspective of a market organizer, e.g., an independent system operator (ISO) or regional transmission organization (RTO). Thus, market-simulation-based methods require detailed information about the network and market structure, which is usually released untimely and incompletely. As we are discussing short-term LMP forecasting for market participants who could only obtain some historical information publicized by the ISO/RTO, we will focus on the latter category of methods in this chapter.

The historical-data-based methods are based on data mining models, which can be further subdivided into statistical models and artificial intelligence (AI) models. The commonly-used statistical models are linear regression and autoregressive (AR) models. Karakatsani et al. discussed the impacts of economic, technical, and strategic factors on intraday prices and the dynamics of these effects over time with a time-varying regression (TVR) model in [6]. Numerical experiments based on data collected in the British market proved the performance of the TVR model among various alternatives, including linear regression and autoregressive models. In [7], Taylor introduced five new univariate exponentially weighted methods based on discount weighted regression and singular value decomposition for forecasting the

Q. Chen et al., *Data Analytics in Power Markets*,
https://doi.org/10.1007/978-981-16-4975-2_7

intraday time series with both intraday and intraweek seasonal cycles. The application of AR model with exogenous variables (ARX) in electricity price forecasting is discussed in [8]. Pircalabu et al. proposed a regime-switching AR model to forecast pairs of day-ahead electricity prices in European markets [9]. Statistical models perform well in explaining the forecasting results, but their ability to model complex nonlinear behaviors of price series and price spikes is limited.

The major strength of AI methods is their ability to handle complexity and nonlinearity [5], making AI suitable for building generalized forecasting models from a number of features. Ma et al. proposed a neural-network (NN)-based method for forecasting zonal LMPs [10]. An NN trained by the Decoupled Extended Kalman Filter (DEKF) was capable of forecasting the LMPs at different zones and the difference between them (*a.k.a.* the congestion component of the LMP). In [11], an NN with Levenberg-Marquardt learning algorithm was used to forecast the electricity price. A novel algorithm iteratively searching the minimum set of features that minimizes the overall redundancy and maximizes the synergy among the features was proposed. In [12], a random forest (RF) model was proposed to predict the California ISO (CAISO) day-ahead market price. Calendar data and ancillary service (AS) requirements were considered as input features in addition to the historical hourly prices. Huang et al. considered the problem of electricity price forecasting as a price classification problem based on certain price thresholds [13]. Classifiers including decision trees, multilayer perceptrons, and k-nearest neighbors are employed. In [14, 15], extreme learning machine (ELM) is applied to estimate the point forecasts of market clearing prices, and its performance is validated with real price data from the Australian electricity market. A deep neural network (DNN) with two hidden layers was used in day-ahead electricity price forecasting in [16]. Bayesian optimization and functional analysis of variance were adopted in the feature selection procedure. Luo et al. used machine learning models including DNN and support vector regression (SVR) to forecast real-time market prices [17]. It was found that selecting wind power generation as an intermediate variable in forecasting could improve the overall accuracy. In [18], the performance of deep learning approaches including long-short term memory (LSTM) are compared with traditional AR models in spot electricity price forecasting. In [19] a hybrid price prediction model of LSTM and wavelet transform was proposed. Hubicka et al. applied the non-linear counterpart of the ARX model—a NARX neural network in day-ahead electricity price forecasts and discussed on the selection of calibration windows in [20]. Despite the advantages, there are many choices in models and hyperparameters, and it is difficult to find an optimal AI method.

The idea of ensemble methodology is to build a predictive model by integrating multiple models [21]. The combined results from individual forecasts (*a.k.a.* sister forecasts) are more accurate or at least equivalent to the individual forecasts in electricity price forecasting and load forecasting [22, 23]. In [22], a linear weighted average was used to combine individual models of ARMIA, TVR, and a Markov regime-switching model. In [23], stacking methods of linearly, exponentially, as well as polynomially weighted average were tested. Lee et al. applied performance-based averaging in stacking multiple forecasting machines for wind power and real-time electricity price [24]. Ensemble model output statistics (EMOS) was used to stack collective ELM-based price forecasting models in [25]. The performance of a stacked or combined method relies on the accuracy, heterogeneity, and complementarity of the individual models. If the results given by multiple sister forecasts are very similar, little improvement could be achieved through model stacking.

In this chapter, we reconsider the problem of short-term electricity price forecasting in a componential way. During the SCED process, the LMP is generated with three components, i.e., marginal cost of energy (MCE), congestion (MCC), and loss (MCL). Most studies in the literature focus on building a direct model for the whole LMP forecasting procedure and ignore certain properties of the LMP. They attempt to link external variables with the LMP, but do not utilize the features behind the three components. In the rest of this chapter, the phrase "direct forecasting" is used to present LMP forecasting with such direct models.

We believe that the respective physical and economic meanings of these marginal components (which will be further introduced in Sects. 7.2 and 7.3) indicate that different forecasting methods should be adopted for each component. To improve the forecasting accuracy with these specific characteristics of the LMP and its components, we fine-tune the forecasters for each component and obtain a novel LMP forecasting method through the result of a summation. We also propose a framework that stacks the composition forecast with the direct LMP forecast. The efficacy of the framework is demonstrated with publicly available market data from CAISO [26] over a wide range of time scales. The main contributions of this chapter are three-fold:

1. It considers the problem of LMP forecasting in a componential approach, in which three individual forecasting models are trained for the corresponding components. The characteristics of each component are analyzed and taken into consideration for feature engineering.
2. It proposes a framework that stacks the novel composition forecasts and the direct forecasts to improve the overall performance with the heterogeneity and complementarity between the individual models.
3. It conducts comprehensive numerical experiments to show the effectiveness of the proposed forecasting method and stacking framework with real market data collected from CAISO.

The remainder of this chapter is structured as follows. Section 7.2 presents the basic formulation of componential LMP forecasting and model stacking. Section 7.3 presents the proposed framework and details the forecasting models. Section 7.4 describes the settings and results of the case study. Finally, Sect. 7.5 draws the conclusion.

7.2 Problem Formulation

7.2.1 Decomposition of LMP

7.2.1.1 Energy Market Model

In an energy-only market, the operator solves the SCED problem of (7.1) to determine the optimal generation schedule. Note that the network loss has been embedded into the problem in a linearized form, which is generally conducted by most ISOs/RTOs [27, 28].

$$
\begin{aligned}
&\min_{\mathbf{P}_G} \sum_i c_i(P_{Gi}) && && (7.1a)\\
s.t.\quad &\mathbf{1}^\top(\mathbf{P}_G - \mathbf{P}_D) - P_{loss} = 0 && :\lambda && (7.1b)\\
&P_{loss} = P_0 + LF^\top(\mathbf{P}_G - \mathbf{P}_D) && :\sigma && (7.1c)\\
&|\mathbf{S}(\mathbf{P}_G - \mathbf{P}_D - P_{loss}\cdot \mathbf{d})| \le \mathbf{f}^{line} && :\mu^{+/-} && (7.1d)\\
&\mathbf{P}_{\min} \le \mathbf{P}_G \le \mathbf{P}_{\max} && :\gamma_{\min/\max} && (7.1e)
\end{aligned}
$$

where $c_i(\cdot)$ is the generation cost function for node i, and $\mathbf{1}$ denotes the all-one vector. $\mathbf{P}_G$ and $\mathbf{P}_D$ are the active power vectors of the generators and loads, respectively. $\mathbf{S}$ is the shift factor matrix, and $\mathbf{f}^{line}$ is the vector of line capacities. $\mathbf{P}_{\min}$ and $\mathbf{P}_{\max}$ are the vectors of the generation output limit. P_{loss} is the active power loss, which is linearized by the factor LF, $\mathbf{d}$, and intercept P_0. The Greek letters are dual variables of the corresponding constraints. The LMP π can be formulated as:

$$\pi = \sigma\mathbf{1} - \sigma LF + (\mathbf{S}^\top - \mathbf{1}\mathbf{d}^\top\mathbf{S}^\top)(\mu^+ - \mu^-) \tag{7.2}$$

For detailed derivation on the LMP, readers can refer to [27, 28]. Note that the entries of $\mu^+ - \mu^-$ are zero unless the corresponding lines are congested and the constraints are binding. So $(\mathbf{S}^\top - \mathbf{1}\mathbf{d}^\top\mathbf{S}^\top)(\mu^+ - \mu^-)$ is considered as the congestion component of π. It represents the price spread among different nodes caused by network congestion. LF measures the marginal proportion of network loss to nodal active power injection, which is determined by the system operating point and updated frequently in the spot market [28, 29]. Thus, σLF is regarded as MCL, and the three decomposed components of the MCE π^E, MCC π^C, and MCL π^L can be formulated as:

$$
\begin{aligned}
\pi^E &= \sigma\mathbf{1} && (7.3a)\\
\pi^C &= (\mathbf{S}^\top - \mathbf{1}\mathbf{d}^\top\mathbf{S}^\top)(\mu^+ - \mu^-) && (7.3b)\\
\pi^L &= -\sigma LF && (7.3c)
\end{aligned}
$$

As indicated by (7.3), the MCC is related to the network congestion conditions, and the MCL is related to network power flow conditions. The MCE has the same value at all nodes and is determined by generation bids and the demand level. They ought to have different patterns and be influenced by different variables.

Usually, the data of the three components are released together with the LMP after the market settlement, which enables us to develop individual forecasting models for these data.

7.2.1.2 Joint Energy and Ancillary Market

In the joint energy and ancillary market where the generation cost and AS cost are co-optimized, there are more items in the objective function and more reserve-related constraints in (7.1). Hence, the ancillary requirements, including spin/non-spin reserve and regulation up/down, can have certain effects on the optimal solution of (7.1) and influence the energy prices. A detailed formulation of the modified SCED in the joint market is beyond the scope of this chapter, and interested readers can refer to [30].

7.2.2 *Short-Term Forecast for Each Component*

We aim to present a general forecasting formulation for market participants with certain historical and forecasting information released by the ISO/RTO. When the current time index is t, we need to forecast the values of certain time series at $t+1$. Let Y denote the target random variable and y denote its ground truth value. Given the publicized information set $\mathcal{I}_t$ at t, the point forecast $\tilde{y}_{t+1}$ of the object y_{t+1} can be given by:

$$\tilde{y}_{t+1} = f_t(\mathcal{F}_t),\ \mathcal{F}_t \subset \mathcal{I}_t \tag{7.4}$$

where $\mathcal{F}$ is the selected feature set. $f_t(\cdot)$ is our forecasting model trained by the historical features and object value $(\mathcal{F}_i, y_{i+1}),\ i \in \{0, 1, \cdots, t-1\}$. $f(\cdot)$ could be either a statistical model or an AI model. In most instances, the selection of $\mathcal{F}$ from $\mathcal{I}$ is conducted through a search of some highly related features and the lagged values of y when there is periodicity in the time series of y.

Taking day-ahead price forecasting as an example, Fig. 7.1 shows the timeline of the day-ahead market process in CAISO. All the information publicized before 10:00 can be utilized. Table 7.1 lists the market information potentially related to price forecasting that can be accessed through CAISO's data API [26]. $\mathcal{F}$ could consist of these data and some customized features. Note that the detailed information of submitted bids is never posted before the completion of the market process, which is also true for other ISOs/RTOs.

Equation (7.4) is the formula for a one-step-ahead forecast. In the case of day-ahead price forecasting, the prices of 24 h in the trading day (TD) should be forecasted

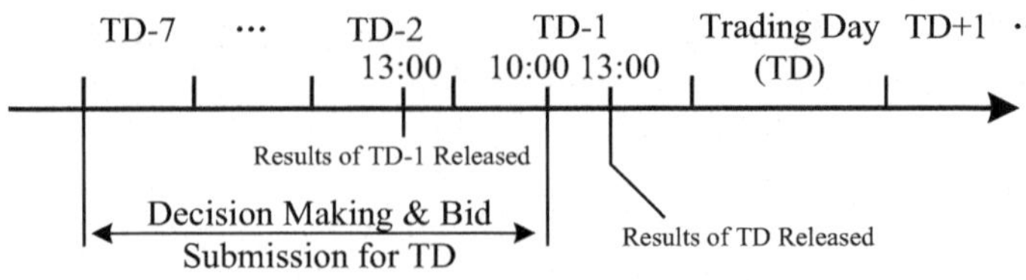

Fig. 7.1 Day-ahead market timeline in CAISO [31]

Table 7.1 Public market information potentially related to LMP forecasting in CAISO

Released data	Description
Energy prices	Data of LMP, MCE, MCC, and MCL at every pricing node; Released at 13:00 before the trading day
Load Forecast	The 5-min, 15-min, and hourly demand (TAC Area); forecast data by transmission access charge area Released before the markets run
Renewable Forecast	The 5-min, 15-min, and hourly wind/solar forecast data aggregated by trading hub; Released before the markets run
Ancillary service requirements	Ancillary service capacity minimum and Maximum per AS region; Released before the markets run
Generation outage schedule	Generator de-rates and outages considered in the day-ahead market; Aggregated by trading hub; Updated every day at 5:00 am;

together. Practically, a multistep ahead forecast is needed for the bidding and offering. When short-term information such as y_t is included in $\mathcal{F}_t$, rolling forecasts should be used to estimate the value of y_{t+1} to y_{t+24} one by one: y_{t+1} is first predicted with y_t, then y_{t+2} to y_{t+24} are predicted using the point forecast $\tilde{y}_{t+1}$ to $\tilde{y}_{t+23}$ through recursion [11].

7.2.3 Summation and Stacking of Individual Forecasts

The LMP π is the summation of π^E, π^C and π^L. For each of the components, we construct an individual forecasting model, and $\tilde{\pi}$ is given by:

$$\tilde{\pi} = \tilde{\pi}^E + \tilde{\pi}^C + \tilde{\pi}^L = f^E(\mathcal{F}^E) + f^C(\mathcal{F}^C) + f^L(\mathcal{F}^L) \tag{7.5}$$

where $f^x(\cdot)$ and $\mathcal{F}^x$ are the selected model and features for component $x \in \{E, C, L\}$, respectively. Note that in (7.3), we use the bold Greek letter $\boldsymbol{\pi}$ as the price symbol to present the price vector at different nodes of the wholesale market. Since π^E is shared by the whole network, the vectorized formula of (7.5) is given as:

$$\tilde{\pi} = \tilde{\pi}^E + \tilde{\pi}^C + \tilde{\pi}^L = f^E(\mathcal{F}^E)\mathbf{1} + \mathbf{f}^C(\mathcal{F}^C) + \mathbf{f}^L(\mathcal{F}^L) \tag{7.6}$$

The idea of summation forecasting with individual sub-models comes from the aggregate and hierarchical forecasting of electricity load [32, 33]. When the data of load in a subregion can be collected and predicted separately, the summed forecasts achieve higher accuracy than traditional ways of forecasting.

It is expected that through proper feature engineering and model parameter tuning, the accuracy of the composition forecast should outperform the traditional direct forecast in non-extreme situations. However, the electricity market is becoming increasingly changeable. Figure 7.2 presents the change in two important statistics, the daily average and intraday variation of the LMP at node PGAE, which is a main aggregated load node in CAISO. As shown in the figure, in 2015 and 2016, the price spikes are within an acceptable range (below 200 $/MWh). In 2017 and 2018, extreme spikes over 500 occurred more frequently, which might be due to special pricing schemes under electricity scarcity. On these occasions, the variation in all three components will become larger than usual, especially for the MCE and MCC. In the summation procedure, the forecast error will accumulate, and the results might be worse than those of the direct forecast.

In model stacking, a metaregressor/classifier gives a combined prediction $\tilde{y}^s$ based on the results produced by lower-level models [24, 25]:

$$\tilde{y}^s = g(\tilde{y}^{(1)}, \cdots, \tilde{y}^{(n)}) \tag{7.7}$$

where $\tilde{y}^{(1)}, \cdots, \tilde{y}^{(n)}$ are the predicted values of y from different models, and $g(\cdot)$ is the metamodel. Generally, a metamodel can take the advantages of its base models and achieve a better forecasting accuracy. The adoption of model stacking could utilize the complementarity of the summation forecasting model and the direct model, thus avoiding the large loss of the summation forecasting model in extreme situations.

7.3 Methodology

In this section, the overall forecasting framework is provided with details for each step.

7.3.1 Framework

The proposed framework of the componential LMP forecasting and model stacking is shown in Fig. 7.3. The basic idea is that, due to the difference between the

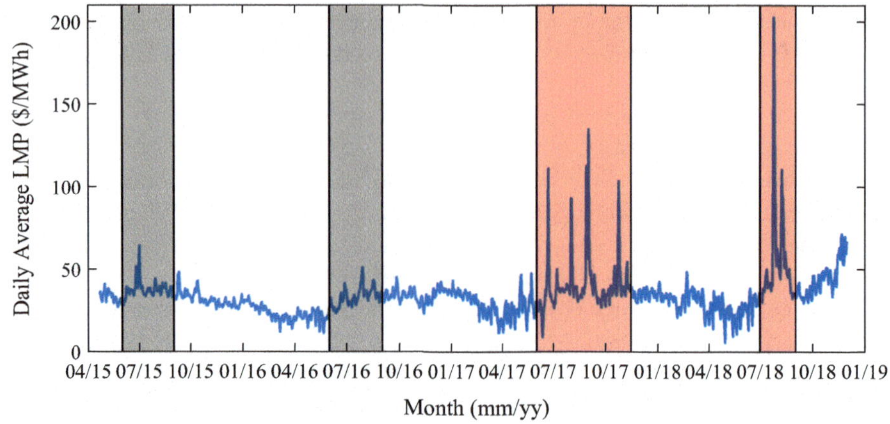

(a) The daily-averaged LMP.

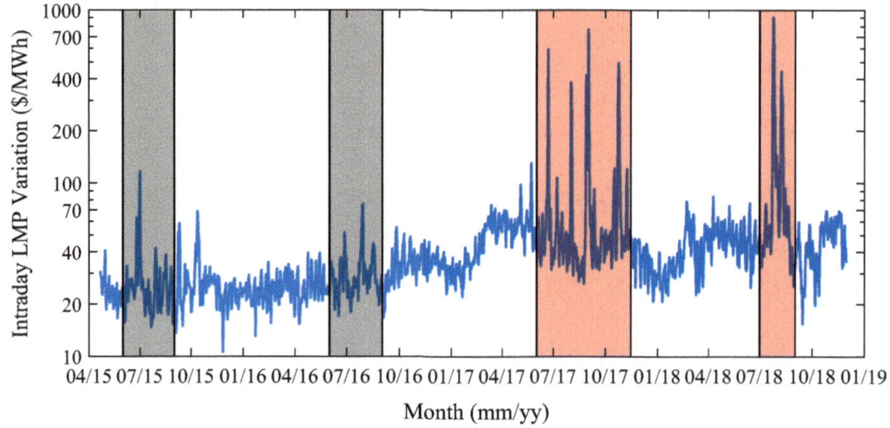

(b) The intraday absolute price variation (logarithmic scaled).

Fig. 7.2 Historical price series and price variation at node PGAE from 2015 to 2018

characteristics of the three components, the features and models should be selected individually to improve the forecasting accuracy. The overall framework involves the following procedures:

1. Data acquisition and cleaning. The historical data of the MCE, MCC, and MCL are required.[1]

[1] The component data can be achieved from the ISO/RTO's website. In some special cases when the ISO/RTO does not release the component data separately, a generator node with a high capacity could be chosen as the reference node, and its LMP is regarded as the MCE. The MCC is regarded as the difference in the LMP between the nodes because the MCL is usually small and can be neglected.

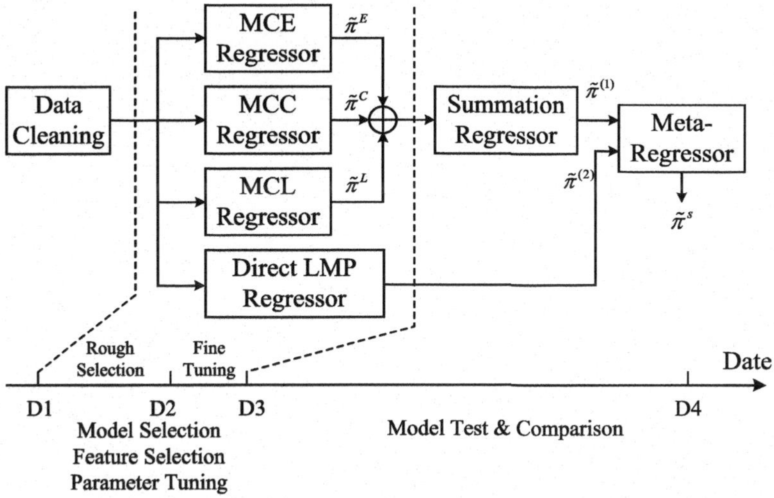

Fig. 7.3 The overall framework of componential and ensemble approach for LMP forecasting. D1 to D4 are specific dates for different procedures

2. Feature and model rough selection for individual models.
3. Parameter fine-tuning for individual models.
4. Aggregate the individual models to get a summation regressor.
5. Model stacking with robust regression.

7.3.2 Feature Engineering

Table 7.2 lists some common features adopted in day-ahead electricity price forecasting [10, 12, 13, 17] which can be achieved from CAISO's website. Basically, auto regressive features are widely used in time series forecasting. It includes certain lagged prices that contain important information for the forecast. This feature is especially useful in MCE forecasting because there are regular patterns in the bidding strategy of the market generators that directly impact the MCE value but cannot be captured by exogenous variables such as network conditions.

The network conditions specify the parameters in the constraints of the SCED problem in the joint energy and ancillary market. Since the MCC and MCL are determined by the system operation conditions, these features should be considered to be the most important features for MCC/MCL forecasting. In the feature selection procedure, the optional feature combinations $\mathcal{F}^E$ for the MCE are AUTOR, (AUTOR+TIME), and (AUTOR+TIME+COND), while the feature combinations $\mathcal{F}^{C/L}$ for the MCC/MCL are COND, (COND+TIME), and (COND+TIME+AUTOR).

Table 7.2 Basic features $\mathcal{F}$ for day-ahead price forecasting

Feature categories		Format	Description
Auto- regressive feature (AUTOR)	Very short term	Numeric	Last 24 h prices
	Short term	Numeric	Previous 24 to 48 h prices
	Long term	Numeric	Price at the same hour last week
Calendar (TIME)	Weekday/month/year	Binary	One-hot encoded
Network condition (COND)	Regional load renewable generation AS requirements outage schedule	Numeric	The latest forecast data released by ISO

7.3.3 *Regression Model Selection and Parameter Tuning*

7.3.3.1 Models

We choose three widely adopted tree-based regression models, decision tree (DT) regressors, random forest (RF) regressors, and extremely randomized tree (ET) regressors for individual forecasting.

DT is a tree-like flowchart with conditional statements on its internal nodes. It is a well-known supervised learning model because it is simple to understand and interpret. RF and ET are ensemble learning methods where each estimator is a decision tree, but they have different ways of implementing tree splitting. RF is based on tree bagging with a randomized way of training each tree. For each tree, a subset of random features is selected, and certain numbers of random samples are selected for training. After tree bagging, it usually has a better and more stable performance than a single DT. ET is similar to RF in randomized feature sets for each tree. But there are two main differences. In the training of each tree in ET, all samples from the training set are used. In addition, in the splitting procedure of tree learner, ET uses a random cut-point for each feature in the considered feature set, while RF computes the locally optimal cut-point. The hyperparameters of these models are easy to tune, and they can achieve high performance in the task of supervised learning [34].

7.3.3.2 Rough Selection of Features and Parameters

In forward forecasting of time series, the well-known k-fold cross-validation cannot be deployed due to the sequentiality of $\mathcal{I}_t$. Since the structure of the electricity market and the behavior of the participants are constantly changing, the forecasting model should be updated appropriately. Thus, we use an adaptive training procedure, i.e., the forecasting model for the next TD being trained by the data of the recent K TDs.

K is considered a hyperparameter for all the models, and the number of estimators N of RF and ET is regarded as the main parameter. As shown in Fig. 7.3, we partition the timeline into three parts: rough selection, fine-tuning, and model testing.

In the rough selection, we set the range of N between 10 and 160 with a step of 30 and the range of K as $\{10, 30, 60, 90\}$. First, the model parameters are selected by grid search. Then, the best feature set is chosen.

7.3.3.3 Fine-Tuning

After the rough selection, the feature sets and the parameters of K are fixed at their best values, and the step of N is changed to 10. Though fine-tuning, the best set of $(\mathcal{F}, K, N)$ is determined for every model.

7.3.4 Model Stacking with Robust Regression

Since the variation in the LMP is large, we use two robust regression algorithms, Huber regression and ℓ_1 regression as the metaregressors for their robustness with high-variance data and outliers [35]. Let $e_t^{(i)}$ denote the forecast error from the ith model at time t, and e^s denote the error from the stacked model. Compared with the traditional least-square approach that tries to minimize the ℓ_2 loss,

$$Loss_{LS} = \sum_t (e_t^s)^2 = \sum_t (w_1 e_t^{(1)} + \cdots + w_n e_t^{(n)})^2 \tag{7.8}$$

Huber regression minimizes the Huber loss function:

$$Loss_H = \sum_t \rho(e_t^s, h) \tag{7.9}$$

where

$$\rho(x, h) = \begin{cases} \frac{1}{2}x^2, & |x| \le h \\ h|x| - \frac{1}{2}h^2, & |x| > h \end{cases}. \tag{7.10}$$

And ℓ_1 regression minimizes the ℓ_1 loss:

$$Loss_{\ell_1} = \sum_t |w_1 e_t^{(1)} + \cdots + w_n e_t^{(n)}| \tag{7.11}$$

As we can see from (7.10) and (7.11), since the ℓ_2 loss is replaced by the ℓ_1 loss, the effect of extreme data on the regression results can be reduced. In comparison, linear regression and SVR are also tested in this chapter. For each stacking algorithm, four situations are considered:

- Stacking the best models (Method I): Only the best direct forecaster and the best summation forecaster are stacked (2 models).
- Stacking the direct models (Method II): For each of the DT, RF, and ET, find the best model in direct forecasting and stack them (3 models).
- Stacking the summation models (Method III): For each of the DT, RF, and ET, find the best model in summation forecasting and stack them (3 models).
- Stacking all the models (Method IV): Stack the 6 models in Method II and Method III.

Method II to Method IV try to improve the overall accuracy by increasing model diversity. However, as shown in Sect. 7.4, more does not mean better in LMP forecasting. Similar to Sect. 7.3.3, the stacking models are trained adaptively, with K fixed at 30.

7.3.5 Metrics

To conduct a comprehensive evaluation, two metrics, the mean absolute error (MAE) and the symmetric mean absolute percentage error (sMAPE) [36], are used in the test of point forecasting. The MAE is defined as:

$$MAE = \frac{1}{|\mathcal{T}|}\sum_{t\in\mathcal{T}}|e_t| = \frac{1}{|\mathcal{T}|}\sum_{t\in\mathcal{T}}|y_t - \tilde{y}_t| \tag{7.12}$$

where $\mathcal{T}$ is the time interval set to be evaluated. The sMAPE is defined as

$$sMAPE = \frac{1}{|\mathcal{T}|}\sum_{t\in\mathcal{T}}\frac{|y_t - \tilde{y}_t|}{|y_t| + |\tilde{y}_t|} \times 100(\%) \tag{7.13}$$

A smaller MAE or sMAPE indicates a better method.

7.4 Case Study

We test the proposed method with day-ahead market data from CAISO [26]. The time range of the data is from April 21, 2015, to November 29, 2018, over three years in total. The numerical experiments are conducted on all four main load aggregation nodes in CAISO, i.e., PGAE, SCE, SDGE, and VEA. The data between April 21, 2015, and January 25, 2016, are used in model selection. Thus, D1 to D4 in Fig. 7.3 are 2015/04/29, 2015/10/27, 2016/01/25, and 2018/11/29, correspondingly.

The training and testing of the models are conducted in a rolling way with a look-ahead time of 24 h as mentioned in Sect. 7.2.2.

7.4.1 Model Selection Results

Since there are four targets (the LMP, MCE, MCC, and MCL) in the model selection procedure, in some steps, the results of the MCE are presented as an example of rough tuning. Other results are omitted due to space limitation. The first step is to roughly tune K and N, while $\mathcal{F}^E$ is kept as the full set of (AUTOR+TIME+COND). The MAE is used as the performance indicator. Table 7.3 shows the rough parameter tuning results. Note that only the best two sets of parameters for each model are shown. The ET with $N = 160$ and $K = 30$ achieves a minimum MAE of 1.749 $/MWh. After this step, the best models for the LMP, MCE, MCC and MCL are the ET ($N = 70$ and $K = 60$), ET ($N = 160$ and $K = 30$), ET ($N = 160$ and $K = 30$), and ET ($N = 160$ and $K = 10$).

The next step is to determine the best feature set for each component. The feature combinations in Sect. 7.3.2 are tested for the four targets. The selection results are shown in Table 7.4. The best feature set for the LMP and MCE is AUTOR+TIME+COND, while COND is the best feature set for the MCC and MCL. This result is consistent with our analysis in Sect. 7.3.2 that network condition is the key factor for the MCC and MCL. The patterns of the LMP and MCE series are more complicated because they rely not only on their shifted values but also on the calendar feature and network conditions.

The final step is the fine-tuning of N. The step of N is changed to 10, and data from another time range (October 27, 2015, to January 25, 2016, 90 days) are used to enhance the reliability. The results for the MCE are shown in Fig. 7.4, indicating that the performance of the selected ET model is stable when N changes within a

Table 7.3 Rough tuning results (MCE) for parameter K and N

Model name	N	K	Average MAE ($/MWh)
ET	160	30	1.749
ET	130	30	1.752
RF	160	90	2.032
RF	130	90	2.041
DT	–	60	2.880
DT	–	30	2.937

Table 7.4 The average MAE ($/MWh) for different feature sets with the best K and N

$\mathcal{F}$	LMP	MCE	$\mathcal{F}$	MCC	MCL
AUTOR	2.534	2.511	**COND**	**0.376**	**0.256**
AUTOR+TIME	2.465	2.395	COND+TIME	0.380	0.268
AUTOR+TIME +COND	**1.999**	**1.749**	COND+TIME +AUTOR	0.4139	0.3060

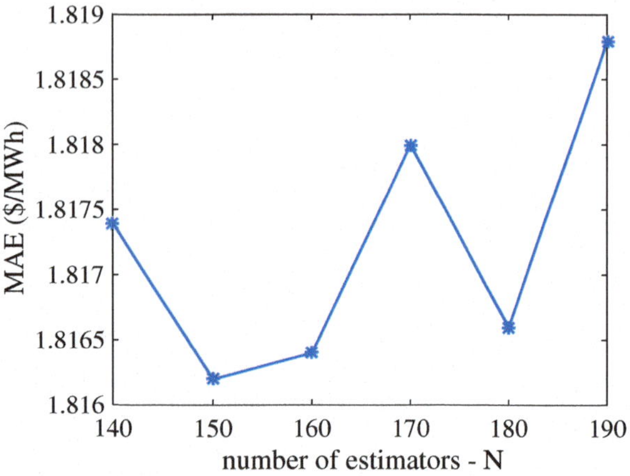

Fig. 7.4 The MAE for the MCE with respect to different *N* during fine-tuning

small range. After the fine-tuning, the best *N* values for the MCE, MCC, and MCL components are 150, 180, and 180, respectively, and *N* for direct LMP forecasting is 90. The hyperparameters for the four targets are different.

7.4.2 Componential Results

Figure 7.5 presents comparisons between the predicted results of the direct forecast or componential forecast for our four targets during a week in February 2016. As shown in the figure, the prediction of the MCC is not as accurate as that of the other three targets. This is reasonable because the occurrence of network congestion is rather random, and the series of the MCC are non-smooth and exhibit high variation. In this chapter, we only obtain regional load forecast data because CAISO does not release nodal load data. We believe that if more detailed information for the load distribution among the nodes and historical generation data are given, the MCC forecast accuracy would be improved. The low accuracy of the MCC forecast will reduce the accuracy of the summation forecast in peak and valley hours but barely have any effect in other hours. As mentioned in Sect. 7.2.3, the summation regressor is not expected to be a perfect method, and we try to compensate for its drawback in these high-variation hours via model stacking.

Table 7.5 lists the nodal average MAE and sMAPE for the four targets during the test period. The LMP is predicted by the direct forecast and the summation forecast. The sMAPE values for MCC/MCL seem quite large, because their ground truth values are small in most circumstances. The MAE values of the LMP, MCE, and MCC during the test period are much larger than the values in Tables 7.3 and 7.4 during the model selection period. This is caused by the high variations during summer times in 2017 and 2018 shown in Fig. 7.2.

Fig. 7.5 Direct and componential forecasts for the four targets of node PGAE on February 4 to 10, 2016

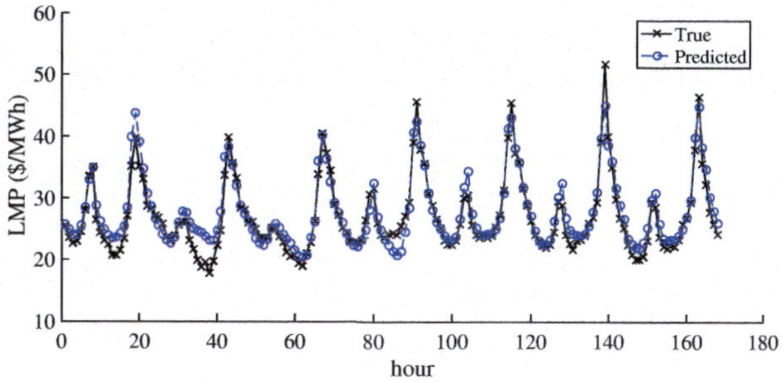

(a) The best direct forecasting results for the LMP

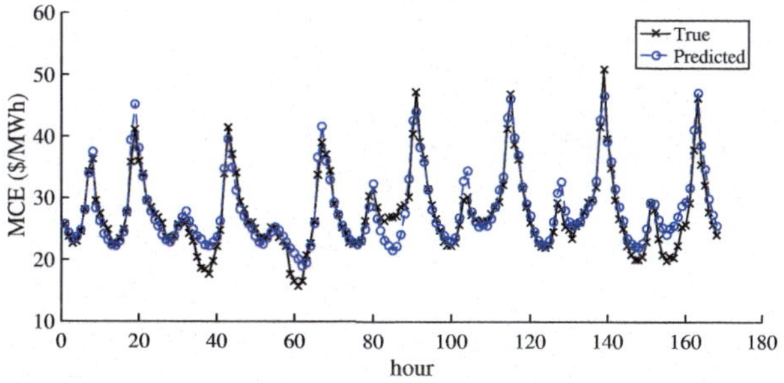

(b) The best componential forecasting results for the MCE

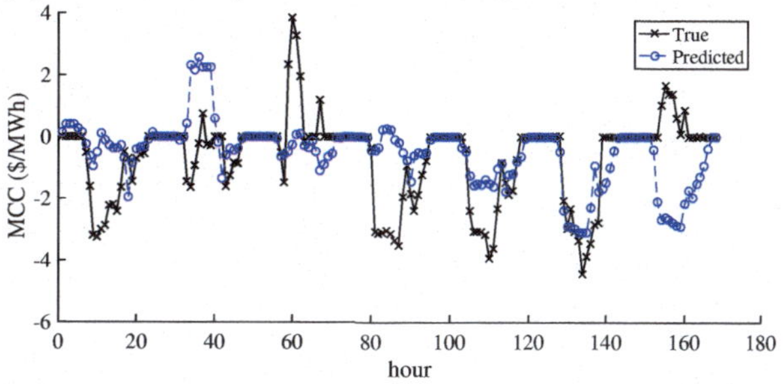

(c) The best componential forecasting results for the MCC

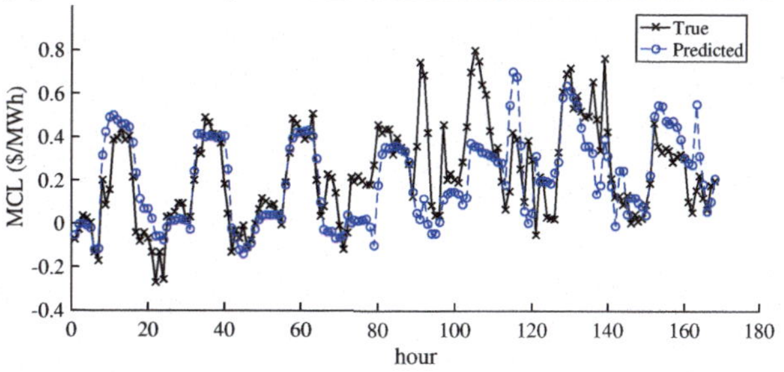

(d) The best componential forecasting results for the MCL

Table 7.5 The average MAE and sMAPE for the targets in the test period

	PGAE	SCE	SDGE	VEA
MCE	4.502 & 7.29%			
MCC	1.618 77.4%	1.116 79.5%	2.918 76.2%	1.904 72.17%
MCL	0.265 37.4%	0.195 34.4%	0.316 43.13%	0.258 13.48%
LMP—direct	4.589 7.51%	5.727 8.84%	6.637 9.26%	5.266 9.16%
LMP—summation	4.292 7.40%	5.217 8.26%	6.255 9.03%	5.040 8.99%

7.4.3 Stacking Results (Overall Improvements)

We test the four stacking methods mentioned in Sect. 7.3.4 and present comparisons between the methods and the individual forecast benchmarks in Table 7.6. For the Huber regression, the threshold h in (7.10) is selected as $1.35 \times \sigma$, where σ is the standard deviation of the training data. For SVR, the radial basis function kernel is used.

The best results (bold) are from stacking Method I (stacking the best models) with robust regression, which achieves improvements to a certain degree in all four nodes. Huber regression and ℓ_1 regression yields similar results, ℓ_1 regression performs slightly better for most occasions. We can see that the summation forecast has about 4 to 9% improvement in the MAE, and 1.4 to 6.6% in the sMAPE. The stacking procedure ensures a further 0.75 to 5.06% improvement in the MAE. As for the sMAPE, there is an improvement for three nodes but a slight drop in the accuracy for the node SCE. In general, the adoption of the summation forecast and stacking model proves to increase the precision and robustness of LMP forecasting. Figure 7.6 shows the actual and predicted LMPs on December 5, 2017. Through model stacking, the forecasts become slightly closer to the ground truth values during most hours, including the valley (approximately 12 o'clock) and the peak (approximately 18 o'clock) hours.

The impact of Huber threshold h on the model performance is also analyzed for Method-I. The detailed numerical results are given in Table 7.7. Meta-regressor named as "Huber—ε" indicates the value of h is $\varepsilon \times \sigma$. It is shown that the overall model performance is not sensitive to h.

To verify the robustness of ℓ_1 regression and Huber regression in handling LMP spikes/outliers, define that an LMP spike occurs in a day when the maximum LMP of the day exceeds 120 \$/MWh. We use the spike-influenced days to denote those days whose training phase contains one or more LMP spikes. And the uninfluenced days denote those whose training phase has no LMP spike. Table 7.8 shows the average MAE at node PGAE using different stacking algorithms. The robust regressors perform much better than the direct forecast and summation forecast during the spike-

Table 7.6 Comparison of the MAE and sMAPE between different stacking models in the test period[1]

Stacking method	Meta-regressor	MAE				sMAPE			
		PGAE	SCE	SDGE	VEA	PGAE	SCE	SDGE	VEA
Method-I	Linear	4.528	5.586	6.780	5.261	7.56%	8.62%	9.34%	9.12%
	Huber	**4.075**	5.216	6.161	4.943	**7.17%**	8.33%	8.97%	8.83%
	ℓ_1	4.080	**5.178**	**6.133**	**4.924**	**7.17%**	8.30%	**8.94%**	**8.80%**
	SVR	4.759	5.957	6.967	5.523	7.80%	9.06%	9.61%	9.55%
Method-II	Linear	5.265	6.950	8.061	6.680	8.13%	10.15%	10.70%	11.09%
	Huber	4.663	5.951	6.953	5.658	7.69%	9.18%	9.65%	9.78%
	ℓ_1	4.645	5.912	6.918	5.626	7.66%	9.17%	9.64%	9.78%
	SVR	5.098	6.518	7.403	6.025	8.29%	9.81%	10.14%	10.24%
Method-III	Linear	5.085	6.291	7.368	6.460	8.22%	9.24%	9.99%	10.58%
	Huber	4.412	5.494	6.476	5.530	7.65%	8.62%	9.32%	9.68%
	ℓ_1	4.401	5.466	6.442	5.511	7.64%	8.61%	9.32%	9.69%
	SVR	5.079	6.234	7.190	5.891	8.23%	9.34%	9.81%	10.01%
Method-IV	Linear	5.046	6.884	8.338	6.828	7.97%	9.66%	10.59%	10.89%
	Huber	4.223	5.616	6.649	5.441	7.36%	8.80%	9.45%	9.59%
	ℓ_1	4.208	5.617	6.731	5.526	7.36%	8.84%	9.59%	9.70%
	SVR	5.037	6.327	7.256	5.917	8.14%	9.47%	9.90%	10.04%
Benchmark	Direct	4.589	5.727	6.637	5.266	7.51%	8.84%	9.26%	9.16%
	Summation	4.292	5.217	6.255	5.040	7.40%	**8.26%**	9.03%	8.99%
Relative improvement 1		6.47%	8.91%	5.76%	4.29%	1.46%	6.56%	2.48%	1.86%
Relative improvement 2		5.06%	0.75%	1.95%	2.30%	3.11%	-0.48%	1.00%	2.11%
Overall improvement		11.20%	9.59%	7.59%	6.49%	4.53%	6.11%	3.46%	3.93%

The bold numerics stand for the best results. Relative Improvement 1 is achieved by the summation forecast against the direct forecast. Relative Improvement 2 is a further improvement achieved by stacking Method I with Huber regression. Overall Improvement is achieved by Method I with Huber regression against the benchmark of direct forecast

influenced days, while they are a little worse than the direct forecast and summation forecast during the uninfluenced days. The linear regression has poor performance during spike-influenced days, as the squared loss is likely to be affected by the spikes.

7.4.4 Error Distribution Analysis

To provide our readers with a comprehensive view on how the proposed method improves forecasting accuracy, the error distributed at different hours and different months is illustrated. Figure 7.7 shows the average MAE of PGAE at 24 h. The error of the stacking model is significantly lower than that of the direct forecasting model

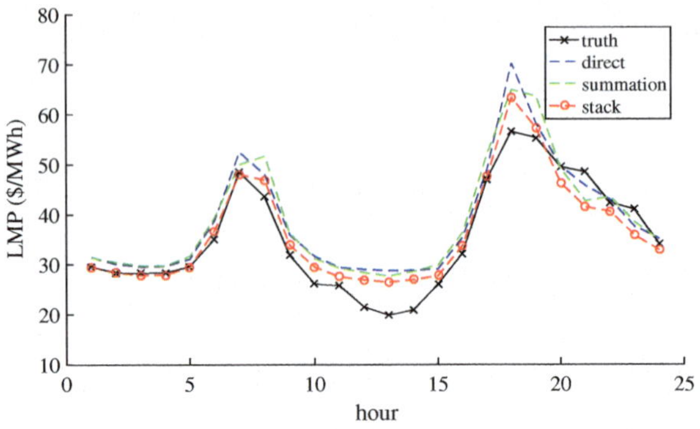

Fig. 7.6 The actual and predicted LMP at node PGAE on December 5, 2017

Table 7.7 Sensitivity analysis results of the threshold h

Meta-regressor	MAE			
	PGAE	SCE	SDGE	VEA
ℓ_1 regression	4.080	5.178	6.133	4.924
Huber—1.2	4.072	5.199	6.145	4.934
Huber—1.35	4.075	5.216	6.161	4.943
Huber—1.5	4.083	5.225	6.180	4.958
Huber—2.0	4.0115	5.257	6.235	5.006
Huber—2.5	4.162	5.278	6.279	5.033
Huber—3.0	4.215	5.294	6.320	5.057
Huber—5.0	4.355	5.442	6.570	5.193
Linear	4.528	5.586	6.780	5.261
Direct	4.589	5.727	6.637	5.266
Summation	4.292	5.217	6.255	5.040

Table 7.8 Average MAE for different stacking methods during spike-influenced and uninfluenced days

Meta-regressor	Spike-influenced days	Uninfluenced days
ℓ_1 regression	5.532	3.202
Huber	5.527	3.199
Linear	6.711	3.209
Direct	6.563	3.090
Summation	5.680	3.094

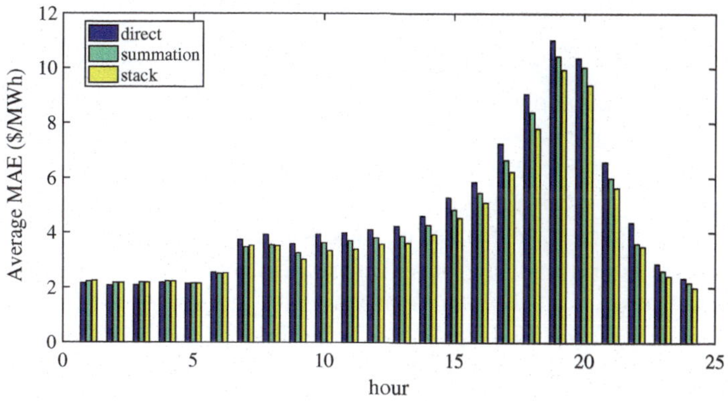

Fig. 7.7 The average absolute forecast error (MAE) of node PGAE at different hours

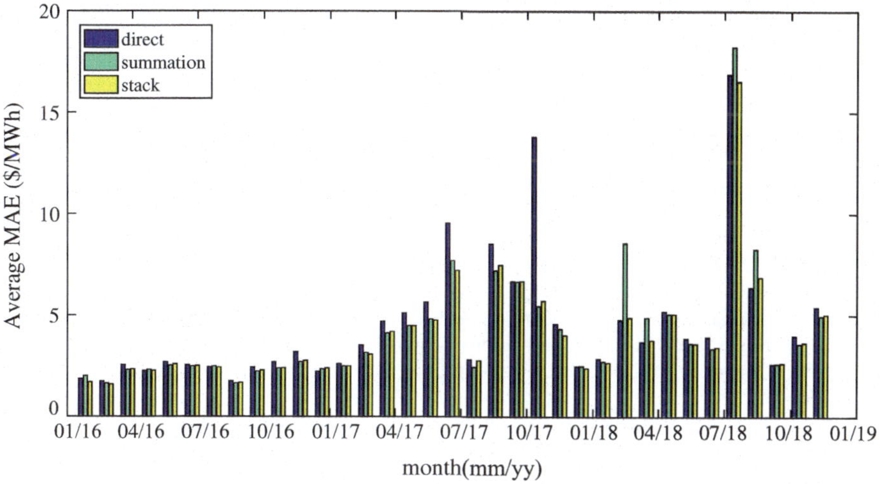

Fig. 7.8 The monthly average absolute forecast error (MAE) of node PGAE

and the summation forecasting model. The MAE during peak hours between 17 and 20 o'clock is larger than usual because the variation in the peak LMP is much larger than that during non-peak hours.

Figure 7.8 shows the monthly average MAE of PGAE from January 2016 to November 2018. In these months, the MAE bar of the stacking model is lower or close to the bar of the direct model and the summation model. Recall the pink boxes in Fig. 7.2 and the errors in October 2017 and July 2018 will not be surprising. The LMP in CAISO has become more extreme during the last several years. Special forecasting models should be adapted to handle these extreme data, which is beyond the scope of this chapter.

7.5 Conclusion

This chapter proposes a novel short-term LMP forecasting method based on componential decomposition and model stacking. Starting from the original SCED formulation of the LMP, we analyze the factors behind the three components of the LMP and formulate the traditional LMP forecasting problem as individual forecasting of its components. Specialized feature engineering and model selection are adopted to improve the accuracy of the individual models. Next, an ensemble framework that stacks the novel componential approach with the traditional direct forecast is proposed. Numerical results based on realistic market data over three years show that the proposed method achieves good accuracy and robustness in day-ahead LMP forecasting.

References

1. Rahimiyan, Morteza, and Luis Baringo. 2016. Strategic bidding for a virtual power plant in the day-ahead and real-time markets: A price-taker robust optimization approach. *IEEE Transactions on Power Systems* 31 (4): 2676–2687.
2. Toubeau, Jean-François., Jérémie. Bottieau, François Vallée, and Zacharie De Grève. 2019. Deep learning-based multivariate probabilistic forecasting for short-term scheduling in power markets. *IEEE Transactions on Power Systems* 34 (2): 1203–1215.
3. Baltaoglu, Sevi, Lang Tong, and Qing Zhao. 2019. Algorithmic bidding for virtual trading in electricity markets. *IEEE Transactions on Power Systems* 34 (1): 535–543.
4. Nicholson, Emma. 2014. Price formation in organized wholesale electricity markets. Technical Report Docket No. AD14-14-000, Federal Energy Regulatory Commission, Washington, DC 20426.
5. Weron, Rafał. 2014. Electricity price forecasting: A review of the state-of-the-art with a look into the future. *International Journal of Forecasting* 30 (4): 1030–1081.
6. Karakatsani, Nektaria V, and Derek W Bunn. 2008. Forecasting electricity prices: The impact of fundamentals and time-varying coefficients. *International Journal of Forecasting* 24 (4): 764–785.
7. Taylor, James W. 2010. Exponentially weighted methods for forecasting intraday time series with multiple seasonal cycles. *International Journal of Forecasting* 26 (4): 627–646.
8. Chitsaz, Hamed, Payam Zamani-Dehkordi, Hamidreza Zareipour, and Palak P Parikh. 2018. Electricity price forecasting for operational scheduling of behind-the-meter storage systems. *IEEE Transactions on Smart Grid* 9 (6): 6612–6622.
9. Pircalabu, Anca, and Fred Espen Benth. 2017. A regime-switching copula approach to modeling day-ahead prices in coupled electricity markets. *Energy Economics* 68: 283–302.
10. Ma, Yaming, Peter B Luh, Krishnan Kasiviswanathan, and Eman Ni. 2004. A neural network-based method for forecasting zonal locational marginal prices. In *IEEE Power Engineering Society General Meeting*, 296–302. IEEE.
11. Abedinia, Oveis, Nima Amjady, and Hamidreza Zareipour. 2016. A new feature selection technique for load and price forecast of electrical power systems. *IEEE Transactions on Power Systems* 32 (1): 62–74.

12. Sadeghi-Mobarakeh, Ashkan, Mahdi Kohansal, Evangelos E Papalexakis, and Hamed Mohsenian-Rad. 2017. Data mining based on random forest model to predict the California ISO day-ahead market prices. In *2017 IEEE Power Energy Society Innovative Smart Grid Technologies Conference (ISGT)*, 1–5.
13. Huang, Dongliang, Hamidreza Zareipour, William D Rosehart, and Nima Amjady. 2012. Data mining for electricity price classification and the application to demand-side management. *IEEE Transactions on Smart Grid* 3 (2): 808–817.
14. Wan, Can, Zhao Xu, Yelei Wang, Zhao Yang Dong, and Kit Po Wong.2014. A hybrid approach for probabilistic forecasting of electricity price. *IEEE Transactions on Smart Grid* 5 (1): 463–470.
15. Wan, Can, Ming Niu, Yonghua Song, and Xu. Zhao. 2017. Pareto optimal prediction intervals of electricity price. *IEEE Transactions on Power Systems* 32 (1): 817–819.
16. Lago, Jesus, Fjo De Ridder, Peter Vrancx, and Bart De Schutter. 2018. Forecasting day-ahead electricity prices in Europe: The importance of considering market integration. *Applied Energy* 211: 890–903.
17. Luo, Shuman, and Yang Weng. 2019. A two-stage supervised learning approach for electricity price forecasting by leveraging different data sources. *Applied Energy* 242: 1497–1512.
18. Lago, Jesus, Fjo De Ridder, and Bart De Schutter. 2018. Forecasting spot electricity prices: Deep learning approaches and empirical comparison of traditional algorithms. *Applied Energy* 221: 386–405.
19. Chang, Zihan, Yang Zhang, and Wenbo Chen. 2019. Electricity price prediction based on hybrid model of Adam optimized LSTM neural network and wavelet transform. *Energy* 187: 115804.
20. Hubicka, Katarzyna, Grzegorz Marcjasz, and Rafa L. Weron. 2018. A note on averaging day-ahead electricity price forecasts across calibration windows. *IEEE Transactions on Sustainable Energy* 10 (1): 321–323.
21. Rokach, Lior. 2010. Ensemble-based classifiers. *Artificial Intelligence Review* 33 (1–2): 1–39.
22. Bordignon, Silvano, Derek W Bunn, Francesco Lisi, and Fany Nan. 2013. Combining day-ahead forecasts for British electricity prices. *Energy Economics* 35: 88–103.
23. Nowotarski, Jakub, Bidong Liu, Rafał Weron, and Tao Hong. 2016. Improving short term load forecast accuracy via combining sister forecasts. *Energy* 98: 40–49.
24. Lee, Duehee, Hunyoung Shin, and Ross Baldick. 2018. Bivariate probabilistic wind power and real-time price forecasting and their applications to wind power bidding strategy development. *IEEE Transactions on Power Systems* 33 (6): 6087–6097.
25. Chai, Songjian, Xu. Zhao, and Youwei Jia. 2019. Conditional density forecast of electricity price based on ensemble ELM and logistic EMOS. *IEEE Transactions on Smart Grid* 10 (3): 3031–3043.
26. California ISO. 2017. California ISO Market Notice: Interface specification for OASIS 2017 release. http://www.caiso.com/Documents/OASIS-InterfaceSpecification_v4_3_5Clean_Spring2017Release.pdf.
27. Litvinov, Eugene, Tongxin Zheng, Gary Rosenwald, and Payman Shamsollahi. 2004. Marginal loss modeling in LMP calculation. *IEEE Transactions on Power Systems* 19 (2): 880–888.
28. Eldridge, Brent, Richard P O'Neill, and Anya Castillo. 2017. Marginal loss calculations for the DCOPF. *Federal Energy Regulatory Commission, Technical Report*. https://www.ferc.gov/sites/default/files/2020-05/marginallosscalculations.pdf.
29. Zechun, Hu., Haozhong Cheng, Zheng Yan, and Furong Li. 2010. An iterative lMP calculation method considering loss distributions. *IEEE Transactions on Power Systems* 25 (3): 1469–1477.
30. Deng, Weisi, Yuting Ji, and Lang Tong. Probabilistic forecasting and simulation of electricity markets via online dictionary learning. In *Proceedings of the 50th Hawaii International Conference on System Sciences*.
31. California ISO. 2018. California ISO business practice manual for market operations. https://bpmcm.caiso.com/Pages/BPMDetails.aspx?BPM=Market.
32. Wang, Yi., Qixin Chen, Mingyang Sun, Chongqing Kang, and Qing Xia. 2018. An ensemble forecasting method for the aggregated load with subprofiles. *IEEE Transactions on Smart Grid* 9 (4): 3906–3908.

33. Hong, Tao, Jingrui Xie, and Jonathan Black. 2019. Global energy forecasting competition 2017: Hierarchical probabilistic load forecasting. *International Journal of Forecasting* 35 (4): 1389–1399.
34. Geurts, Pierre, Damien Ernst, and Louis Wehenkel. 2006. Extremely randomized trees. *Machine Learning* 63 (1): 3–42.
35. Chun, Yu., and Weixin Yao. 2017. Robust linear regression: A review and comparison. *Communications in Statistics - Simulation and Computation* 46 (8): 6261–6282.
36. Tofallis, Chris. 2015. A better measure of relative prediction accuracy for model selection and model estimation. *Journal of the Operational Research Society* 66 (8): 1352–1362.

Chapter 8
Economic Impact of Price Forecasting Error

8.1 Introduction

In a modern deregulated electricity market, market participants including load serving entities (LSEs), generators, and virtual bidders need short-term price forecasts (STPF) for bidding curve constructing and operation scheduling [1, 2]. Sometimes, the market organizer, usually an independent system operator (ISO) or regional transmission organization (RTO), releases its STPF. However, there is no guarantee on the accuracy of the public forecast from ISO. In recent years, private companies providing accurate STPF results or market information services have appeared, including Genscape and EBW analytics group [3, 4]. Market participants would purchase private forecasts or employ researchers to conduct their own STPF, in order to achieve a higher accuracy, which will cause an additional cost. Even though a higher forecasting accuracy indicates higher incomes and lower risks, a trade-off between the marginal benefit of STPF accuracy and its cost is required in the market participant's decision framework. Therefore, the impact of STPF errors on market bidding should be studied and quantified to support the decision.

In a general case of data purchasing or third-party electricity forecasting service, the quality of an STPF product or software is usually and most intuitively described by the error metrics such as mean absolute error (MAE) and root mean square error (RMSE). It is important for the buyers of such forecasting products to evaluate the economic impact of reducing STPF error for its specific bidding model. For example, how much cost could be saved if a 50-MW LSE improves its STPF from an MAE of 8 \$/MWh to 5 \$/MWh by purchasing a 3000 \$ (yearly) software? Or how much revenue could be improved for a certain generation company (GenCo) with 2 GW of thermal units? In addition to purchasing third-party STPF service, some companies would conduct projects on their own research & development (R&D) on STPF methods. It would be necessary to evaluate the potential revenue in advance.

The market bidding problem of market participants is a kind of optimization problem minimizing the total cost or maximizing the overall profit. There are two essential types of market participants, price-takers and price-makers. The price-makers can

Q. Chen et al., *Data Analytics in Power Markets*,
https://doi.org/10.1007/978-981-16-4975-2_8

influence the electricity price since they have enough market power, while the price-takers accept the prices as exogenous variables. In the bidding of price-markers, certain market clearing rules can be integrated into the decision model and the optimization problem is usually transformed into a mathematical program with equilibrium constraints (MPEC) [5, 6]. Such a problem is complex and can only be formulated by mixed-integer linear programming (MILP). The price-makers' model is beyond the scope of this chapter, and we suggest referring to [5, 6] for further reading.

In the literature, the bidding problem of price-takers is usually formulated as linear programming or MILP. In [7], an optimal bidding strategy for generation companies is proposed considering piece-wise staircase offer curves. A parametric linear programming model is used to handle incomplete information. Song et al. considered the bidding problem of a retailer with flexible loads in the spot market in [8]. A linear programming model was used to reduce low-profit risk. A bidding strategy for LSE considering price-responsive loads was modeled and solved by MILP in [9]. Zhao et al. developed a bidding model for an electric vehicle aggregator with linear programming in [10]. Compared with MILP-based models, linear models are much more computational effective, and they have been widely adopted in the forming of bidding curves in electricity spot markets.

The economic impact of electricity price forecasting errors was first studied by Zareipour et al. in [11]. A cost-related index named the forecasting inaccuracy economic impact (FIEI) was proposed, which is the relative percentage change in the expected cost attributable to price forecasting error. The same impact on self-scheduling of generation companies was then studied in [12]. Chazarra et al. discussed the economic impact of forecasting errors in residual reserve curves on a pumped-storage hydropower plant participating in the day-ahead (DA) energy market and the secondary regulation reserve market [13]. However, the simulation in [11–13] are based on predefined single inaccuracy forecast series, which lacks theoretical analysis on the impact of errors with different characteristics (e.g. error levels and distributions).

The traditional methods mainly solve the bidding model by the original STPF and the improved STPF (some predefined forecasts) and compare the historical cost/revenue of the two STPF results. We think that such settings can be further improved for the following reasons:

- Calculating the economic impact by just comparing two STPF series is not comprehensive enough. The randomness is unavoidable for single time series while using metrics like MAE and RMSE is more robust.
- The historical STPF results provided by a third-party seller for the buyer's evaluation might be tuned or embellished so that it would look better.

Thus, it would be better to quantify the economic impact of STPF with only the bidding model, the widely-adopted metrics describing the error level, and the historical ground truth prices.

In this chapter, we will study the impact of STPF errors on bidding from a price-taker's perspective. First, a general bidding model is formulated in both deterministic and stochastic situations. The forecasted prices are considered as one of the exogenous parameters for the model. Then, the theory of multiparametric linear programming (MPLP) is used to analyze the impact of forecasting error. A numerical method is proposed to quantify such impact with respect to different error levels. Finally, numerical results and discussions are conducted.

In comparison to the existing methods in the literature, we are the first to consider both deterministic and stochastic bidding models. Moreover, our method does not rely on any specific predefined forecasts, and it is able to give the expected loss caused by STPF errors defined by certain widely-adopted evaluation metrics, e.g. MAE and RMSE. Also, we reveal that when the STPF error falls in the same polytope (later defined as the Critical Region), the cost and revenue of the bidder would not be affected.

The contribution of this chapter is three-fold:

1. It studies the impact of STPF error on the bidding cost in a model-based approach. Both deterministic and stochastic linear bidding models are considered.
2. It proposes a novel MPLP-based method to quantify the aforementioned economic impact. Once the bidding model of a market participant is determined, the percentage loss of revenue can be calculated with respect to the error level.
3. It conducts three numerical experiments to illustrate the application of the methods in both deterministic and stochastic bidding models. The obtained results are instructive for market participants.

The rest of this chapter is organized as follows. Section 8.2 presents the basic deterministic and stochastic bidding models for market participants. Section 8.3 introduces the MPLP-based methodology and the overall analysis framework. The case study and the numerical results are shown in Sect. 8.4. Finally, Sect. 8.5 draws conclusions.

8.2 General Bidding Models

The bidding models are used to solve the optimal bidding curve of a certain market participant. The bidding curves are then collected by the market organizer (ISO/RTO) to perform market clearing.

Two basic kinds of bidding models are described in this section. Without loss of generality, the market participant is assumed to be an LSE with flexible demand in the cost analysis part of this chapter. Such demand flexibility can be achieved by demand response or energy storage systems (ESS) [14]. For details of some concrete bidding models, readers can refer to [8, 15].

Figure 8.1 shows an example of the bidding curve submitted to the ISO. The curve consists of n endpoints for each time interval t. For an LSE, when the price is higher, less electricity will be purchased. p_t^* is the ground truth price, q_t^* is the actual consumed electricity volume, and q_t^{Δ} is the load imbalance at t.

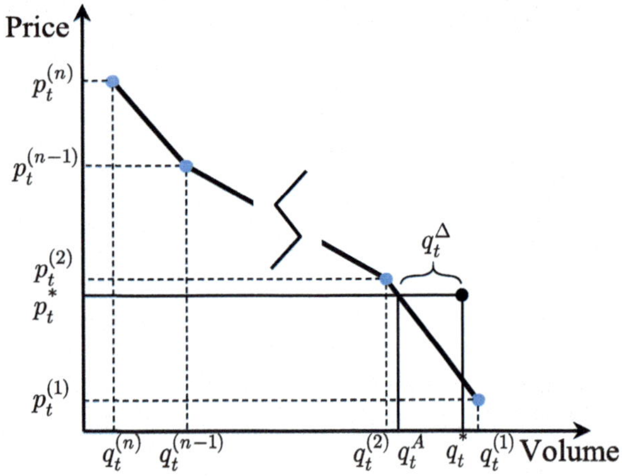

Fig. 8.1 An example of the bidding curve of an LSE. n is the number of endpoints. q_t^A is the purchased electricity volume

8.2.1 Deterministic Bidding Model

In the deterministic bidding model, the market participant submits an inelastic bidding curve with volume fixed at $\mathbf{q}$. Let $\mathbf{x}$ denote the decision variables of the market participant apart from the bidding volume. The objective is to maximize the total profit with certain operation constrains:

$$\max \sum_{t\in\mathcal{T}} p^S q_t^* - \left((p_t^* + \theta_t)q_t + \bar{p}q_t^{\Delta+} - \underline{p}q_t^{\Delta-}\right)$$

where $T = |\mathcal{T}|$ is the number of time intervals (usually 24 in the DA market), and θ_t is the point forecasting error of electricity price. In some cases the retail price p^S is time-dependent determined by the retail contract signed between LSE and consumers. For simplification, p^S is considered as a fixed value. Thus, the income of the LSE is constant and omitted in the optimization problem:

$$\begin{aligned}
\min_{\mathbf{q},\mathbf{x}}\ & (\mathbf{p}^* + \text{`})^\top \mathbf{q} + \bar{p}\,\mathbf{1}^\top \mathbf{q}^{\Delta+} - \underline{p}\,\mathbf{1}^\top \mathbf{q}^{\Delta-} \\
\text{s.t.}\ & \mathbf{A}_1\mathbf{q} + \mathbf{A}_2\mathbf{x} \le \mathbf{b},\ \mathbf{q} \ge 0, \\
& \mathbf{q}^{\Delta+} - \mathbf{q}^{\Delta-} = \mathbf{q}^* - \mathbf{q}, \\
& \mathbf{q}^{\Delta+/-} \ge 0.
\end{aligned} \tag{8.1}$$

For more complicated p^S (e.g. time-of-use pricing), a linear term of income can be added into the objective function. In (8.1) $\mathbf{p}^*$ is the ground truth price vector, $\boldsymbol{\theta} \in \Theta \subset \mathbb{R}^\top$ is the STPF error following a certain predefined distribution. The incumbent electricity retail price $\bar{p}$ and the feed-in tariff $\underline{p}$ are used to address the load imbalance (q_t^{Δ} in Fig. 8.1) between the real demand and the purchased electricity. Note that if $\bar{p}$ is high enough and $\underline{p}$ is low enough (i.e., $\mathbf{p}^* + \boldsymbol{\theta} \in [\underline{p}, \bar{p}]$ holds), the bidding demand $\mathbf{q}$ of a deterministic bidder who has no flexibility would not be affected by the STPF results. It would be rational to bid at the real demand $\mathbf{q}^*$ and make sure that $\mathbf{q}^{\Delta+/-} = 0$. Thus, we can simplify (8.1) to omit the load imbalance:

$$\begin{aligned} \min_{\mathbf{q},\mathbf{x}} \ & (\mathbf{p}^* + \boldsymbol{\theta})^\top \mathbf{q}, \\ \text{s.t.} \ & \mathbf{A}_1\mathbf{q} + \mathbf{A}_2\mathbf{x} \leq \mathbf{b}, \ \mathbf{q} \geq 0, \\ & \boldsymbol{\theta} \in \Theta. \end{aligned} \tag{8.2}$$

The decision variable $\mathbf{x}$ may seem confusing at the moment. It indicates the internal decision variables of the market participant and usually contains the schedule of flexible resources such as demand response or ESS. Some examples of $\mathbf{x}$ are given later in (8.16)–(8.17) in Sect. 8.4.

8.2.2 Stochastic Bidding Model

The stochastic bidding model has been widely adopted since it considers the uncertainties of the decision framework in a proper way. Assume that the DA price uncertainty is described by a T-dimensional joint distribution $G(\mathbf{p})$ from which a scenario set Ω can be sampled. For each scenario $\omega \in \Omega$, $\mathbf{p}_\omega$ is a realization of price from $G(\mathbf{p})$ with probability π_ω. Thus, the objective function of expected cost can be approximated in a scenario-based way:

$$\mathbb{E}\left(\mathbf{p}^\top \mathbf{Q}(\mathbf{p}) \middle| \mathbf{p} \sim G\right) \approx \sum_{\omega \in \Omega} \pi_\omega \cdot \mathbf{p}_\omega^\top \mathbf{Q}(\mathbf{p}_\omega) \tag{8.3}$$

The forecasting error in this model is different from the one in the deterministic model as the forecasting result is a distribution $G(\mathbf{p})$ rather than a price vector. It is still an open question on the definition of forecasting error, as the divergence between two T-dimensional joint distribution is very complex. To simplify the bias in the objective function caused by STPF error, we consider it as the bias in the scenario probability π_ω.

For each time interval t, a piecewise linear bidding curve $Q_t(p)$ is determined to minimize the expected cost (similar to Sect. 8.2.1, the retail income is considered as a constant value and omitted in the objective function):

$$\begin{aligned} \min_{\mathbf{Q}(p),\mathbf{Y},\mathbf{X}_\omega,\forall\omega} \ & \sum_{\omega,t} (\pi_\omega^* + \theta_\omega) \cdot \Big(p_{t,\omega} \cdot Q_t(p_{t,\omega}) \\ & + \bar{p} \cdot q_{t,\omega}^{\Delta+} - \underline{p} \cdot q_{t,\omega}^{\Delta-} \Big), \\ \text{s.t.} \ & \mathbf{A}_1\mathbf{Q}(\mathbf{p}_\omega) + \mathbf{A}_2\mathbf{X}_\omega + \mathbf{A}_3\mathbf{Y} \leq \mathbf{b}, \ \forall \omega \in \Omega, \\ & \mathbf{q}_\omega^{\Delta+} - \mathbf{q}_\omega^{\Delta-} = \mathbf{q}^* - \mathbf{Q}(\mathbf{p}_\omega), \ \mathbf{q}_\omega^{\Delta+/-} \geq 0. \end{aligned} \tag{8.4}$$

where $\omega \in \Omega$ stands for a certain scenario, $p_{t,\omega}$ is the price at time t for scenario ω, and π_ω^* is its ground truth probability. $\mathbf{q}_\omega^{\Delta+/-}$ is the electricity imbalance in the day-ahead market, and θ_ω is the bias in scenario probability due to forecasting error. The market participant need to solve two kinds of variables in the scenario-based stochastic programming, i.e., the wait-and-see variables and the here-and-now variables. The former ones are related to scenarios and are denoted as $\mathbf{X}_\omega$, while the latter ones are irrelevant to scenarios and are denoted as $\mathbf{Q}(p) = [Q_t(p)]_t$ and $\mathbf{Y}$.

Equation (8.4) is not yet a linear programming problem because Q_t is a function. A common way to linearize it is to parametrize Q_t with pairs of $(q_t^{(j)}, p_t^{(j)})$ in Fig. 8.1. Note that:

$$\forall t, \omega,\ \exists j \in \{1, \ldots, n-1\},\ \text{s.t. } p_t^{(j)} \le p_{t,\omega} \le p_t^{(j+1)}. \tag{8.5}$$

Practically, values of prices $p_t^{(j)}$ can be empirically fixed at certain points $p^{(1)}, \ldots, p^{(n)}$ [8], whereas the volume values $q_t^{(j)}$ need to be solved in the model. Thus, the linearized form of (8.4) is:

$$\begin{aligned}
\min_{\mathbf{q}^{(j)}, \mathbf{Y}, \mathbf{X}_\omega, \forall \omega} \ & \sum_{\omega,t} (\pi_\omega^* + \theta_\omega) \cdot \left(p_{t,\omega} \cdot q_{t,\omega}^A + \bar{p} \cdot q_{t,\omega}^{\Delta+} - \underline{p} \cdot q_{t,\omega}^{\Delta-} \right) \\
\text{s.t. } & \mathbf{A}_1 \mathbf{q}_\omega^A + \mathbf{A}_2 \mathbf{X}_\omega + \mathbf{A}_3 \mathbf{Y} \le \mathbf{b},\ \forall \omega \in \Omega, \\
& \mathbf{q}_\omega^{\Delta+} - \mathbf{q}_\omega^{\Delta-} = \mathbf{q}^* - \mathbf{q}_\omega^A,\ \mathbf{q}_\omega^{\Delta+/-} \ge 0, \forall \omega \in \Omega, \\
& q_{t,\omega}^A = q_t^{(j)} + (q_t^{(j+1)} - q_t^{(j)}) \frac{p_{t,\omega} - p^{(j)}}{p^{(j+1)} - p^{(j)}},\ \forall t, \omega, \\
& \text{where } j \text{ satisfies: } p^{(j)} \le p_{t,\omega} \le p^{(j+1)}.
\end{aligned} \tag{8.6}$$

Later in this chapter, $[\theta_\omega]_{\omega \in \Omega}$ is denoted as $\boldsymbol{\theta}_\Omega$ to present the scenario probability forecasting errors in (8.4) and (8.6). The vector form of the objective function in (8.6) is:

$$(\boldsymbol{\pi}_\Omega^* + \boldsymbol{\theta}_\Omega)^\top \left(\mathbf{p}_{\mathcal{T},\Omega} \circ \mathbf{q}_{\mathcal{T},\Omega}^A + \bar{p} \cdot \mathbf{q}_{\mathcal{T},\Omega}^{\Delta+} - \underline{p} \cdot \mathbf{q}_{\mathcal{T},\Omega}^{\Delta-} \right)^\top \mathbf{1}_T, \tag{8.7}$$

where $\mathbf{1}_T$ is the T-dimensional all-one vector, and $\circ$ is the Hadamard product operator. Note that the linearization of $q_{t,\omega}^A$ in (8.6) is for piecewise linear bidding curves of Fig. 8.1. Sometimes, the step-wise bidding curve is also used. In this occasion, the linearization of $q_{t,\omega}^A$ would have a different formulation, but it can still be obtained using the aforementioned method by fixing $p^{(1)}, \ldots, p^{(n)}$ at certain values.

Both deterministic and stochastic models of (8.1) and (8.6) can be analyzed using the MPLP theory.

Remark 8.1 In real world probabilistic STPF, the forecasted price scenario set of $\mathbf{p}_\omega (\omega \in \Omega_1)$ is usually biased from the ground truth price scenario set of $\mathbf{p}_\omega^* (\omega \in \Omega^*)$. In (8.6), we could construct a whole set $\Omega = \Omega_1 \cup \Omega^*$ containing both ground truth and biased scenarios, and $\pi_\omega = 0$ $(\forall \omega \in \Omega_1)$. The vector $\mathbf{ß}_\Omega$ would have several zero entries in correspondence to those biased scenarios. Through this transform, the STPF error is modeled as the bias in scenario probability π_ω rather than the price value error.

8.3 Methodology and Framework

In this section, the detailed methodology and framework for quantifying the economic impact of forecasting errors on the linear bidding models of Sect. 8.2 is introduced.

8.3.1 *Forecasting Error Modeling*

The first step is to estimate the distribution of the forecasting error given by a certain forecasting model, i.e., to model the distribution of $\boldsymbol{\theta}_N$ or θ_ω based on historical forecasts and ground truth price values. There are many statistical approaches to model such multivariate distributions, including the Cholesky decomposition method and Kernel Density Estimation (KDE) [16, 17]. After error modeling, the distribution of $\boldsymbol{\theta}$ is presented as a joint cumulative distribution function (CDF) and a probabilistic density function (PDF).

The detail of error modeling is beyond the scope of this chapter. Even though it is included in our framework, we focus more on calculating the general economic impact under certain STPF error level. In the case study, $\boldsymbol{\theta}$ is assumed to be uniformly distributed on some high-dimensional surface. However, if a bidder wants to quantify the error impact under certain distribution, our method in this section is still applicable.

8.3.2 *Multiparametric Linear Programming (MPLP) Theory*

In the theory of multiparametric programming [18], the optimization problem is solved as function of one or multiple parameters:

$$\begin{aligned} J(\boldsymbol{\theta}) &= \min_{\mathbf{x}\in\mathbb{R}^n} f(\mathbf{x}, \boldsymbol{\theta}) \\ &\text{s.t. } \mathbf{g}(\mathbf{x}, \boldsymbol{\theta}) \leq 0, \\ &\text{where } \boldsymbol{\theta} \in \Theta \subset \mathbb{R}^m. \end{aligned} \tag{8.8}$$

Besides the optimal objective function value $J(\boldsymbol{\theta})$, the optimal decision variable $\mathbf{x}^* = \mathbf{x}^*(\boldsymbol{\theta})$ is also regarded as a function of the parameters $\boldsymbol{\theta}$. Assume that $\mathbf{g}(\cdot, \cdot)$ is linear, and denote $\mathcal{A}$ as the set of all the constraints defined by $\mathbf{g}$. Since there are inequalities in $\mathcal{A}$, for any $\boldsymbol{\theta}$ and the corresponding optimal $\mathbf{x}^*(\boldsymbol{\theta})$, $\mathcal{A}$ can be split into two subsets, the binding set $\mathcal{B}(\boldsymbol{\theta})$ and the non-binding set $\mathcal{N}(\boldsymbol{\theta})$. The constraints that are binding at $\mathbf{x}^*(\boldsymbol{\theta})$ are contained in $\mathcal{B}$, while others are in $\mathcal{N}$.

A critical region (CR) is defined as a convex polytope that keeps the same binding constraints for all $\boldsymbol{\theta}$ inside it. In other words, if $\boldsymbol{\theta}$ changes within the same CR, $\mathcal{B}$ and $\mathcal{N}$ remain unchanged. An example of CRs is given in Fig. 8.2.

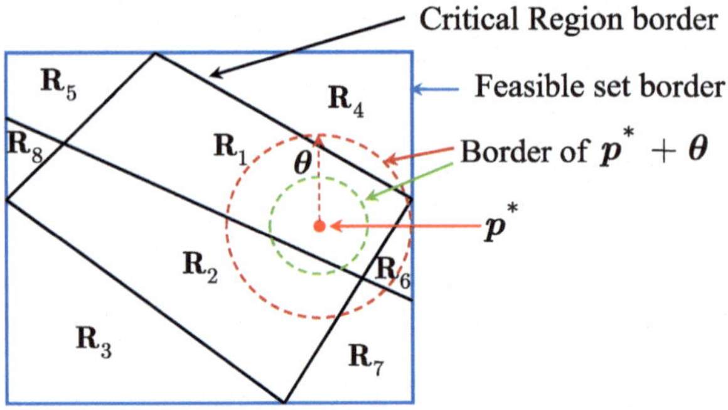

Fig. 8.2 An example of critical regions (CRs) in the MPLP theory. $\mathbf{R}_1$ to $\mathbf{R}_8$ are CRs in Θ space. Dashed color lines indicate the potential border of the biased parameters

Theorem 8.1 *In a general MPLP problem of (8.8), when the parameters are only contained in the objective function, the optimal decision variable* $\mathbf{x}^*$ *holds the same value for each CR.*

Corollary 8.1 *The minimum cost function* $J(\boldsymbol{\theta})$ *is a piecewise affine function, and the optimal decision variable* $\mathbf{x}^*(\boldsymbol{\theta})$ *is a piecewise constant function in* Θ.

For details of the proof of Theorem 8.1 and Corollary 8.1, readers can refer to Chap. 1 in [18].

The borders of CRs are determined once the MPLP problem is specified. Functions of $J(\boldsymbol{\theta})$ and $\mathbf{x}^*(\boldsymbol{\theta})$ can be easily achieved using the Multi-Parametric Toolbox (MPT) [19] when f and $\mathbf{g}$ are given.

Sometimes the high dimensionality of $\boldsymbol{\theta}$ would cause computational burden, and reducing its dimension can help in getting faster results of $J(\boldsymbol{\theta})$ and $\mathbf{x}^*(\boldsymbol{\theta})$ as well as CRs. Such dimensional reduction should be performed in a linear way to make sure the MPLP theory is still applicable:

$$\boldsymbol{\theta} = \mathbf{B}\boldsymbol{\theta}_N \tag{8.9}$$

where $\mathbf{B} \in \mathbb{R}^{T \times N}$ represents the error basis vectors, and N is the reduced error dimension.

8.3.3 Error Impact Formulation

If we consider the ground truth price $\mathbf{p}^*$ as a point in Θ space, the forecasting error $\boldsymbol{\theta}_N$ shifts the point from its original position to a biased position, as shown in Fig. 8.2. The shift caused from an accurate forecasting model is expected to be more moderate than that from an inaccurate one. Assuming that the market participant knows the ground truth price $\mathbf{p}^*$ or price scenario probability π^*_ω in advance, the biding volume would be $\mathbf{q}^*(0)$ in the deterministic model and $q_t^{(i)}(0)$ in the stochastic model. So the loss due to the forecasting errors is formulated as:

$$J_\ell^d(\boldsymbol{\theta}_N) = \mathbf{p}^* \cdot \Big(\mathbf{q}^*(\boldsymbol{\theta}_N) - \mathbf{q}^*(0)\Big), \tag{8.10a}$$

$$J_\ell^s(\boldsymbol{\theta}_\Omega) = \sum_{t,\omega} \pi_\omega^* p_{t,\omega} \left(q_{t,\omega}^A(\boldsymbol{\theta}_\Omega) - q_{t,\omega}^A(0)\right), \tag{8.10b}$$

where

$$q_{t,\omega}^A(\boldsymbol{\theta}_\Omega) = q_t^{(j)}(\boldsymbol{\theta}_\Omega) + \left(q_t^{(j+1)}(\boldsymbol{\theta}_\Omega) - q_t^{(j)}(\boldsymbol{\theta}_\Omega)\right) \cdot \frac{p_{t,\omega} - p^{(j)}}{p^{(j+1)} - p^{(j)}}$$

In (8.10), $J_\ell^d(\boldsymbol{\theta}_N)$ and $J_\ell^s(\boldsymbol{\theta}_\Omega)$ are the loss in the deterministic model and the stochastic model, respectively. To fully present the loss with the error distribution model calculated in Sect. 8.3.1, the expected loss in cost for the two models is calculated:

$$\begin{aligned} E_\Theta^d(\mathbf{P}) = \mathbb{E}_\Theta\big[J_\ell(\boldsymbol{\theta}_N)\big|\boldsymbol{\theta}_N \sim \mathbf{P}\big] = \int_\Theta \mathbf{p}^*\Big(\mathbf{q}^*(\boldsymbol{\theta}_N) - \\ \mathbf{q}^*(0)\Big)\, d\mathbf{P}(\boldsymbol{\theta}_N) = \int_\Theta \mathbf{p}^*\Big(\mathbf{q}^*(\boldsymbol{\theta}_N) - \mathbf{q}^*(0)\Big)\rho(\boldsymbol{\theta}_N)\, d\boldsymbol{\theta}_N, \end{aligned} \tag{8.11a}$$

$$\begin{aligned} E_\Theta^s(\mathbf{P}) = \mathbb{E}_\Theta\big[J_\ell^s(\boldsymbol{\theta}_\Omega)\big|\boldsymbol{\theta}_\Omega \sim \mathbf{P}\big] = \int_\Theta \sum_{t,\omega} \pi_\omega^* p_{t,\omega} \cdot \\ \Big(q_{t,\omega}^A(\boldsymbol{\theta}_\Omega) - q_{t,\omega}^A(0)\Big)\rho(\boldsymbol{\theta}_\Omega)\, d\boldsymbol{\theta}_\Omega, \end{aligned} \tag{8.11b}$$

where $\mathbf{P}(\cdot)$ is the joint CDF and $\rho(\cdot)$ is the PDF of $\boldsymbol{\theta}$ or $\boldsymbol{\theta}_\Omega$. For arbitrary function ρ, it is hard to calculate the high-dimensional integrations of (8.11a) and (8.11b) explicitly even though $\mathbf{q}^*(\boldsymbol{\theta}_N)$ and $q_{t,\omega}^A(\boldsymbol{\theta}_\Omega)$ are piecewise constant functions. Still, they can be numerically calculated using the Monte Carlo method or quasi-Monte Carlo methods [20]. And the percentage loss (α) is defined as:

$$\alpha^d = \frac{E_\Theta^d(\boldsymbol{P})}{Cost_0 - \mathbf{p}^* \cdot \mathbf{q}^*(0)} \times 100\,(\%) \tag{8.12a}$$

$$\alpha^s = \frac{E_\Theta^s(\mathbf{P})}{\sum_\omega \pi_\omega^* \mathbf{p}_\omega \cdot \mathbf{q}_\omega^A(0)} \times 100\,(\%) \tag{8.12b}$$

where $Cost_0$ is the cost when LSE has no flexibility. The maximum revenue by intergrating flexibility is $Cost_0 - \mathbf{p}^* \cdot \mathbf{q}^*(0)$ when there is no forecasting error. α^d is used to quantify the percentage loss of the revenue from flexibility resources invested by the LSE, while α^s is used to quantify the percentage loss of the overall electricity purchasing cost. The former one is used for deterministic bidders and the latter one is used for stochastic bidders.

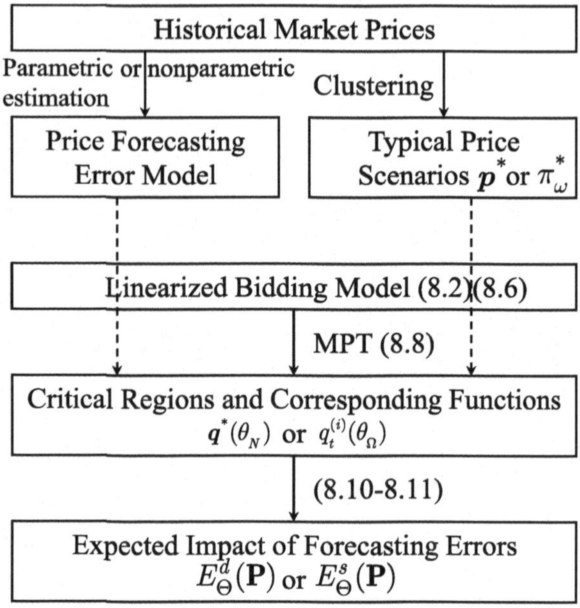

Fig. 8.3 The overall forecasting error impact analysis framework based on MPLP

Since $E_\Theta(\mathbf{P})$ in (8.11a) and (8.11b) is related to the selection of $\mathbf{p}^*$ and π^*_ω, it is recommended that some typical values of $\mathbf{p}^*$ and π^*_ω should be tested. Clustering techniques like k-means can be used to select such typical values.

8.3.4 Overall Framework

The overall framework of the proposed electricity price forecasting error impact analysis methodology is shown in Fig. 8.3. It has the following four steps:

1. Use statistic methods to estimate the price forecasting error model from historical market data or predefined error levels.
2. Find the typical price scenarios $\mathbf{p}^*$ or π^*_ω with clustering techniques.
3. Obtain the border of the critical regions and the function of $\mathbf{q}^*(\boldsymbol{\theta}_N)$ or $q_t^{(i)}(\boldsymbol{\theta}_\Omega)$ with MPT according to the market participant's linearized bidding model.
4. Substitute the distribution of $\boldsymbol{\theta}_N$ or $\boldsymbol{\theta}_\Omega$ into $\mathbf{q}^*(\boldsymbol{\theta}_N)$ or $q_t^{(i)}(\boldsymbol{\theta}_\Omega)$ and calculate the integration with typical values of $\mathbf{p}^*$ or π^*_ω to obtain the final expected impact.

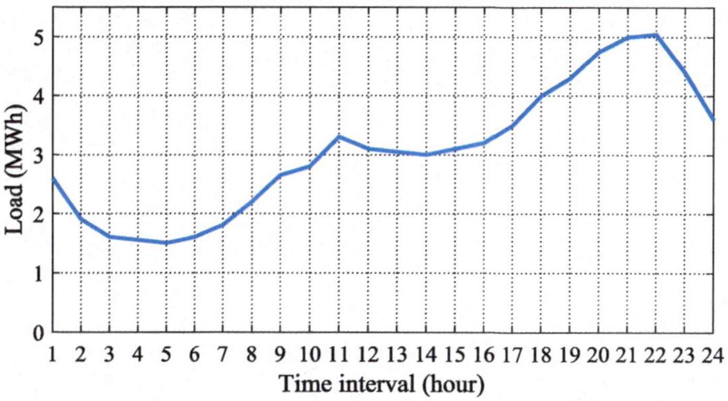

Fig. 8.4 The aggregated load curve in the case study

8.4 Case Study

In this section, we give three numerical cases to analyze the error impact on bidding. The market participant is assumed to be an LSE who is in charge of serving the load of 10000 typical residential households. We use the typical residential load curve given in [21], and the aggregated load demand is given as Fig. 8.4. The day-ahead price data are collected from the California ISO (CAISO [22]) who is the operator of California's bulk power system and electricity market. The one-year price data of an aggregated pricing node Pacific Gas and Electric (PG&E) which is one of the largest pricing nodes in CAISO are collected and clustered using k-means as shown in Fig. 8.5. Five representative price curves are generated to test our method.

In Case 1 and Case 2, the deterministic bidding model is analyzed and the LSE is assumed to have demand response programs or ESS. Especially, we discussed the impact of incorrect prediction of price spikes in Case 1. In Case 3, we focus on the stochastic model, and the impact of STPF error on load purchasing cost is analyzed.

All the numerical cases are conducted on an Intel Core i7-10510U@1.80GHz laptop with MPT [19] and Gurobi in MATLAB R2020a.

8.4.1 *Measurement of STPF Error Level*

8.4.1.1 Deterministic Models

The MAE and RMSE are adopted as the evaluation metrics of STPF for their wide application in forecasting. However, MAE is more robust in evaluating price forecasting results, because RMSE is likely to be influenced by extreme values. The formulas of MAE and RMSE are given as:

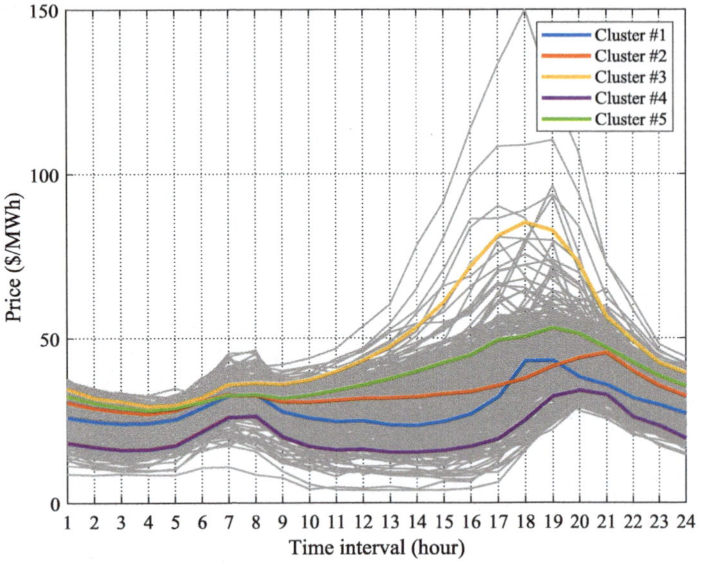

Fig. 8.5 The one-year day-ahead price data of PG&E in CAISO from April 21, 2015 to April 20, 2016. Gray lines: the original price data. Colored lines: five clusters given by k-means

$$
\begin{aligned}
\text{MAE} &= ||\mathbf{p} - \mathbf{p}^*||_{\ell_1}/T = \sum_t |p_t - p_t^*|/T, \\
\text{RMSE} &= \sqrt{||\mathbf{p} - \mathbf{p}^*||_{\ell_2}/T} = \sqrt{\sum_t (p_t - p_t^*)^2/T}.
\end{aligned}
\tag{8.13}
$$

Note that when a price spike occurs, MAE and RMSE cannot address the STPF error properly. While there has not yet been a widely-adopted metric for the spike scenarios, the following indicator of absolute error concentration level is proposed:

$$
I = \frac{\sum_{t \in \mathcal{S}} |p_t - p_t^*|}{\sum_{t \in \mathcal{T}} |p_t - p_t^*|} \tag{8.14}
$$

where $\mathcal{S} \subset \mathcal{T}$ is the set of time intervals with price spikes. I is not specified in the non-spike cases but fixed at 0.8 in the spike cases.

8.4.1.2 Stochastic Models

As we mentioned in Remark 8.1, the error in this case is modeled as the bias in scenario probabilities, and we use the ℓ_1 norm to quantify the error level:

Table 8.1 Daily Cost and Revenue ($) in Case 1

Price cluster	Cost without flexibility	Cost ($\theta = 0$)	Max revenue
#1	2245.8	2169.6	76.2
#2	2593.5	2537.6	55.9
#3	3871.8	3716.5	155.3
#4	1651.0	1575.5	75.5
#5	2971.1	2903.7	67.4

$$\mathrm{ERR}_{\ell_1} = ||\boldsymbol{\theta}_{\Omega}||_{\ell_1} = \sum_{\omega \in \Omega} |\theta_{\omega}| \tag{8.15}$$

8.4.2 *Case 1: LSE with Demand Response Programs*

The demand response programs enable the shift or reduction of load during peak hours. In this case, 40% of the LSE's load at each hour are assumed to be flexible, and can be shifted to the hour right before or after the original hour. The decision variables $\mathbf{x}$ in (8.2) contains three parts $\mathbf{x} = [\mathbf{x}^+, \mathbf{x}^0, \mathbf{x}^-]$, where $\mathbf{x}^+$ and $\mathbf{x}^-$ are the vectors of the forward-shifted and backward-shifted load, respectively, and $\mathbf{x}^0$ is the unshifted part. Part of the constraints in (8.2) are specified as:

$$\begin{aligned} q_t = q_t^* + x_{t+1}^+ + x_{t-1}^- - x_t^+ - x_t^-, \quad \forall t \\ \mathbf{x}^+ + \mathbf{x}^0 + \mathbf{x}^- = 40\% \times \mathbf{q}^*, \\ \mathbf{q}, \mathbf{x}^+, \mathbf{x}^0, \mathbf{x}^- \geq 0. \end{aligned} \tag{8.16}$$

The MPLP can be formulated according to the above constraints, and in order to simplify the model complexity, the dimension of basis vectors N in (8.9) is reduced to 4. $\mathbf{B}$ is constructed by considering the temporal dependence between adjacent time intervals. The STPF errors with MAE and RMSE from 0 to 15 are tested, respectively. Table 8.1 shows the daily cost and revenue in Case 1. Due to the assumption that the flexible demand can only be shifted one hour before or after, the maximum revenue with perfect STPF (i.e., $\boldsymbol{\theta} = 0$) is about 3–5% of the total purchasing cost. Figure 8.6 shows the percentage loss α under different MAEs and RMSEs, and the error bar in the figure is the standard deviation of α under different price clusters. As is shown in the figure, the α values under the same level of MAE and RMSE are slightly different, as the MAE line is higher than the RMSE lines. This is because $\boldsymbol{\theta}$ defined by the MAE is geometrically located on a ℓ_1 norm surface which has a diamond structure, and the $\boldsymbol{\theta}$ defined by the RMSE (of the same level) is on a ℓ_2 norm surface which is also the inscribed sphere of the ℓ_1 diamond.

The MAE of STPF in the literature generally ranges from 4 to 8 and the RMSE ranges from 6 to 10. When the error level (MAE) increases from 8 to 15, α^d increases

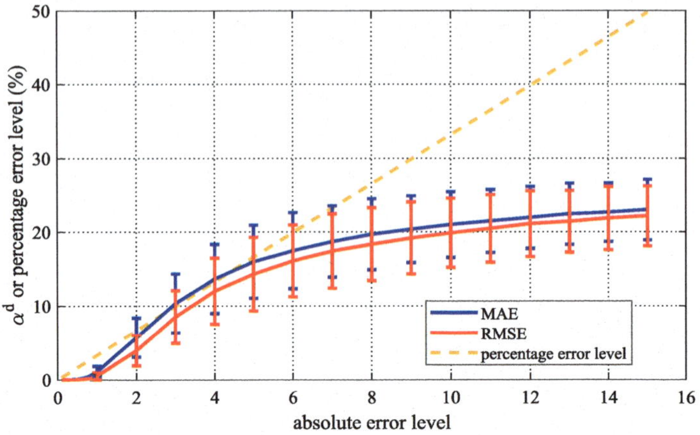

Fig. 8.6 Case 1: change of the percentage loss α^d with respect to the error level. The dashed brown line is the expected percentage error level of MAE

from 19.7% to 23.0%, while the expected percentage error level (the dashed brown line in Fig. 8.6) increases from 26.6% to 49.8%. If the error level (MAE) is reduced from 5 to 3, α^d improves about 5%. We can conclude that, bad and worse forecasters do not differ much, but good forecasters get much more revenue than average forecasters.

To study the impact of incorrect prediction of price spikes, an extra 57.6 $/MWh-spike is added onto price cluster #1 at hour 16, 17, 18, 19, and 20, respectively. Two situations are considered, with their examples shown in Fig. 8.7: Situation 1 is underestimating the price spike; Situation 2 is misplacing the price spike in a different but adjacent time interval.

The results are shown in Fig. 8.8. The level of underestimation in Fig. 8.8a is tuned by MAE with $I = 0.8$: 1 means that the price spike is not predicted at all, and 0 means that the price spike is predicted perfectly. For the price spike at hour 18 and 19, the underestimation does not have a significant economic impact since in the original price cluster #1 the prices at these two hours are already the largest. It is shown that, if the order of price values at different time intervals is predicted correctly, it is economically acceptable to underestimate the spike values in STPF. Figure 8.8b presents the average α^d for misplacing the price spike at an adjacent time interval. Such failure in spike forecasting has a very large impact on the cost.

8.4.3 Case 2: LSE with ESS

Compared with demand response, the ESS is much more flexible and can shift limited demand to any other time within its capacity. In this case, the LSE is assumed to have an ESS with 4 MWh capacity and 1.0 MW maximum charge/discharge

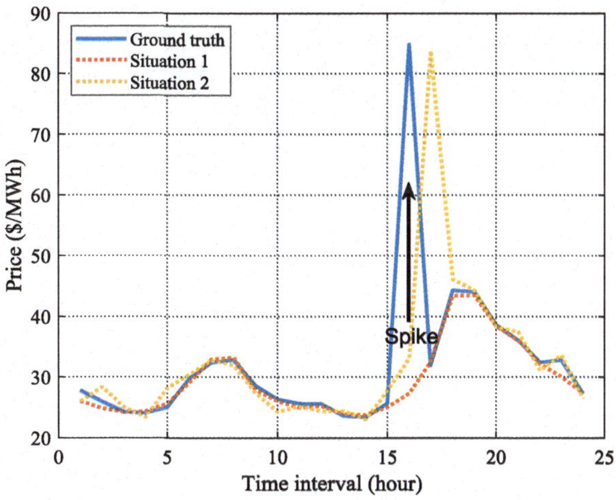

Fig. 8.7 Example on the two situations in price spike forecasting

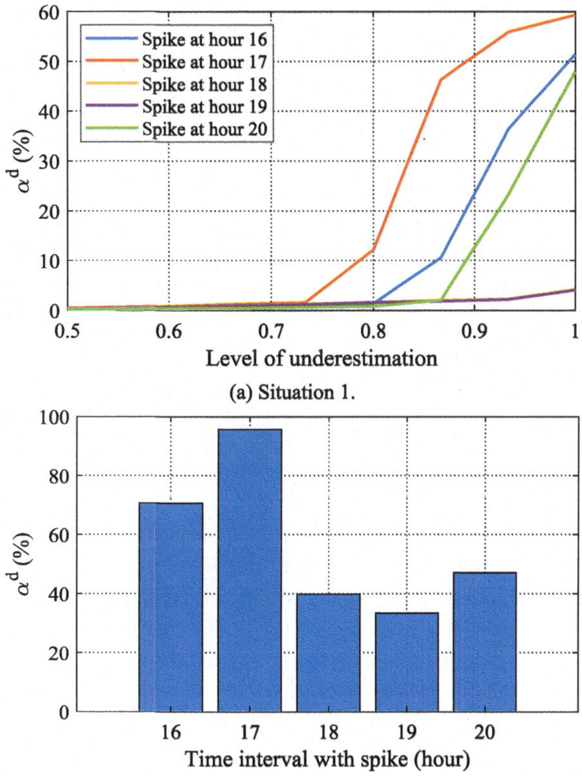

(a) Situation 1.

(b) Situation 2.

Fig. 8.8 Case 1: α^d in the two situations of incorrect prediction of price spikes

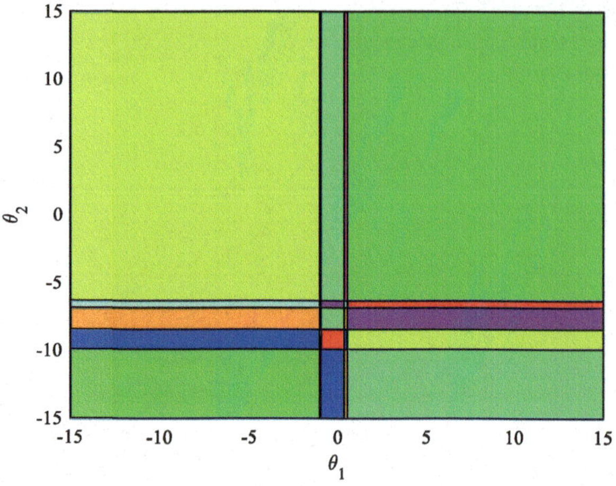

Fig. 8.9 A simplified view of the CRs in Case 2. θ_1 is the error associated with price at hour 10, and θ_2 is associated with hour 20

power. The energy efficiency is set as 0.95. The decision vector $\mathbf{x}$ contains three parts $\mathbf{x} = [\mathbf{x}^c, \mathbf{x}^d, \mathbf{Soc}]$, where $\mathbf{x}^c$ and $\mathbf{x}^d$ are the charging and discharging power, respectively, and $\mathbf{Soc}$ is the state of charge of the ESS. The constraints in (8.1) are specified as:

$$\begin{aligned} \mathbf{q} - \mathbf{x}^c + \mathbf{x}^d &= \mathbf{q}^*, \\ Soc_{t+1} &= Soc_t + 0.95 \times x_t^c - x_t^d, \; t = 1, \ldots, 24, \\ 0 \le \mathbf{x}^c, \mathbf{x}^d \le 1.0, \; 0 &\le \mathbf{Soc} \le 4, \; Soc_0 = Soc_{25} = 1.5. \end{aligned} \tag{8.17}$$

The MPLP is then formulated, and in order to present a direct view of the CRs, the price errors of hour 10 and 20 are selected as parameters. The simplified CRs are plotted in Fig. 8.9. The rectangles in the figure represent the CRs, and if $\boldsymbol{\theta}$ falls in the same rectangle, the loss of the LSE keeps the same. We then select the same $\mathbf{B}$ as in Case 1, and solve the α with respect to different MAE and RMSE levels.

Basic statistics of the daily cost and revenue of this case are given in Table 8.2. The revenue of the ESS is highly dependent on the intra-day price variation. Thus, the LSE achieves the maximum revenue with price cluster #3 with perfect STPF. The change of α^d with respect to different MAE or RMSE is shown in Fig. 8.10. Compared with the LSE with demand response programs in Case 1, the revenue of the ESS-LSE is sensitive to STPF errors even when the error level is above 8. As the MAE increases from 8 to 15, α^d increases from 21.9 to 30.9%. The results of this case have shown that bad and worse forecasters differ a lot for ESS. This might be caused by the fact that the ESS has much more flexibility than demand response programs, and that it deviates more from the optimal points when the error is large.

Table 8.2 Daily cost and revenue ($) in case 2

Price cluster	Cost without flexibility	Cost ($\theta = 0$)	Max revenue
#1	2245.8	2145.8	100.0
#2	2593.5	2514.0	79.5
#3	3871.8	3648.7	223.1
#4	1651.0	1557.3	93.7
#5	2971.1	2864.2	106.9

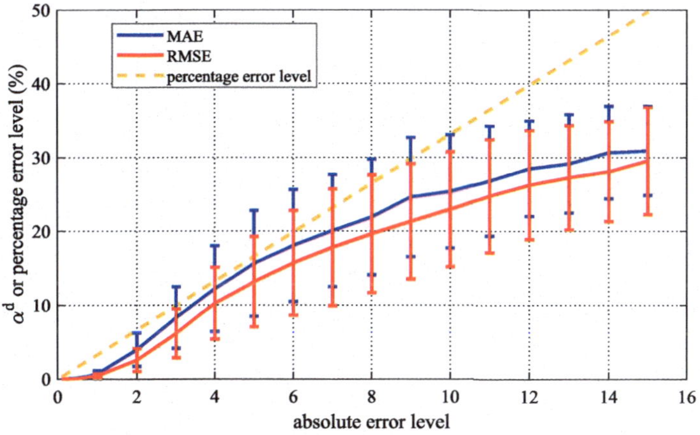

Fig. 8.10 Case 2: change of the percentage loss α^d with respect to the error level

8.4.4 Case 3: Stochastic LSE Bidding Model

In this case, we focus on the impact of STPF error on the bidding curve, because the endpoints on the bidding curve determine how much electricity will be purchased in the spot market and thus influence the cost of LSE as the ground truth prices differ from the forecasted prices. The LSE has the same demand response program as in Case 1. According to the linearization of (8.6), the values of prices $p_t^{(j)}$ are fixed at [10, 20, 30, 40, 50, 60] \$/MWh, and $0 \leq q_t^{(j+1)} \leq 0.95 q_t^{(j)}$. The prices $\bar{p}$ and $\underline{p}$ are set at 70 and 5 \$/MWh, respectively.

Assume that the ground truth price scenario set Ω^* is generated by adding a zero-mean Gaussian noise with $\sigma = 2$ on the price cluster #1. Due to STPF errors, the LSE gets a biased scenario set Ω_1 generated from the price cluster #2 with the same Gaussian noise. The MAE between the two price clusters is 4.51. The error level ERR_{ℓ_1} from 0 to 2.0 with a step of 0.01 is tested, and 2.0 means the forecasted scenarios are fully biased from the ground truth ones. In this case, the expected cost with perfect STPF is 2232 \$.

Figure 8.11 shows the change of α^s w.r.t. ERR_{ℓ_1}. As shown in the figure, α^s increases sharply when the error level is above 1.8. A fully biased scenario set will

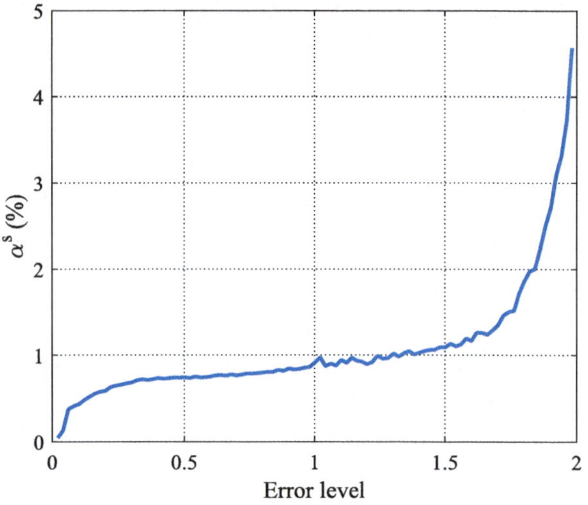

Fig. 8.11 Case 3: change of the percentage loss α^s with respect to ERR_{ℓ_1}

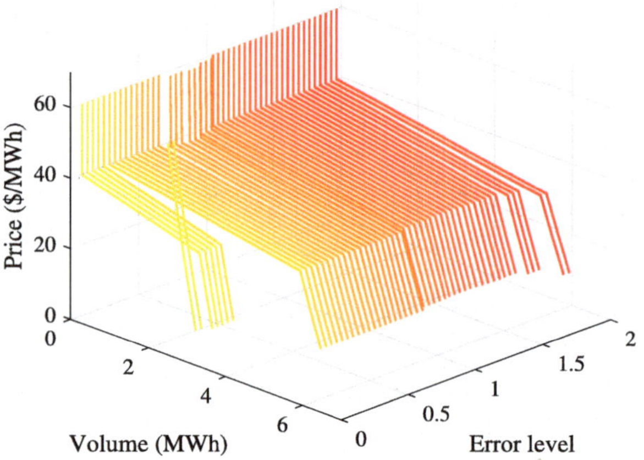

Fig. 8.12 Case 3: the change of the bidding curve at hour 12 under different ERR_{ℓ_1}

cause a nearly 5% loss on the total purchasing cost, which is rather considerable for an LSE. To present an intuitive result, we restrict θ_Ω on the line of identical values for $\theta_\omega(\omega \in \Omega_2)$ and plot the series of bidding curves. Figure 8.12 shows the change of the bidding curve at hour 12 under different ERR_{ℓ_1}. We can see that at some points of ERR_{ℓ_1}, the bidding curve shifts from one pattern to another abruptly, indicating that the STPF error has a large impact on the bidding curve. This is consistent with the property of critical region and Corollary 8.1 as the bidding curve's parameter $q_t^{(j)}$ is piece-wise constant in Θ.

Table 8.3 Average Time Consumption on the Calculation of CRs

Case	Case 1	Case 2	Case 3
Time (second)	18.61	1373.47	10600.00

8.4.5 *Time Consumption*

Table 8.3 lists the average time consumption of step 3 of the proposed framework in Sect. 8.3.4, which is the most time-consuming part of the method. It is noteworthy that the method would take hours or even days to finish the computation for large bidding models. However, this is acceptable because the evaluation of the impact of STPF error is a long-term task rather than a daily routine.

8.5 Conclusions and Future Work

In this chapter, we proposed a novel MPLP-based framework to quantify the impact of electricity price forecasting errors on the cost and revenue of market bidders from a price-taker's perspective. Both the deterministic and stochastic bidding models are considered. For the deterministic models, the STPF error is formulated as a vector added to the ground truth prices. For the stochastic models, the STPF error is formulated as the bias in the scenario probabilities. Then the error impact formulations under both conditions are given as high-dimension integration with the piece-wise constant function solved from the MPLP. Such integration can be calculated using Monte Carlo or quasi-Monte Carlo methods. Finally, three case studies are conducted to show the application of the method for three typical LSEs. The proposed framework and method can be used in the bidder's decision-making on improving STPF accuracy.

Future work includes improving computation efficiency. Also, some bidding models with ESS or electric vehicles are based on MILP, and the impact of integer variables is to be considered.

References

1. da Silva Catalão, João Paulo, Sílvio José Pinto Simões Mariano, VMF Mendes, and LAFM Ferreira. 2007. Short-term electricity prices forecasting in a competitive market: A neural network approach. *Electric Power Systems Research* 77 (10): 1297–1304.
2. Panapakidis, Ioannis P, and Athanasios S Dagoumas. 2016. Day-ahead electricity price forecasting via the application of artificial neural network based models. *Applied Energy* 172 :132–151.
3. Genscape. 2020. PowerIQ^{TM}, electric power data and intelligence for regional power market visibility. https://www.genscape.com/products/power/poweriq. Accessed 10 September 2020.

4. EBW Analytics Group. 2020. Energy market research and analysis. https://ebwanalytics.com/. Accessed 10 September 2020.
5. Ruiz, Carlos, and Antonio J Conejo. 2009. Pool strategy of a producer with endogenous formation of locational marginal prices. *IEEE Transactions on Power Systems* 24 (4): 1855–1866.
6. Tómasson, Egill, Mohammad Reza Hesamzadeh, and Frank A Wolak. 2020. Optimal offer-bid strategy of an energy storage portfolio: A linear quasi-relaxation approach. *Applied Energy* 260 :114251.
7. Gao, Feng, Gerald B. Sheble, Kory W. Hedman, and Yu. Chien-Ning. 2015. Optimal bidding strategy for GENCOs based on parametric linear programming considering incomplete information. *International Journal of Electrical Power and Energy Systems* 66: 272–279.
8. Song, Meng, and Mikael Amelin. 2017. Purchase bidding strategy for a retailer with flexible demands in day-ahead electricity market. *IEEE Transactions on Power Systems* 32 (3): 1839–1850.
9. Saez-Gallego, Javier, Mahdi Kohansal, Ashkan Sadeghi-Mobarakeh, and Juan M Morales. 2018. Optimal price-energy demand bids for aggregate price-responsive loads. *IEEE Transactions on Smart Grid* 9 (5): 5005–5013.
10. Zhao, Yuxuan, Changsen Feng, Zhenzhi Lin, Fushuan Wen, Chuan He, and Zhemin Lin. 2018. Development of optimal bidding strategy for an electric vehicle aggregator in a real-time electricity market. In *2018 IEEE Innovative Smart Grid Technologies - Asia (ISGT Asia)*, 288–293. IEEE.
11. Zareipour, Hamidreza, Claudio A Canizares, and Kankar Bhattacharya. 2010. Economic impact of electricity market price forecasting errors: A demand-side analysis. *IEEE Transactions on Power Systems* 25 (1): 254–262.
12. Mohammadi-Ivatloo, B., H. Zareipour, M. Ehsan, and N. Amjady. 2011. Economic impact of price forecasting inaccuracies on self-scheduling of generation companies. *Electric Power Systems Research* 81 (2): 617–624.
13. Chazarra, Manuel, Juan I Pérez-Díaz, Javier García-González, and Arild Helseth. 2017. Economic impact of forecasting errors in residual reserve curves in the day-ahead scheduling of pumped storage plants. In *2017 IEEE Manchester PowerTech*, 1–6. IEEE.
14. Huang, Wujing, Ning Zhang, Chongqing Kang, Mingxuan Li, and Molin Huo. 2019. From demand response to integrated demand response: Review and prospect of research and application. *Protection and Control of Modern Power Systems* 4 (1): 1–13.
15. Rashidizadeh-Kermani, Homa, Mostafa Vahedipour-Dahraie, Amjad Anvari-Moghaddam, and Josep M. Guerrero. 2019. Stochastic risk-constrained decision-making approach for a retailer in a competitive environment with flexible demand side resources. 2019, 29 (2): e2719.
16. Wang, Jin, and Chunlei Liu. 2006. Generating multivariate mixture of normal distributions using a modified cholesky decomposition. In *Proceedings of the 38th Conference on Winter Simulation*, 342–347. IEEE.
17. Scott, David W. 2015. *Multivariate density estimation: Theory, practice, and visualization*. Wiley Inc.
18. Pistikopoulos, Efstratios N, Michael C Georgiadis, and Vivek Dua. 2007. *Multi-parametric programming: Theory, algorithms and applications*, vol 1. Wiley-VCH Verlag GmbH & Co. KGaA.
19. Herceg, Martin, Michal Kvasnica, Colin N Jones, and Manfred Morari. 2013. Multi-parametric toolbox 3.0. In *Proceedings of the European Control Conference*, 502–510. Zürich, Switzerland. http://control.ee.ethz.ch/~mpt.
20. Dick, Josef, Frances Y Kuo, and Ian H Sloan. 2013. High-dimensional integration: The quasi-monte Carlo way. *Acta Numerica* 22 :133.
21. Intelligent Energy Europe. 2008. Residential monitoring to decrease energy use and carbon emissions in Europe, (REMODECE). Technical report. https://ec.europa.eu/energy/intelligent/projects/sites/iee-projects/files/projects/documents/remodece_publishable_report_en.pdf.
22. California ISO. 2017. California ISO market notice: Interface specification for OASIS 2017 release. http://www.caiso.com/Documents/OASIS-InterfaceSpecification_v4_3_5Clean_Spring2017Release.pdf.

Chapter 9
LMP Forecasting and FTR Speculation

9.1 Introduction

Financial transmission rights (FTRs) or congestion revenue rights are financial instruments implemented in many electricity markets to hedge the risk of price spread across different nodes caused by network congestion [1]. It can be acquired in an auction-based primary market or a secondary market built by the same system operator in charge of the day-ahead (DA) market. Detailed market structures can be found in [2]. An FTR is generally specified with a *from* node, a *to* node, an *MW* amount, and a holding period. The revenue of the FTR comes from the cumulative DA congestion price spread of the two nodes in the holding period. Although there are discussions on lossy FTR [3, 4] which can hedge full price spreads (including those caused by network loss), most electricity markets still implement the traditional FTR; thus, we will focus on the traditional FTR scheme.

There are two kinds of FTR investors, the hedgers, and the speculators. The hedgers are usually electricity market participants such as generation companies (GenCos) or load serving entities (LSEs). These companies buy or sell a large amounts of electricity in the wholesale market through bilateral or spot contracts. They would face the risk of congestion in the DA or real-time market and are instinctively willing to hold FTRs for hedging purposes. The speculators are not limited to GenCos or LSEs and have more diversified identities such as banks, commodity traders, and private funds. They regard FTR as a pure financial derivative and hope to optimize utility through proper pricing and investing. Speculators play an important role in a deregulated market, as they help in price discovery.

The system operator usually organizes the auctions annually, quarterly, and monthly in the primary market, and FTRs with the corresponding holding period are granted to bid winners. FTRs can be reconfigured in terms of size and time, e.g., a 100 MW FTR could be sold as two 50 MW FTRs, and a monthly FTR could be split so that one portion covers next week only, or only the hours from 6 pm to 8 pm (page 210 of [5]), which makes it much more flexible to be purchased in the secondary market. Basically, the FTR tool is categorized as either an obligation or an option type. An FTR obligation requires the holder to pay an offsetting congestion charge, while an FTR option exempts the right holder from the obligation to pay offsetting congestion charge so that it would only yield nonnegative returns [6]. In this chapter,

Q. Chen et al., *Data Analytics in Power Markets*,
https://doi.org/10.1007/978-981-16-4975-2_9

we focus on FTR obligations, because FTR options could be modeled in a similar way and will be discussed in a future work.

Researchers have conducted analyses on the bidding strategy and portfolio construction of FTRs since the early 2000s [7]. Li et al. proposed a risk-constrained FTR bidding strategy in the primary market [8]. Then, a bilevel optimization framework with the upper subproblem representing bidders and the lower subproblem representing auctions is used to simulate the market outcomes. In [9], a two-tier matrix game theoretic modeling approach was developed that can be used to obtain joint bidding strategies for GenCos and LSEs in FTR and energy markets. In [10], Apostolopoulou et al. proposed an FTR portfolio construction framework based on the identification of congested network elements. Detailed network topology was assumed to be accessible, and the power transfer distribution factor (PTDF) matrices under different contingency cases were used to form an optimization problem with a sparsity objective. The numerical experiment was conducted on a PJM-based large-scale test system. In [11], Fang et al. proposed a bidding strategy for LSEs considering coupon-based demand response programs and congestion compensation from FTR. In [12], the cross-product manipulation problem of FTR and virtual transactions was studied. The three-stage equilibrium in the FTR auction, DA, and real-time market was analyzed, and a two-node numerical case was given. Many articles focus on FTR bidding for hedgers, while the problem of FTR portfolio construction for speculators has not yet been studied properly. For FTR speculations, there are some difficulties since the decision space contains almost all pairs of the whole network nodes. It would be an onerous task to model the forecasting and uncertainty of so many targets in large-scale systems. The aim of this chapter is to propose a stochastic optimization framework for FTR speculation with a clustering method applied to reduce the decision space and a forecasting-based conditional price distribution modeling method to improve the excess return.

The techniques used in the proposed FTR speculation framework of this chapter are related to pricing node clustering and mid-term electricity price forecasting. A brief literature review on these two fields is given below.

The clustering of pricing nodes in the network was mainly studied for system operators to identify potential load pockets and prevent market power exercising [13, 14]. The Laplacian structure of the power network was used for clustering in [13], while the sensitivity matrix of generator dispatch and price perturbations was clustered in [14]. Vaskovskaya et al. also performed a series of studies [15, 16] on finding congestion zones with the PTDF matrix and LMP data. The clustering techniques used in these articles include the max-min distance algorithm, hierarchical clustering, k-d tree, k-means, etc. To the best of the authors' knowledge, existing studies mainly focus on network partitioning and decomposition, and there has been little research on the clustering to select key pricing nodes targeting FTR investment.

Because the holding period of an FTR usually varies from days to months, its pricing relies on mid-term forecasting of the joint distribution of nodal prices in the market. There is a comprehensive review by Ziel et al. in this field [17], whose main focus is to forecast the price density hour by hour from weeks ahead to years ahead. In [18], a market model for the Turkish electricity market was proposed to consider the impact of supply and demand on prices. A generalized additive model was then used to obtain daily demand forecasts. This model can generate average prices and hourly price forecasts in the mid-term. In [19], a hybrid nonlinear regression and support vector machine (SVM) model was used to predict the month-ahead average daily electricity price profile. In [20], deep learning approaches were tested and compared with traditional forecasting methods. Existing research has mainly focused on price forecasting for a single node. For FTR speculation, the price distribution of multiple nodes should be modeled and analyzed.

The payoff of FTRs is directly determined by the congestion price spread, which is highly related to the network's PTDF matrix. However, FTR investors are market participants and usually do not have detailed network information. In other words, their portfolios of FTRs are decided under uncertainty and imperfect information. As the existing studies focus on FTR hedging or speculation with perfect information, studies on FTR speculation under imperfect information need to be conducted to bridge this gap. Additionally, the problem of price forecasting in FTR speculation is slightly more complicated since the prices at multiple pricing nodes should be forecasted and their dependence should be modeled. In this chapter, we proposed a stochastic optimization framework for FTR speculation and portfolio construction with clustering and probabilistic mid-term price forecasting. K-means clustering with frequent congestion patterns as input features is first used to find representative pricing nodes and reduce the problem scale. Quantile regression forests (QRFs) are used to predict the probabilistic density of average LMPs during the holding period, and a copula-based model is used to model the dependence of LMPs. The methods in our framework require only public market data, including LMP data, and do not rely on confidential network information. The contributions of this chapter are three-fold:

1. It proposes a framework for FTR speculation with forecasting on the joint distribution of the average prices of multiple nodes during a mid-term holding period. The problem is formulated as a stochastic optimization problem and solved using a scenario-based approach.
2. K-means clustering with network congestion patterns is used to select nodes with representative price profiles and reduce the problem scale. Quantile regression forests with copula models are used to generate the conditional joint distribution of prices. The methods are applicable for decision-making under imperfect information.
3. It conducts comprehensive numerical experiments with public market data from the southwest power pool (SPP). The results demonstrate the performance of clustering and probabilistic forecasting methods.

The rest of this chapter is structured as follows. Section 9.2 presents the stochastic optimization model for FTR speculation and portfolio construction. Section 9.3 gives an overview of the proposed data-driven framework. Section 9.4 details the data-driven methodology of clustering and forecasting. Section 9.5 shows the case study with SPP data. Finally, Sect. 9.6 draws the conclusion.

9.2 Stochastic Optimization Model

The model is based on linear programming (LP). It assumes that the conditional joint distribution of prices is given.

9.2.1 Model of FTR Portfolio Construction Problem

Let N denote the number of pricing nodes for the construction of FTR portfolio, and p_i denote the average LMP at node i $(i = 1, 2, \ldots, N)$ during the holding period. According to the decomposition of LMP [21, 22]:

$$p_i = p_i^E + p_i^C + p_i^L \tag{9.1}$$

where p_i^E, p_i^C, p_i^L are the LMP components of energy, congestion, and loss, respectively. p_i^E is identical at all pricing nodes, while p_i^C and p_i^L may differ by node. We then denote an FTR obligation with the ordered triplet Γ:

$$\Gamma = \{i, j, \gamma_{i,j}\} \tag{9.2}$$

where i is the *from* node, j is the *to* node $(1 \leq i, j \leq N, i \neq j)$, and $\gamma_{i,j} \geq 0$ is the *MW* amount. According to the definition, the hourly revenue r of Γ (wherein the DA market is assumed to clear hourly without loss of generality) is

$$r = (p_j^C - p_i^C)\gamma_{i,j} \tag{9.3}$$

Given the probabilistic forecast of the average LMP vector denoted as a high-dimensional joint distribution function $\hat{G}(p_1^C, p_2^C, \ldots, p_N^C)$, the decision variable for the weight of FTR assets would be a $N(N-1)$-dimensional vector $\boldsymbol{\gamma}$ with elements $\gamma_{i,j}$. The objective function of the speculator can be formulated as

$$\max_{\gamma} \mathbb{E}\Big(\sum_{1 \leq i,j \leq N, i \neq j} (p_j^C - p_i^C) \cdot \gamma_{i,j} \Big| \mathbf{p}^C \sim \hat{G} \Big) - \sum_{1 \leq i,j \leq N, i \neq j} \gamma_{i,j} q_{i,j} + \beta \cdot \mathrm{CVaR} \tag{9.4}$$

where $q_{i,j}$ is the cost for procuring the FTR, CVaR is the conditional value-at-risk, and β is the risk factor. The cost analysis on procuring a certain FTR is a very complex problem in both primary auctions and secondary markets [10], and we will try to consider the uncertainty in the cost of FTR in future work. Currently, we use a deterministic budget constraint instead:

$$\sum_{1 \leq i,j \leq N, i \neq j} \gamma_{i,j} q_{i,j} \leq K \tag{9.5}$$

To simplify the formulation of the optimization problem, we introduce an $N \times 1$ financial nodal injection vector $\mathbf{a}$ with elements $a_i = \sum_{1 \le i,j \le N, i \ne j} (\gamma_{i,j} - \gamma_{j,i})$ to represent the original decision variable γ. There exists an $N \times [N(N-1)]$ matrix $\mathbf{M}$ that satisfies $\mathbf{a} = \boldsymbol{M}\gamma$:

$$\mathbf{M} = \left\{ \begin{bmatrix} 0 \\ \vdots \\ 1 \\ \vdots \\ -1 \\ \vdots \\ 0 \end{bmatrix} \begin{matrix} \\ \\ i \\ \\ j \\ \\ \\ \end{matrix} \right\}_{i,j} = \begin{bmatrix} 1 & 1 & \cdots 0 & 0 \\ -1 & 0 & \cdots 0 & 0 \\ 0 & -1 & \cdots 0 & 0 \\ \vdots & \vdots & \ddots \vdots & \vdots \\ 0 & 0 & \cdots 0 & -1 \\ 0 & 0 & \cdots 1 & 1 \end{bmatrix} \tag{9.6}$$

The optimization problem of (9.4)–(9.5) can be transformed as follows:

$$\begin{aligned} \max_{\mathbf{a},\gamma} \; & \mathbb{E}\Big(-(\mathbf{p}^C)^\top \mathbf{a} \Big| \mathbf{p}^C \sim \hat{G} \Big) - \mathbf{q}^\top \gamma + \beta \cdot \text{CVaR} \\ \text{s.t.} \; & \mathbf{a} = \boldsymbol{M}\gamma, \\ & \mathbf{q}^\top \gamma \le K \end{aligned} \tag{9.7}$$

where $\mathbf{q} = \left[q_{i,j} \right]_{i \ne j}$ is the vector of cost.

9.2.2 *Scenario-Based Stochastic Optimization Model*

The problem of (9.7) with a general distribution function $\hat{G}$ and CVaR can be well represented in a scenario-based stochastic optimization model:

$$\begin{aligned} \max \; & \sum_{\omega \in \Omega} -\pi_\omega (\mathbf{p}^C_\omega)^\top \mathbf{a} - \mathbf{q}^\top \gamma + \beta \cdot \text{CVaR} \\ \text{s.t.} \; & \mathbf{a} \in \boldsymbol{M}\gamma, \\ & \mathbf{q}^\top \gamma \le K \end{aligned} \tag{9.8}$$

where Ω is the set of certain number of realizations from $\hat{G}$ and π_ω is the probability of some scenario $\omega \in \Omega$. Let α denote the degree of confidence. CVaR is defined as the expected value of the profit smaller than the $(1 - \alpha)$-quantile of the profit distribution. It can be embedded into the scenario-based optimization problem with preservation of convexity by introducing auxiliary variables [23, 24]:

$$\begin{aligned} \text{CVaR} = \max \zeta - \frac{1}{1-\alpha} \sum_{\omega \in \Omega} \pi_\omega \eta_\omega \\ \text{s.t.} \quad \zeta + (\mathbf{p}_\omega^C)^\top \mathbf{a} \le \eta_\omega, \ \forall \omega \in \Omega \\ \eta_\omega \ge 0, \qquad \forall \omega \in \Omega \end{aligned} \tag{9.9}$$

where ζ and η are auxiliary variables. Thus, the final optimization problem is formulated as

$$\begin{aligned} \max_{\mathbf{a},\zeta,\eta_\omega} & \Big(\sum_{\omega \in \Omega} -\pi_\omega (\mathbf{p}_\omega^C)^\top \mathbf{a} \Big) - \mathbf{q}^\top \gamma + \beta \Big(\zeta - \frac{1}{1-\alpha} \sum_{\omega \in \Omega} \pi_\omega \eta_\omega \Big) \\ \text{s.t.}\ & \zeta + (\mathbf{p}_\omega^C)^\top \mathbf{a} \le \eta_\omega,\ \eta_\omega \ge 0,\ \forall \omega \in \Omega, \\ & \mathbf{a} = \boldsymbol{M}\gamma, \\ & \mathbf{q}^\top \gamma \le K \end{aligned} \tag{9.10}$$

which is a linear programming problem and can be solved efficiently by commercial or open-source solvers.

9.3 Data-Driven Framework

In a typical LMP market, there are usually thousands of pricing nodes, and it would still be difficult to forecast the conditional joint distribution of thousands of LMPs. Actually, the payoff of FTR is directly related to price spread $\mathbf{p}^C$ across different nodes, which is caused by network binding constraints (e.g., congested lines). Once some lines become congested, the congestion component of LMP is formed, and there will be congestion zones such that the intra-zone price spreads are smaller than the inter-zone price spreads. It is rational to select a small number of representative nodes according to the congestion zones for FTR investment. Thus, the data-driven framework includes two main procedures: clustering and forecasting. An overview of the proposed data-driven framework is shown in Fig. 9.1. The overall framework includes the following steps:

1. *Data Collection*: Collect the historical LMP and congestion price data for all pricing nodes of the system.
2. *Congestion Pattern Estimation*: Recover the missing data, and estimate the basis matrix $\Delta\mathbf{S}$ for a congestion price spread with (9.14). $\Delta\mathbf{S}$ is highly related to the PDFT matrix, and this relationship will be further discussed in Sect. 9.4.
3. *Node Selection*: Cluster using a distance matrix constructed from $\Delta\mathbf{S}$, and find the key pricing nodes with representative price profiles.
4. *Probabilistic Forecasting*: Train a quantile regression model for each key node, and fit copulas to model the dependence of prices across nodes. The probabilistic forecasts can be obtained with (9.20) and (9.22).

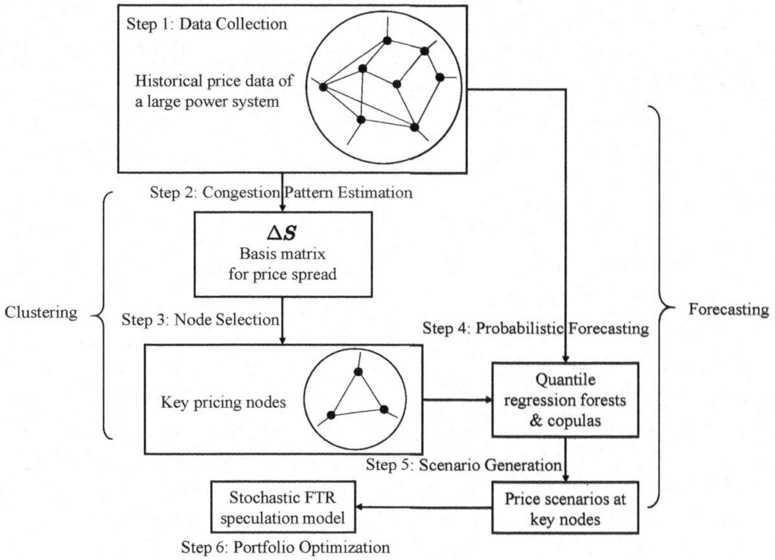

Fig. 9.1 Overview of the proposed data-driven framework

5. *Scenario Generation*: Generate the price scenarios from the fitted models with (9.24) and (9.25).
6. *Portfolio Optimization*: Solve the stochastic optimization problem of (9.10) in Sect. 9.2.

9.4 Methodology

The detailed clustering and price distribution forecasting methods are introduced in this section.

9.4.1 Clustering

Since historical LMP data and network binding constraints are usually available online, these data can be used to construct the features for clustering. In former studies [14–16], the PTDF matrix was usually assumed to be accessible, as their methods were designed for system operators. However, such data are usually confidential to others. Thus, we will try to estimate the PTDF matrix (or network congestion patterns) from historical data. First, let us recall the decomposition of LMP [21, 22]:

$$\mathbf{p} = \mathbf{p}^E + \mathbf{p}^C + \mathbf{p}^L$$
$$\mathbf{p}^E = \sigma\mathbf{1},\ \mathbf{p}^C = \underbrace{\mathbf{S}^\top\boldsymbol{\mu}}_{\mathbf{p}_A^C} - \underbrace{\mathbf{1}\mathbf{d}^\top\mathbf{S}^\top\boldsymbol{\mu}}_{\mathbf{p}_B^C},\ \mathbf{p}^L = -\sigma\mathbf{l} \tag{9.11}$$

where $\mathbf{S}$ is the PTDF matrix, $\mathbf{l}$ is the linearized factor for network loss, $\mathbf{d}$ is the loss distribution factor, and σ and $\boldsymbol{\mu}$ are shadow prices in the lossy security-constrained economic dispatch (SCED) model. Since the component $\mathbf{p}_B^C$ is identical at all nodes, the price spread in $\mathbf{p}^C$ is mainly caused by $\mathbf{p}_A^C$. Each element in $\boldsymbol{\mu}$ corresponds to a transmission constraint, and $\mathbf{p}_A^C$ is spanned by column vectors of $\mathbf{S}$. It is reasonable that $\boldsymbol{\mu}$ accounts for most of the variance in the price spread, and its impact can be described by matrix $\mathbf{S}$.

$\mathbf{S}$ is a constant matrix unless there are line outages or network parameter changes. However, the impact of the error of $\mathbf{S}$ on FTR applications was proven to be in an acceptable range [25]. Therefore, $\mathbf{S}$ is considered as constant in this chapter, and we will also demonstrate the efficacy of using $\mathbf{S}$ in the case study.

The proposed clustering method takes the following steps.

9.4.1.1 Missing Data Recovery

Although the data of $\mathbf{p}^C$ and $\boldsymbol{\mu}$ are public data, there are some times when $\boldsymbol{\mu} = 0$ but $\mathbf{p}^C \neq 0$. For example, $\mathbf{p}^C$ for 11:00 to 12:00 on Jan 16, 2015 is nonzero, while no binding constraint is released for the same hour.[1] This is because the real $\boldsymbol{\mu}$ is corrupted or not publicized due to internal errors. Thanks to the sparsity of $\boldsymbol{\mu}$, we can use an orthogonal matching pursuit [26] (OMP)-based method to recover the original $\boldsymbol{\mu}$. Since the number of intervals with missing data is small, we do not go through the details of this step.

9.4.1.2 Estimating Congestion Patterns

As $\mathbf{p}_B^C$ is identical at all nodes, we can eliminate it from $\mathbf{p}^C$ by subtracting the value of $\mathbf{p}^C$ at a certain node i from the $\mathbf{p}^C$ at every node[2]:

[1] The DA LMP data can be obtained at https://marketplace.spp.org/file-browser-api/download/da-lmp-by-bus?path=%2F2015%2F2015.zip, and the binding constraints can be obtained at https://marketplace.spp.org/file-browser-api/download/da-binding-constraints?path=%2F2015%2F2015.zip.

[2] Without loss of generality, node 1 is selected as node i.

$$\Delta \mathbf{P}^C = \mathbf{P}^C - \mathbf{p}_1^C \mathbf{1} = \begin{bmatrix} p_{1,t_1}^C & p_{2,t_1}^C & \cdots & p_{N,t_1}^C \\ p_{1,t_2}^C & p_{2,t_2}^C & \cdots & p_{N,t_2}^C \\ \vdots & \vdots & \ddots & \vdots \\ p_{1,t_T}^C & p_{2,t_T}^C & \cdots & p_{N,t_T}^C \end{bmatrix} - \begin{bmatrix} p_{1,t_1}^C & \cdots & p_{1,t_1}^C \\ p_{1,t_2}^C & \cdots & p_{1,t_2}^C \\ \vdots & \ddots & \vdots \\ p_{1,t_T}^C & \cdots & p_{1,t_T}^C \end{bmatrix} \tag{9.12}$$

The congestion pattern matrix $\Delta\mathbf{S}$ is defined as:

$$\begin{bmatrix} \boldsymbol{\mu}_{t_1}^\top \\ \vdots \\ \boldsymbol{\mu}_{t_T}^\top \end{bmatrix} \Delta\mathbf{S} = \Delta\mathbf{P}^C \tag{9.13}$$

which can be estimated using ordinary least squares (OLS) if $\Delta\mathbf{P}^C$ and $\boldsymbol{\mu}$ are given. Let $\mathbf{U} = [\boldsymbol{\mu}_{t_1}, \ldots, \boldsymbol{\mu}_{t_T}]^\top$. The estimation of $\Delta\mathbf{S}$ can be calculated using the Moore-Penrose pseudo inverse:

$$\hat{\Delta\mathbf{S}} = \mathbf{U}^\dagger \Delta\mathbf{P}^C \tag{9.14}$$

Note that the congestion pattern matrix $\Delta\mathbf{S}$ is different from the PTDF matrix $\mathbf{S}$. Let $\Delta S_{i,j}$ and $S_{i,j}$ denote the elements of $\Delta\mathbf{S}$ and $\mathbf{S}$, respectively. We have:

$$\Delta S_{i,j} = S_{i,j} - S_{1,j} \tag{9.15}$$

The row vector of $\Delta S_{i,j}$ is N-dimensional, and its j-th element represents the impact of the corresponding constraints on the price spread between p^C at node j and node 1. Thus, $\Delta\mathbf{S}$ contains important information on the congestion pattern of different constraints.

9.4.1.3 Feature Construction

According to the aforementioned characteristic of the $\Delta\mathbf{S}$ matrix, we can perform an ℓ_2 row normalization on $\Delta\mathbf{S}$ and then use its columns as the feature vector for each node. Another straightforward way is to use historical LMP data as the feature vector. We refer to these two ways as **congestion pattern-based** and **price-based** approaches, and the latter is used as benchmark.

9.4.1.4 Clustering Algorithm

We use the classic k-means algorithm to obtain the price zones. Once the partition is finished, the node that has the median LMP profile in a certain price zone is selected as a representative.

9.4.2 Mid-Term Probabilistic Forecasting

9.4.2.1 What Are We Forecasting

To generate price scenarios for FTR investment, the forecasting target is the average congestion component p^C of LMP at different nodes during the holding period (e.g., weeks to months). Since $p^C = 0$ when there is no congestion in the network, p^C is a sparse time series which makes it difficult to forecast with traditional forecasting techniques. Since p^E is identical in all nodes, we can simply replace $\mathbf{p}^C$ in (9.10) with $\mathbf{p}^C + \mathbf{p}^E$ without changing the payoff. Thus, we will forecast the average value of $p^C + p^E$ for a certain holding period.

9.4.2.2 Single Quantile Regression

Let the vector $\mathbf{x}_t$ denote the explanatory variables, and let y_t denote the response variable. A single QR model is used to describe the conditional distribution with $y_t(q)$:

$$\Pr(y_t \leq y_t(q)|\mathbf{x}_i) = q \tag{9.16}$$

where $q \in (0, 1)$ is the quantile level. $y_t(q)$ can be obtained by some probabilistic forecasting engine, e.g., LP-based models and machine learning models trained with quantile loss or pinball loss function.

9.4.2.3 Tree-Based Forecasting Engine

The focus of this chapter is not on the development or introduction of any novel forecasting engines. Since tree-based algorithms have been widely used in machine-learning tasks and have shown good generalization ability, we use the quantile regress forests (QRF) [27] as the forecasting engine which is based on random forest (RF) models.

Assume that m samples are used to train a random forest with s trees. Each tree j assigns a weight $w_i(\mathbf{x}, \boldsymbol{\theta}_j)$ with given explanatory variables $\mathbf{x}$ for the i-th response variable y_i, where $\boldsymbol{\theta}$ is the tree's parameter vector. The ensemble weight of the forests is the average of tree weight:

$$w_i(\mathbf{x}) = \frac{1}{s}\sum_{j=1}^{s} w_i(\mathbf{x}, \boldsymbol{\theta}_j),\ i = 1, 2, \ldots, m \tag{9.17}$$

In a typical RF model, one would only care for the conditional mean of the response variable, which is predicted as

$$\hat{\mathbb{E}}(y|\mathbf{x}) = \sum_{i=1}^{m} w_i(\mathbf{x}) y_i \tag{9.18}$$

In the QRF, the conditional cumulative distribution function (CDF) is further calculated by

$$\hat{F}(y|\mathbf{x}) = \Pr(Y \le y|\mathbf{x}) = \sum_{i=1}^{m} w_i(\mathbf{x}) \cdot 1_{\{y_i \le y\}}, \ \forall y \in \mathbb{R} \tag{9.19}$$

where $1_{\{\cdot\}}$ is the Boolean function for the input condition. Once $\hat{F}(y|\mathbf{x})$ is obtained, the q-quantile $y(q)$ that satisfies (9.16) can be predicted:

$$\hat{y}(q) = \inf\{y : \hat{F}(y|\mathbf{x}) \ge q\}, \ \forall q \in (0, 1) \tag{9.20}$$

The computational burden of QRF does not differ from RF in training. It can predict the conditional quantiles efficiently and consistently, i.e., the annoying quantile crossing problem in other forecasting engines is not an issue.

9.4.2.4 Selection of Features

Three kinds of basic features are considered in $\mathbf{x}_t$: the lagged prices y_{t-l}, the grid-level and zonal-level demand forecasts, and the grid-level generation mix, as they are highly related to the target series [28, 29]. Having access to more details about the system data is helpful in forecasting but is not discussed in this chapter. Recursive feature elimination [30] with RF and cross-validation is applied in selecting a subset of features for final prediction.

9.4.3 Copulas for Dependence Modeling

To model the dependence among single QR results, copula theory is adopted in this chapter. According to Sklar's theorem [31], any multivariate CDF $H(u_1, u_2, \ldots, u_N)$ can be expressed in terms of its marginals $F_i(u_i) = \Pr(U <= u_i)$ with a copula function C:

$$H(u_1, u_2, \ldots, u_N) = C\Big(F_1(u_1), F_2(u_2), \ldots, F_N(u_N)\Big) \tag{9.21}$$

Thus, for the conditional joint distribution $G_t(p_1, p_2, \ldots, p_N)$ at time t, we can model it with certain families of copulas:

$$\hat{G}_t(p_1, p_2, \ldots, p_N) = C\Big(F_1(p_1|\mathbf{x}_{1,t}), F_2(p_2|\mathbf{x}_{2,t}), \ldots, F_N(p_N|\mathbf{x}_{N,t})\Big) \tag{9.22}$$

Let $v_{i,t} = F_i(p_i|\mathbf{x}_{i,t})$ $(t = 1, \ldots, T)$ is the historical quantile level. The parameters of C can be trained with historical data of $v_{i,t}$. There are different families and structures for a high-dimensional copula, and we choose the simplest Gaussian copulas, which can be parametrized by an N-dimensional correlation matrix $\mathbf{R}$:

$$C_{\mathbf{R}}^{\text{Gauss}}(u_1, \ldots, u_N) = \Phi_{\mathbf{R}}\Big(\Phi^{-1}(u_1), \ldots, \Phi^{-1}(u_N)\Big) \tag{9.23}$$

where Φ^{-1} is the inverse CDF of a standard norm distribution and $\Phi_{\mathbf{R}}$ is the joint CDF of N norm variables with zeros means and covariance $\mathbf{R}$. $\mathbf{R}$ can be estimated from $v_{i,t}$ using maximum likelihood estimation or Kendall's τ estimation. When Gaussian copulas are applied to high-dimensional data, the parametrization of $\mathbf{R}$ has three basic structures: the full structure, giving freedom to all its nondiagonal elements (**Gaussian-full**), the Toeplitz structure, limiting elements in each descending diagonal from left to right to the same (**Gaussian-toep**), and the exchangeable structure, whose nondiagonal elements are the same (**Gaussian-ex**). The numbers of parameters in these three structures are $N(N-1)/2$, $N-1$, and 1, respectively. Gaussian-toep and Gaussian-ex use fewer parameters under a stronger assumption and can perform better if there are not enough data for the training of parameter-heavy models. In the case study, we will analyze the performance of Gaussian copula structures as well as other complicated copula structures.

9.4.4 Training and Evaluation Timeline

Considering all FTR holding periods with identical length (e.g., one week), the whole historical data set could contain T continuous holding periods. We partition the T periods into three parts: the pretraining set $\mathcal{T}_1 = \{1, 2, \ldots, T_1\}$, the calibration set from $\mathcal{T}_2 = \{T_1 + 1, \ldots, T_2\}$, and the test set $\mathcal{T}_3 = \{T_2 + 1, \ldots, T\}$. Feature selection is conducted on $\mathcal{T}_1$, and the best feature set for each node is then fixed in $\mathcal{T}_2$ and $\mathcal{T}_3$. In $\mathcal{T}_2$, the QRF model for each node is retrained after every period and generates one-step ahead prediction. In the test set, the QRF model is fixed at T_2. Clustering is performed with the features from congestion patterns or price profiles in $\mathcal{T}_1$ and $\mathcal{T}_2$, and the selected key nodes are fixed in $\mathcal{T}_3$. Copulas are estimated using data in $\mathcal{T}_2$. Scenarios are generated, and (9.10) is solved for every period in $\mathcal{T}_3$.

9.4.5 Scenario Generation

Scenario generation from probabilistic forecasting of individual nodes and copula models is essentially a high-dimensional distribution sampling problem. In the specific case of Gaussian copulas, the sampling is simplified compared with other copula structures (e.g., Archimedean copulas). In this chapter, we only introduce how sampling is conducted in the case of Gaussian copulas. For other copulas, one can refer to [32].

Given matrix $\mathbf{R}$, generate a sample $\mathbf{u}_\omega = [u_{\omega,1}, \ldots, u_{\omega,N}] \in \mathbb{R}^N$ following an N-dimensional Gaussian distribution with zero means and covariance $\mathbf{R}$ for scenario ω. Then, the joint quantiles $\mathbf{q}_\omega$ can be obtained by

$$q_{\omega,i} = \Phi^{-1}(u_{\omega,i}), \quad i = 1, \ldots, N \tag{9.24}$$

Then, the QRFs for each individual node can be used to convert the quantile scenario $\mathbf{q}_\omega$ to price scenario $\mathbf{p}_\omega$:

$$p_{\omega,i} = \inf\{p_i : \hat{F}(p_i|\mathbf{x}_i) \geq q_{\omega,i}\} \tag{9.25}$$

9.5 Case Study

To illustrate the application of the proposed stochastic optimization framework and demonstrate the performance of the forecasting method based on QRF and copulas, we conduct a case study with real-world data from SPP, a system operator for the large-scale southwestern electricity market in the U.S. The forecasting and sampling are performed using the R language (version 4.0.0), while the stochastic optimization is solved with Yalmip and Gurobi in MATLAB R2019a. The size of the scenario set Ω for each method is set at 1000, and the degree of confidence α is set at 95%. Since we do not focus on the modeling of the FTR cost, the budget K is set at 100 \$, and $q_{i,j}$ is set as 1 \$/MWh. In the case study, the budget constraint in (9.10) is always binding; thus, we can compare the methods by considering only the cumulated revenue (or payoff) of FTR, as the cost is identical.

9.5.1 Data Description

All the data are obtained from the SPP's website, including the generation mix and demand forecast[3] as well as binding constraints and LMP with its components in the DA market.[4] Although other data from earlier years can be obtained, the binding

[3] https://marketplace.spp.org/groups/operational_data.

[4] https://marketplace.spp.org/groups/day-ahead-market.

constraint data only starts in May 2013. Therefore, the time range of the data covers the full five years from 2014 to 2018, with 1 826 days from 262 weeks. The number of nodes N is 5 041.

We consider the problem of investing in weekly FTR obligations instead of monthly FTR obligations to ensure there's enough data for the training of the QRF and revenue results evaluation. Weekly FTR obligations can be procured in the secondary market. $\mathcal{T}_1$ covers the first 100 weeks, $\mathcal{T}_2$ covers the next 102 weeks, and $\mathcal{T}_3$ covers the remaining 60 weeks.

9.5.2 Comparison Methods

We set four comparison methods in modeling the joint price distribution. The first is a naïve benchmark (further referred to as **naïve**) that uses historical prices as future scenarios without any forecasting. The scenarios are renewed after every period in $\mathcal{T}_3$.

The second one is based on a multivariate truncated skew Student's t distribution [33] (further referred to as **skew-t**), which is basically a parametric unconditional approach. The parameters of skew-t are renewed after every period in $\mathcal{T}_3$. To prevent unrealistic scenarios, the lower truncation point is set at 0 $/MWh, and the upper truncation point is set at 60 $/MWh for all nodes.

The third is copula-GARCH [34], further referred to as **cGARCH**. It is based on the GARCH and copula model to predict the conditional variance of the forecasting error and has been widely used in multivariate financial time series modeling. The Student's t distribution is used in the GARCH model, and the Gaussian copula is used to model the dependence. The model is trained on $\mathcal{T}_2$.

The last is based on QR with the vine copula, which has been widely used in scenario generation [35, 36]. We want to test the performance of other copula structures in addition to the Gaussian copula structures. The famous canonical vine (**C-vine**) structure is used. For each vine (a.k.a. edge), the family of Gaussian, Student's t, Clayton, Gumbel, Frank, and Joe copulas are automatically selected using the AIC criterion with the method in [35]. The model is trained on $\mathcal{T}_2$.

Skew-t and cGARCH are trained on the forecasting error filtered by the point forecast of the same QRF, and C-vine is trained on the same historical quantile $v_{i,t}$ as the proposed method. This is an attempt to avoid any bias caused by the point forecast engine.[5]

[5] But the bias of cGARCH is still unavoidable since we do not renew the model in $\mathcal{T}_3$.

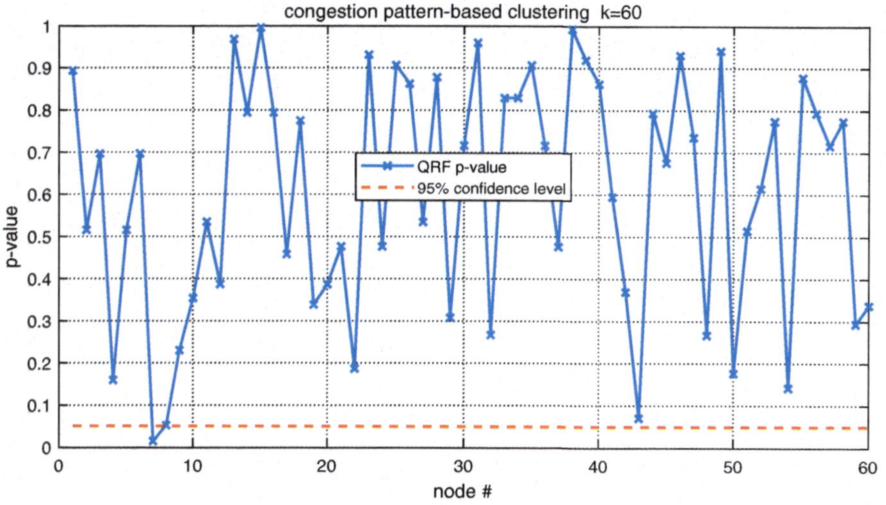

Fig. 9.2 P-value of $v_{i,t}$ in Pearson's chi-squared test

9.5.3 *Statistical Validation of Quantile Regression*

We set the number of clusters k to 60 and use congestion pattern-based clustering. Theoretically, a QR model fits the data well if the historical quantile level $v_{i,t}$ is an i.i.d. random variable that follows a uniform distribution on $(0, 1)$. To test this property, Pearson's chi-squared test is run for each i. The $(0, 1)$ interval is partitioned into ten 0.1-length bins and the confidence level is set at 95%. The resulting p-values are shown in Fig. 9.2. The p-values at all nodes are above 0.05, which means that the QRF can fit the average prices well. If there exist some nodes whose p-value is below the threshold, the forecasting engine fails to fit the nodal price data and produces unreliable results. The speculator is better off not taking a position on such nodes.

9.5.4 *Scenario Quality Evaluation*

To visualize the price scenarios, the results generated by Gaussian-toep as well as the true prices at the first period of $\mathcal{T}_3$ are shown in Fig. 9.3. The node selection settings are set the same as in Sect. 9.5.3. The true prices have been covered by scenarios with a similar variation trend across different key pricing nodes.

To further quantify the performance of scenario generation, we use the energy score Es [37] as the verification metric for multivariate scenario generation:

$$Es_t = \frac{1}{|\Omega|}\sum_{\omega\in\Omega}\left|\left|\mathbf{z}_t - \hat{\mathbf{z}}_{t,\omega}\right|\right|_2 - \frac{1}{2|\Omega|^2}\sum_{\omega_1\in\Omega}\sum_{\omega_2\in\Omega}\left|\left|\hat{\mathbf{z}}_{t,\omega_1} - \hat{\mathbf{z}}_{t,\omega_2}\right|\right|_2 \tag{9.26}$$

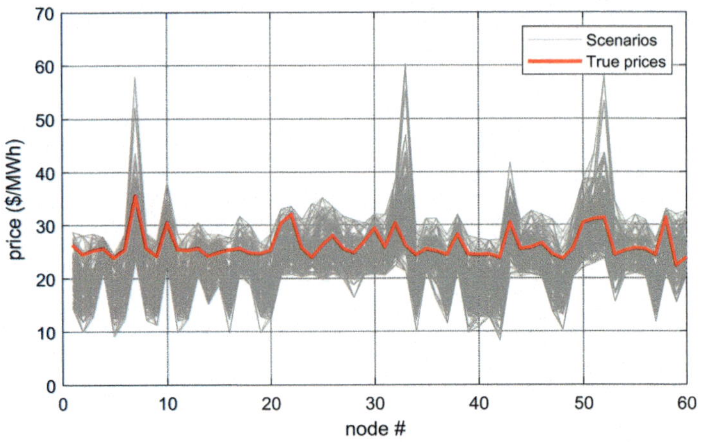

Fig. 9.3 True prices and price scenarios generated by Gaussian-toep at the first period of $\mathcal{T}_3$

Table 9.1 The average normalized Energy score when $k = 60$

Method	Normalized Es_t
naïve	3.876
skew-t	3.293
cGARCH	3.184
C-vine	3.121
Gaussian-full	3.119
Gaussian-toep	3.133
Gaussian-ex	3.121

where $\mathbf{z}_t$ is the true value at period t and $\hat{\mathbf{z}}_{t,\omega}$ is the sample of scenario ω. The Es is minimal when the true distribution is used for generating samples, so it is a negatively oriented score. Since Es is related to the scale of $\mathbf{z}_t$, we consider the normalized energy score $Es_t/\sqrt{k}$.

Table 9.1 presents the average normalized Es_t with $k = 60$, and Fig. 9.4 shows the boxplot of normalized Es_t with $k = 20, 40, 60$ in congestion pattern-based clustering for $t \in \mathcal{T}_3$. The red line in the middle of each box indicates the median value. From the table and the figure, we can clearly see that the scenario quality of the QR-based methods with copula structures are better than the other methods in terms of energy score and regardless of the copula structure. Among the QR-based methods, C-vine and Gaussian-full have the best energy score but the difference is rather small.

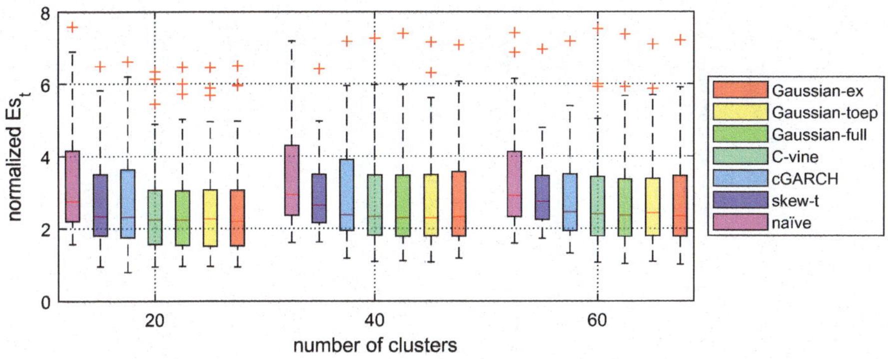

Fig. 9.4 Boxplot of normalized Es_t with different k

9.5.5 *Impact of Node Reduction with Clustering*

The performance of clustering in selecting nodes with high and low prices can be evaluated by sorting the original 5 041 nodes by their average prices in $\mathcal{T}_3$ in ascending order and finding the ranks of the selected nodes. The clustering approach with the rank percentiles of selected nodes closer to 0 or 100% is considered a better one. Here, we calculated the rank percentiles of the three minimum and maximum prices of the selected nodes for each approach with the number of clusters ranging from 10 to 100, and plotted them in Fig. 9.5. It is shown that the congestion pattern-based approach is better in selecting nodes for most of the conditions. An interesting phenomenon is that the ability to select nodes with clustering is asymmetric, as nodes with high prices are easier to be selected. When max-3 exceeds 95%, min-2 and min-3 are still somewhere between 20 and 45%.

9.5.6 *Revenue and Risk Estimation*

The stochastic optimization problem of (9.10) is solved for each period in the test set $\mathcal{T}_3$ and the revenues are recorded. The risk factor β is tuned to yield optimized portfolios with different levels of diversity. The diversity of the portfolio is defined as the average number of nonzero entries in γ:

$$d(\beta) = \frac{1}{|\mathcal{T}_3|} \sum_{t \in \mathcal{T}_3} ||\gamma_t(\beta)||_0 \tag{9.27}$$

Generally, a larger β indicates a more diversified portfolio and a larger $d(\beta)$, but such monotonicity is not strict.

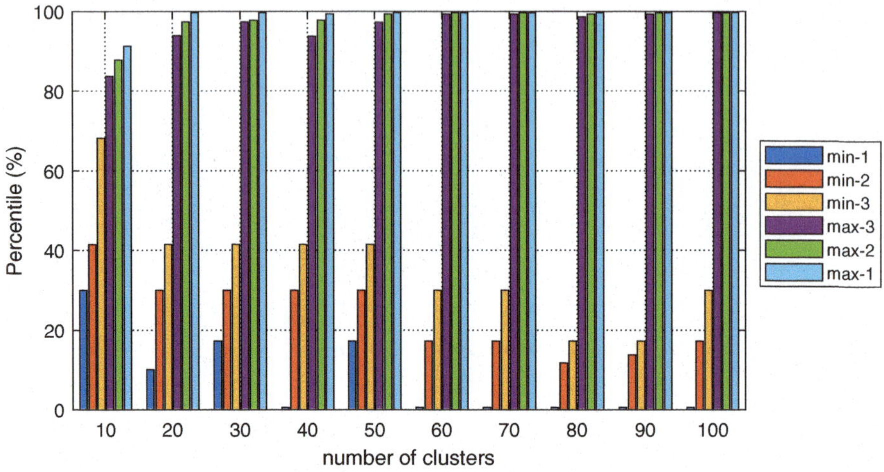

(a) Congestion pattern-based clustering.

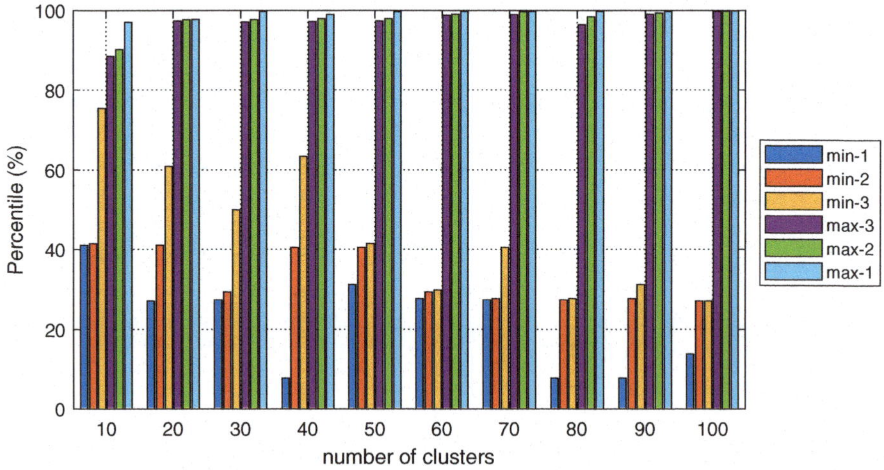

(b) Price-based clustering.

Fig. 9.5 The rank percentiles of the first, second and third minimum and maximum prices of the selected nodes in the original 5 041 nodes. (min-1 indicates the first minimum, etc.)

The comparisons are conducted with $k = 60$ for both congestion pattern-based and price-based clustering approaches.

Figure 9.6 shows the cumulative hourly revenue for different methods in $\mathcal{T}_3$ when the diversity of the portfolio is approximately 4.0. Figure 9.6a shows the results for congestion pattern-based clustering. At the end of $\mathcal{T}_3$, the cumulative revenue of QR-based methods Gaussian-full, Gaussian-toep, Gaussian-ex, and C-vine reach 8.198, 9.813, 9.802, and 8.597 $\times 10^6$ \$, respectively, while naïve, skew-t and cGARCH reach

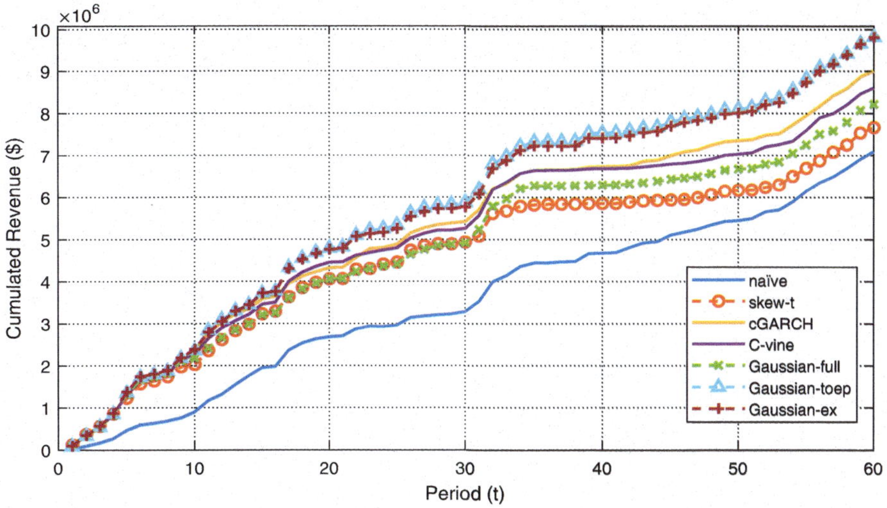

(a) Congestion pattern-based clustering.

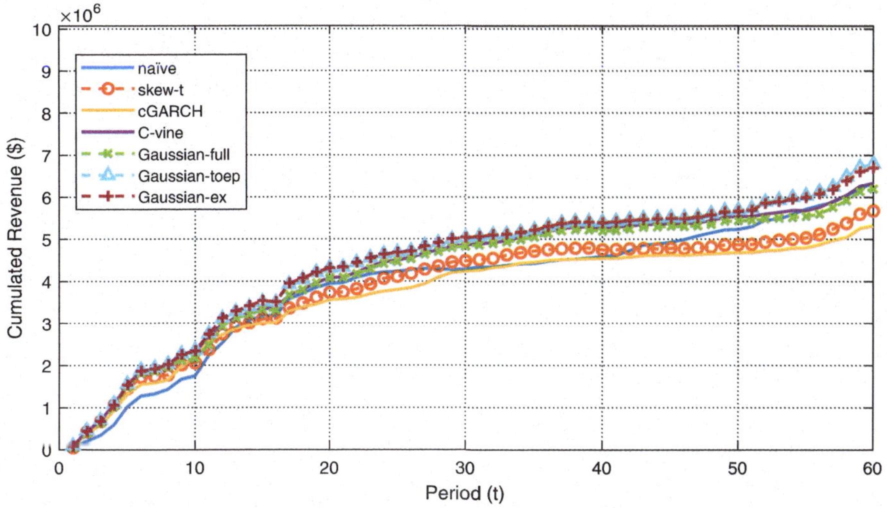

(b) Price-based clustering.

Fig. 9.6 Cumulative revenue (hourly) over the 60 holding periods in $\mathcal{T}_3$ ($d(\beta)$ is approximately 4)

Table 9.2 Cumulative revenue comparison between different methods

Method		Cumulative revenue (congestion pattern-based)	Cumulative revenue (price-based)
Benchmark	naïve	7.085×10^6	6.325×10^6
	skew-t	7.653×10^6	5.664×10^6
	cGARCH	8.995×10^6	5.303×10^6
	Average	7.911×10^6	5.764×10^6
QR and copula-based (parameter-heavy)	C-vine	8.597×10^6	6.330×10^6
	Gaussian-full	8.198×10^6	6.180×10^6
QR and copula-based (parameter-light)	Gaussian-toep	9.813×10^6	6.793×10^6
	Gaussian-ex	9.802×10^6	6.693×10^6
	Average	9.103×10^6	6.499×10^6

7.085, 7.653, and 8.995 $\times 10^6$ \$. The QR and copula-based methods can achieve a significant 15% improvement on average compared with other approaches. Using Gaussian copulas with fewer parameters can improve the revenue by approximately 14%-20% compared with parameter-heavy models.

Figure 9.6b shows the results for price-based clustering. At the end of $\mathcal{T}_3$, the cumulative revenue of QR-based methods Gaussian-full, Gaussian-toep, Gaussian-ex, and C-vine reach 6.180, 6.793, 6.693, and 6.330 $\times 10^6$ \$, respectively, while naïve, skew-t and cGARCH reach 6.325, 5.664, and 5.303 $\times 10^6$ \$. The QR and copula-based methods can achieve an average significant 12.75% improvement compared with other approaches. Using Gaussian copulas with fewer parameters can improve the revenue by approximately 5%-10% compared with parameter-heavy models.

The details of the cumulative revenue of different methods at the end of $\mathcal{T}_3$ are also given in Table 9.2.

Through comparison between the cumulative revenue results for the two clustering approaches, it is shown that all six methods benefit from congestion pattern-based clustering with average revenue improvements of 39.5% compared with price-based clustering. Such large improvements indicate that node selection is of vital importance for FTR speculation.

Figure 9.7 presents the average hourly revenue of different methods and Gaussian copula structures under different portfolio diversity for the two clustering approaches. For some methods, increasing β from 0 to some positive value would improve the resulting revenue. This is because considering the risk or CVaR properly can help investors make more robust decisions by constructing a more diversified portfolio. The results have also shown that the QR-based methods achieve steadier performance than naïve, skew-t, and cGARCH. The parameter-heavy models (i.e. C-vine and Gaussian-full) are worse than the models with fewer parameters (i.e. Gaussian-toep and Gaussian-ex), which is mainly caused by insufficient training data for

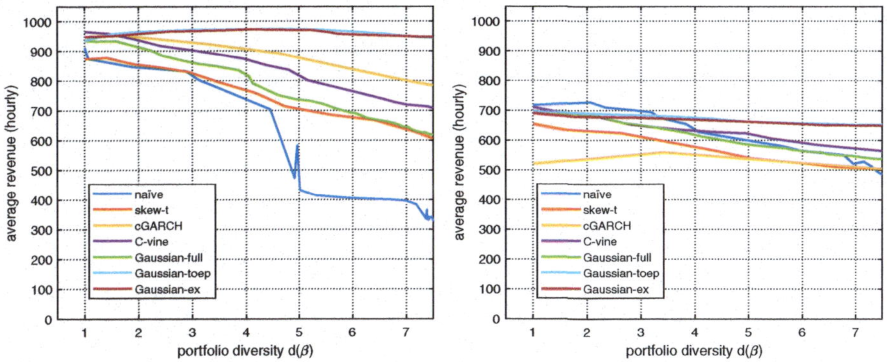

(a) Congestion pattern-based clustering.

(b) Price-based clustering.

Fig. 9.7 Average hourly revenue in $\mathcal{T}_3$ with respect to $d(\beta)$

parameter-heavy models. When the size of calibration set $\mathcal{T}_2$ is 102 and the number of targeted random variables in the copula is 60, parameter-heavy models are more likely to overfit. Additionally, by comparing the results of the two clustering approaches, it is demonstrated that proper node selection can improve the average revenue under different portfolio diversities for almost all the methods. Gaussian-toep and Gaussian-ex have steady and good performance in portfolio construction regardless of node selection.

To illustrate the portfolio construction results, the price forecasting results and absolute nodal injection vectors ($|\mathbf{a}|$ for a better view) from Gaussian-ex are shown in Fig. 9.8. It is natural that $\mathbf{a}$ is nonzero at nodes with the largest or smallest price forecasting results. From Fig. 9.8b, we can see that when β is zero, the constructed portfolio only contains the nodes of 5 and 33. When β is increased to 3, it becomes more diversified and contains several more nodes.

9.5.7 Sensitivity Analysis on the Number of Clusters

Table 9.3 lists the cumulative revenue for different methods with the number of clusters k ranging from 20 to 70 with approximately $d(\beta) = 4$ (i.e., investing approximately 4 FTRs for each period). It is shown that for these k values, Gaussian-toep and Gaussian-ex have good performances, and Gaussian-toep is slightly better, although the difference is negligible. The parameter-heavy models (i.e., Gaussian-full and C-vine) are slightly worse than cGARCH. The naïve method has the worst performance since it relies only on historical scenarios, and the improvement obtained by QRF and proper copula structure exceeds at least 10%.

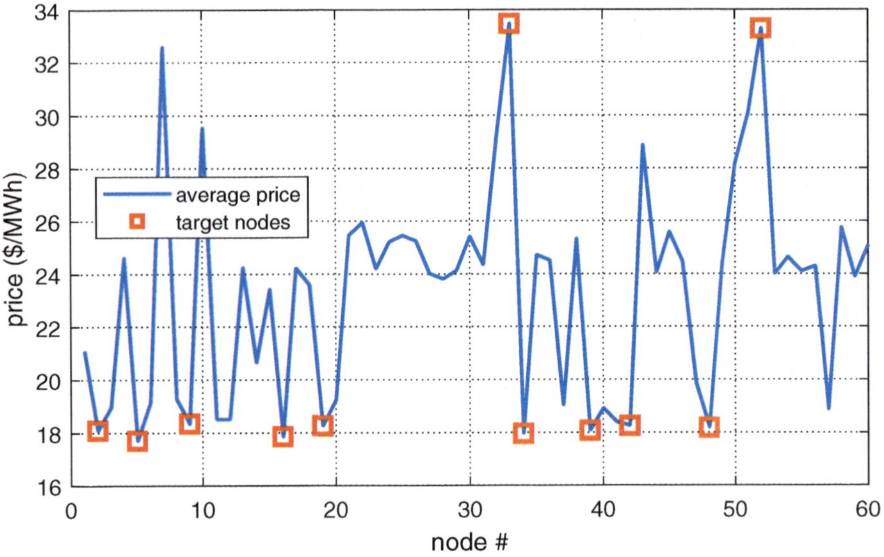

(a) The average nodal price of Ω.

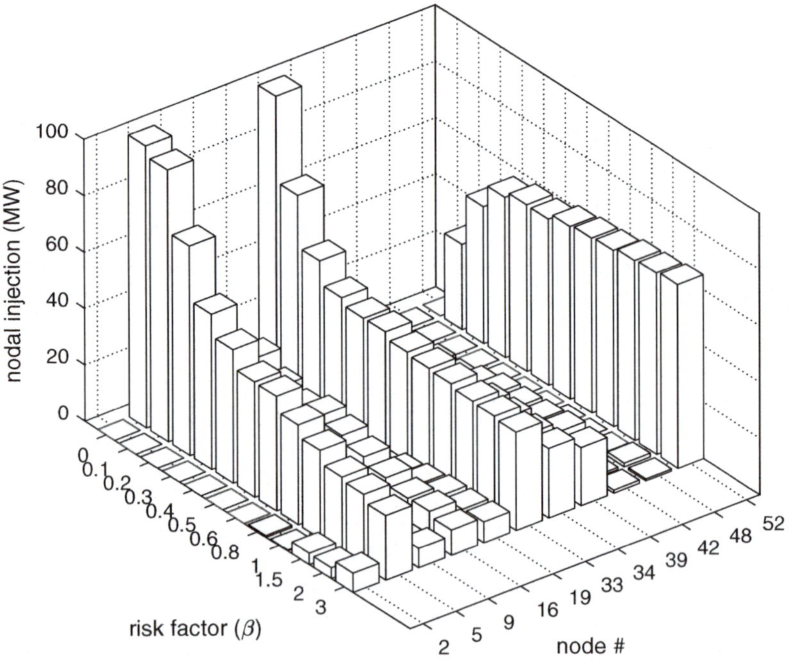

(b) The absolute nodal injection at the target nodes with respect to β.

Fig. 9.8 The constructed portfolio at the first period in $\mathcal{T}_3$. (Congestion pattern-based clustering with $k = 60$)

Table 9.3 The cumulative revenue ($\times 10^6$ \$) with different k in congestion pattern-based clustering. Best values in each column are marked in bold

k value	20	30	40	50	60	70
Method						
naïve	1.596	5.646	6.283	5.463	7.085	7.085
skew-t	6.344	5.587	7.063	5.001	7.653	7.174
cGARCH	6.757	5.974	8.877	5.409	8.995	8.857
C-vine	7.128	5.907	8.231	6.004	8.597	8.612
Gaussian-full	6.964	5.643	7.664	5.745	8.198	7.450
Gaussian-toep	7.726	6.030	**9.814**	**6.526**	**9.813**	9.394
Gaussian-ex	**7.943**	**6.187**	9.737	6.385	9.802	**9.396**

9.6 Conclusion

This chapter proposes a stochastic optimization framework for FTR speculation and portfolio construction. The expected return and risk are modeled in a scenario-based approach and the stochastic optimization problem is transformed into a linear programming problem suitable for real application. A congestion pattern-based clustering method is used to select representative nodes among thousands of pricing nodes and reduce the problem size. Then, QRF with copula models is used to generate high-dimensional joint probabilistic forecasting results. Through a case study with SPP market data, the proposed framework is illustrated and the performance of the method is demonstrated. Congestion pattern-based clustering selects nodes with extreme price values better than the ordinary price-spread approach. QR-copula method is much more powerful than the baseline methods in uncertainty distribution modeling. The selection of the copula structure is also discussed, and Gaussian copula structures with fewer parameters achieve satisfactory performance.

References

1. Hogan, William W. 2013. Financial transmission rights: point-to point formulations. In *Financial transmission rights*, vol. 7, ed. Juan Rosellón, and Tarjei Kristiansen, 1–48., Lecture Notes in Energy Springer.
2. Sarkar, Vaskar, and Shrikrishna A. Khaparde. 2008. A comprehensive assessment of the evolution of financial transmission rights. *IEEE Transactions on Power Systems* 23 (4): 1783–1795.
3. Sarkar, Vaskar, and Shrikrishna A. Khaparde. 2009. Introduction to loss-hedging financial transmission rights. *IEEE Transactions on Power Systems* 24 (2): 621–630.
4. Vaishya, Shri Ram, and Vaskar Sarkar. 2017. Implementation of lossy FTRs for perfect risk hedging under the marginal loss pricing. *IET Generation, Transmission and Distribution* 11 (1): 166–173.
5. Hunt, Sally. 2002. *Making competition work in electricity*. New York: Wiley.

6. Canchi, Devendra. 2010. *Financial transmission rights: Valuation and auction structure*. West Lafayette: Purdue University PhD thesis.
7. Richter, C. W., A. Jayantilal, and J. Kumar. 2001. Improving market participant strategies with FTR options. In *22nd International Conference on Power Industry Computer Applications (PICA 2001)*, 180–185.
8. Li, Tao, and Mohammad Shahidehpour. 2005. Risk-constrained FTR bidding strategy in transmission markets. *IEEE Transactions on Power Systems* 20 (2): 1014–1021.
9. Babayiğit, Cihan, Patricio Rocha, and Tapas K. Das. 2010. A two-tier matrix game approach for obtaining joint bidding strategies in FTR and energy markets. *IEEE Transactions on Power Systems* 25 (3): 1211–1219.
10. Apostolopoulou, Dimitra, George Gross, and Teoman Güler. 2013. Optimized FTR portfolio construction based on the identification of congested network elements. *IEEE Transactions on Power Systems* 28 (4): 4968–4978.
11. Fang, Xin, Fangxing Li, Hu. Qinran, and Yanli Wei. 2016. Strategic CBDR bidding considering FTR and wind power. *IET Generation, Transmission and Distribution* 10 (10): 2464–2474.
12. Prete, Chiara Lo, Nongchao Guo, and Uday V. Shanbhag. 2018. Virtual bidding and financial transmission rights: an equilibrium model for cross-product manipulation in electricity markets. *IEEE Transactions on Power Systems* 34 (2): 953–967.
13. Chéverez-González, Daniel, and Christopher L. DeMarco. 2009. Admissible locational marginal prices via Laplacian structure in network constraints. *IEEE Transactions on Power Systems* 24 (1): 125–133.
14. Rogers, Katherine M, and Thomas J Overbye. 2011. Clustering of power system data and its use in load pocket identification. In *2011 44th Hawaii International Conference on System Sciences*, 1–10.
15. Volodin, Dmitry V., and Tatiana A. Vaskovskaya. 2015. Clustering approach for determination of congestion zones on nodal electricity markets in long term periods. In *2015 IEEE Eindhoven PowerTech*, 1–6.
16. Vaskovskaya, Tatiana, Priyanko Guha Thakurta, and Janusz Bialek. 2017. Identifying congestion zones with weighted decomposition of locational marginal prices. In *2017 IEEE Manchester PowerTech*, 1–6.
17. Ziel, Florian, and Rick Steinert. 2018. Probabilistic mid-and long-term electricity price forecasting. *Renewable and Sustainable Energy Reviews* 94: 251–266.
18. İlseven, Engin, and Murat Göl. 2018. Hydro-optimization-based medium-term price forecasting considering demand and supply uncertainty. *IEEE Transactions on Power Systems* 33 (4): 4074–4083.
19. Ma, Ziming, Haiwang Zhong, Le. Xie, Qing Xia, and Chongqing Kang. 2018. Month ahead average daily electricity price profile forecasting based on a hybrid nonlinear regression and SVM model: an ERCOT case study. *Journal of Modern Power Systems and Clean Energy* 6 (2): 281–291.
20. Lago, Jesus, Fjo De Ridder, and Bart De Schutter. 2018. Forecasting spot electricity prices: Deep learning approaches and empirical comparison of traditional algorithms. *Applied Energy* 221: 386–405.
21. Litvinov, Eugene, Tongxin Zheng, Gary Rosenwald, and Payman Shamsollahi. 2004. Marginal loss modeling in LMP calculation. *IEEE Transactions on Power Systems* 19 (2): 880–888.
22. Zheng, Kedi, Yi. Wang, Kai Liu, and Qixin Chen. 2020. Locational marginal price forecasting: A componential and ensemble approach. *IEEE Transactions on Smart Grid* 11 (5): 4555–4564.
23. Tyrrell Rockafellar, R., and Stanislav Uryasev. 2002. Conditional value-at-risk for general loss distributions. *Journal of Banking and Finance* 26 (7): 1443–1471.
24. Conejo, Antonio J., Miguel Carrión, and Juan M. Morales. 2010. Risk management. In *Decision Making Under Uncertainty in Electricity Markets, volume 153 of International Series in Operations Research & Management Science*, 121–156. Berlin: Springer.
25. Liu, Minghai, and George Gross. 2004. Role of distribution factors in congestion revenue rights applications. *IEEE Transactions on Power Systems* 19 (2): 802–810.
26. Tony Cai, T., and Lie Wang. 2011. Orthogonal matching pursuit for sparse signal recovery with noise. *IEEE Transactions on Information Theory* 57 (7): 4680–4688.

27. Meinshausen, Nicolai. 2006. Quantile regression forests. *Journal of Machine Learning Research* 7: 983–999.
28. Fang, Xin, Hantao Cui, Du. Ershun, Fangxing Li, and Chongqing Kang. 2021. Characteristics of locational uncertainty marginal price for correlated uncertainties of variable renewable generation and demands. *Applied Energy* 282: 116064.
29. Radovanovic, A., T. Nesti, and B. Chen. 2019. A holistic approach to forecasting wholesale energy market prices. *IEEE Transactions on Power Systems* 34 (6): 4317–4328.
30. Pang, Herbert, Stephen L. George, Ken Hui, and Tiejun Tong. 2012. Gene selection using iterative feature elimination random forests for survival outcomes. *IEEE/ACM Transactions on Computational Biology and Bioinformatics* 9 (5): 1422–1431.
31. Sklar, M. 1959. Fonctions de répartition à n dimensions et leurs marges. *Publications de l'Institut Statistique de l'Université de Paris* 8: 229–231.
32. Hofert, Marius. 2008. Sampling Archimedean copulas. *Computational Statistics and Data Analysis* 52 (12): 5163–5174.
33. Li, Rui, and Saralees Nadarajah. 2020. A review of student's t distribution and its generalizations. *Empirical Economics* 58 (3): 1461–1490.
34. Bai, Xiwen, and Jasmine Siu Lee. Lam. 2019. A copula-GARCH approach for analyzing dynamic conditional dependency structure between liquefied petroleum gas freight rate, product price arbitrage and crude oil price. *Energy Economics* 78: 412–427.
35. Dissmann, Jeffrey, Eike C. Brechmann, Claudia Czado, and Dorota Kurowicka. 2013. Selecting and estimating regular vine copulae and application to financial returns. *Computational Statistics and Data Analysis* 59: 52–69.
36. Sun, Mingyang, Jochen Cremer, and Goran Strbac. 2018. A novel data-driven scenario generation framework for transmission expansion planning with high renewable energy penetration. *Applied Energy* 228: 546–555.
37. Pinson, Pierre, and Robin Girard. 2012. Evaluating the quality of scenarios of short-term wind power generation. *Applied Energy* 96: 12–20.

Part III
Market Bidding Behavior Analysis

Chapter 10
Pattern Extraction for Bidding Behaviors

10.1 Introduction

In most deregulated markets, generation companies (GENCOs) are allowed to bid strategically. The flexibility in bidding inevitably provides GENCOs with opportunities to exercise market power or even to manipulate markets, which would disturb the market operations and, furthermore, harm social welfare [1, 2].

As operators of power markets, independent system operators (ISOs) are in charge of ensuring order in power market operations. To prevent potential incidents and to ensure that the markets operate efficiently, ISOs must describe and simulate the market participants' bidding behaviors, forecast the future market operations, and design suitable market rules.

For these tasks, it is fundamental to explicitly understand and describe the bidding behaviors of GENCOs in detail. In recent years, many studies have been conducted on this topic from various perspectives, including calculating the GENCOs' optimal bidding strategies [3, 4], simulating the market equilibrium [5, 6], and designing regulation mechanisms to mitigate market power [7]. Most studies are based on theoretical microeconomic models, especially game theory [8, 9]; hence, they often portray the GENCOs' strategic bids with the idealized assumption that GENCOs always act rationally to maximize their individual payoffs with full utilization of the obtained information [10].

However, there is often a crucial mismatch between the theoretical studies and the actual behaviors of GENCOs, in which the actual biding behaviors deviate from the idealized hypotheses. The mismatch between theory and practice could eventually limit the effectiveness and practicality of the prevalent studies on strategic bidding behaviors. For example, in our preliminary analyses on bidding data, we found that a GENCO's bidding behaviors might not always change with the market scenario but are usually similar among adjacent time periods, which contradicts the simulation results in the theoretical studies [11]. Moreover, since the GENCOs usually only have incomplete information, their bidding strategies may not always be represented as either price-taking or price-making [12] but may be more diversified.

With the development of smart grids, more renewable resources and new electrical equipment are gradually integrated into power systems, thereby increasing the system uncertainty and rendering market simulation even more intractable [6].

Q. Chen et al., *Data Analytics in Power Markets*,
https://doi.org/10.1007/978-981-16-4975-2_10

To more effectively simulate and evaluate the market, it is important to have a more detailed understanding of GENCOs and their bidding behaviors. To identify the GENCOs' bidding behaviors, a suitable analysis method should be based on actual market data.

A few studies have analyzed market data. Doraszelski analyzed the evolution of the UK frequency response market over a 6-year period and showed how prices stabilized and GENCO's bidding strategies changed over time [13]. Hortacsu and Puller used an equilibrium model with detailed data on the Texas electricity market to simulate the GENCOs' optimal bidding strategies and compared the results with their actual bidding behaviors [14]. Kohansal analyzed the price elasticities and considered the renewable energy impacts in California's day-ahead (DA) power market by using the supply bid records for one month with limited data on large generators [15]. In addition, a brief study was conducted on the demand bids in the California energy market, which presents detailed demand bids of several load entities on the same day, discusses future analysis directions, and provides recommendations [16]. However, these studies attempted to either evaluate the accuracy of the theoretical model using data, analyze the price trends of the entire market or simply present one day's bidding data to readers; however, they rarely analyzed the bidding behaviors of the market participants.

In contrast to theoretical analysis methods, data-driven methods allow researchers to understand and analyze phenomena based on data rather than based on intuitions or idealized assumptions. Due to the rapid development of machine learning algorithms and the increasing accessibility of data sources, the application of data-driven analyses to smart grids has become a popular subject of engineering studies in recent years. Common data-driven research has mainly been applied to price [17], load forecasting [18, 19], renewable-output profiles [20], household consumer behavior [21, 22], anomaly detection [23], optimal power flow [24], and power system operation [25] and planning [26].

Few studies attempt to utilize data-driven methods to analyze market participants' bidding behaviors, such as estimating rivals' marginal costs or inferring the aggregate supply curves. Ruiz and Conejo used inverse function methods to identify marginal suppliers' marginal costs based on cleared energy quantities and locational marginal prices [27]. The methods were further developed and applied for price-responsive consumers [28], demand-response loads [29] and optimal bidding decision-making of market suppliers [30]. In addition, Mitridati and Pinson proposed a Bayesian inference approach for inferring the aggregate supply curve in a day-ahead electricity market based on the hidden Markov model [31]. However, these studies mainly evaluated the proposed methods on generated data rather than on real data. In addition, these studies focused on calculating specified variables, such as a rival's marginal costs, in power markets, rather than analyzing the bidding behaviors of market participants. The main reason is that bidding data are usually considered confidential information in most deregulated markets.

Since ISOs have complete information on markets, it is feasible and appropriate for them to analyze the GENCOs' bidding behaviors using a data-driven approach. In addition, certain power markets, such as the Australian Energy Market (AEM) [32] and the Midcontinent Independent System Operator (MISO) in the US [33], have disclosed transaction data openly and transparently, which enable researchers to analyze actual data-driven bidding behaviors.

It is difficult to directly describe the GENCOs' bidding behaviors based on the massive bidding dataset. Since the bids are similar in many time periods and can usually be aggregated to form several patterns, we try to extract individual typical bidding patterns from the large bidding dataset by applying clustering methods. Clustering methods are commonly used for grouping sets of similar data for pattern extraction, which have been widely used for household load classification [34] and scenario reduction [35].

However, in contrast to the typically analyzed time series data, such as market prices and loads, the actual bidding data are in different formats and cannot be directly clustered by the available clustering methods. First, the bidding data are presented in price-quantity pairs, which differ from the commonly analyzed time series formats. Second, the difference in the bidding available capacity affects the length of the analyzed data series and aggravates the analysis process. Third, the wide ranges of bidding prices, which typically range from negative to positive tens of thousands of dollars but are not reflected in market clearing prices, would have a major influence on the analysis results. Thus, a data standardization processing method specifically for the bidding data should be developed.

After addressing the data particularities, many questions remain regarding the effective extraction of the typical bidding patterns of a generator. First, the bidding behaviors are traditionally presented in a stepwise bidding data format, which represents the bidding attitudes of market participants and should be preserved during the clustering. To preserve the features of the bids, the K-medoids method [36] is utilized, which can retain the clustering centers in a stepwise format. Second, no effective distance measurement method is available for bids. Considering the characteristics of bidding data, the Wasserstein distance method [37] is customized and integrated to measure the pairwise distances among the bids. Third, the numbers of typical patterns for each generator's bids are unknown and are likely to differ; hence, the pattern extraction method should be adaptive in the selection of the number of clusters. Thus, an adaptive bidding behavior clustering method is developed, which can adaptively select a suitable number of clusters.

Based on a data standardization processing method and an adaptive clustering method, we propose a data-driven framework for analyzing bidding behaviors. A series of data processing methods, which include bidding curve sampling, data length unification, and extreme price clipping, are used to address the bidding data format particularities and to standardize the original bidding data for further analyses. An adaptive bidding behavior clustering method is proposed for extracting the typical bidding patterns of generators. Finally, empirical analyses are conducted based on AEM bidding data to identify the GENCOs' actual bidding behaviors. The typical bidding patterns of various generators are presented and compared. Three main bidding attitudes are summarized, and the bidding characteristics of generators that differ in terms of fuel type are described. Then, detailed analyses on the discrete aggregation, probability, and time distribution features are conducted. The effectiveness and validity of the proposed framework are also proved.

Since strategic bidding behavior is mainly applied to supply sides and rarely to demand sides, the bidding behavior analyses in this chapter focus on generation assets.

The contributions of this chapter are summarized as follows:

1. A data-driven analysis framework is proposed for extracting typical bidding patterns, which are used to analyze market participants' bidding behaviors based on actual bidding data.
2. A data standardization processing method is integrated to provide a foundation for bidding pattern extraction by addressing the bidding data particularities, with procedures for sampling the bidding curves, unifying the data length, and clipping the extreme bidding prices.
3. An adaptive bidding behavior clustering method is proposed for effectively extracting the generators' typical bidding patterns with stepwise features and adaptive selection that is based on the number of clusters.
4. An empirical analysis is conducted based on the actual AEM data, which identifies the specific characteristics of the actual bidding behaviors in power markets.

Understanding and identifying different GENCOs' bidding behaviors will equip the ISOs with information for advancing not only market simulations and operations but also the design of market rules to promote operational efficiency and prevent market power abuse. The remainder of the chapter is organized as follows. The assumptions and analysis framework are presented in Sect. 10.2. In Sect. 10.3, the data standardization processing procedure is explained. In Sect. 10.4, the adaptive bidding behavior clustering method is developed. In Sect. 10.5, the collected AEM data are comprehensively presented for the analyses. In Sect. 10.6, the results of typical bidding pattern extraction are provided, and three main bidding attitudes are summarized. In Sect. 10.7, the bidding characteristics of the discrete aggregation, probability, and time distributions are discussed. Finally, the conclusions of this study are presented, and future works are discussed in Sect. 10.7.

10.2 Assumptions and Proposed Framework

10.2.1 Model Assumptions

The chapter focuses on the bidding behaviors of suppliers in a centrally optimizing power market, which is a prevailing power market structure that is utilized in many power markets, such as the Pennsylvania-New Jersey-Maryland (PJM) power market, the California electricity market, and the AEM. The ISO is assumed to operate the power market. The dispatchable unit (DU) is assumed to be the supplier unit for bid submission.

Since the DA energy market operates as a fundamental component in a centralized optimizing power market, it is selected as the analysis target of the proposed data-driven framework. In the DA market, each DU submits stepwise bids for the next day, thereby indicating its preferred relationship between price and quantity. After collecting the bids, the ISO will clear the market with consideration of the power system constraints.

According to studies in power markets, the DUs can be roughly categorized as price-takers and price-makers. A price-taker usually does not take into consideration the influences of its bidding behaviors on the market clearing results during its decision-making process on bidding strategies [38]. In addition, the price-taker usually has little influence on the market clearing prices due to its limited market share [39, 40]. In contrast, a price-maker usually has a large market share and considers the impacts of its bids on market prices when it selects its bidding strategy.

In addition, the DUs could also be categorized as nonstrategic participants and strategic participants. The nonstrategic DUs submit the true marginal costs as the price-quantity pairs and their maximum capacities to the ISO. However, the DUs who represent strategic participants can adjust the price-quantity pairs to increase the SMPs or withhold the available capacity to change the supply-demand ratios. The bid adjustments represent strategic bidding behaviors that include physical withholding, financial withholding, or a combination of the two [41].

10.2.2 Bidding Data Format

We apply the price-quantity pair format for the DU stepwise bidding curve in which the available capacity is each DU's dispatchable capacity. Each market-participating DU, i, submits N price-quantity pairs, namely, $\mathbf{B}_{i,t}^{\mathrm{PR}}$, and the available capacity, namely, $Q_{i,t}^{\max}$ for DA scheduling. The detailed format of $\mathbf{B}_{i,t}^{\mathrm{PR}}$ is as follows:

$$\mathbf{B}_{i,t}^{\mathrm{PR}} = \begin{bmatrix} \left(p_{i,t}^{\mathrm{PR}}(1), q_{i,t}^{\mathrm{PR}}(1)\right), \ldots, \left(p_{i,t}^{\mathrm{PR}}(n), q_{i,t}^{\mathrm{PR}}(n)\right) \\ \ldots, \left(p_{i,t}^{\mathrm{PR}}(N), q_{i,t}^{\mathrm{PR}}(N)\right) \end{bmatrix} \tag{10.1}$$

The value of N ranges from 5 to 20. With larger N values, each DU has more degrees of freedom for customizing its bidding curves. t represents the time period in a day. The bidding price, namely, $p_{i,t}^{\mathrm{PR}}(n)$, is limited to a predefined range $\left[P_0^{\mathrm{LC}}, P_0^{\mathrm{UC}}\right]$ in which the absolute values of the upper and lower price caps are usually set to large values with minute influences on the DUs' bidding strategies.

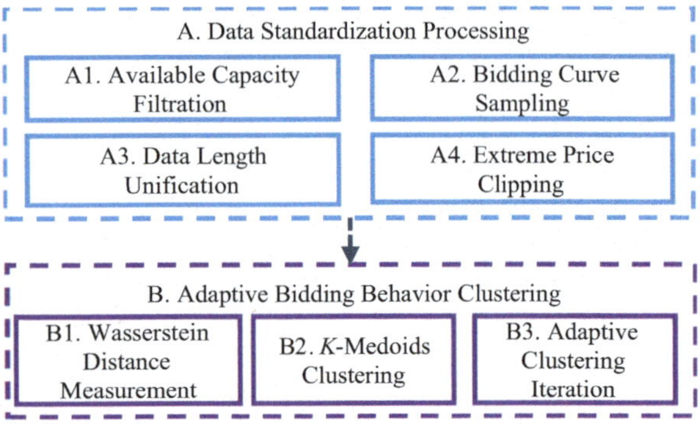

Fig. 10.1 Data-driven analysis framework

10.2.3 Data-Driven Analysis Framework

In this chapter, we propose a data-driven analysis framework for extracting a DU's bidding patterns. The framework includes an adaptive clustering method, which can efficiently and adaptively allocate the bidding curves into clusters that represent the bidding patterns. However, the particularity of the bidding data could hinder the utilization of clustering methods. Thus, several data standardization processing procedures are added to the proposed framework.

The proposed data-driven analysis framework, which is illustrated in Fig. 10.1, consists of two procedures: A. data standardization processing and B. adaptive bidding behavior clustering.

In the *data standardization processing* procedure, several subprocedures (A1-A4) are proposed for addressing the bidding data format particularities. After standardization processing, each bid in the price-quantity pair format is transformed into a series of bidding curves with uniform data resolution and length.

In the *adaptive bidding behavior clustering* procedure, an adaptive clustering method is proposed that is based on the K-medoids clustering method and the Wasserstein distance measure, with modifications that address the bidding data particularity. After adaptive clustering, any DU's bidding curve data can be extracted as several typical bidding patterns.

10.3 Data Standardization Processing

10.3.1 Filtering Available Capacities

Due to planned maintenance or other reasons, DUs in some time periods might submit low available capacity values, which would not be representative of their bidding behaviors. To mitigate the influences of abnormal data in such time periods, a threshold ratio, namely, η_i, on the available capacity value is set for filtering the bidding data. This threshold ratio is generally set to a relatively low value that depends on the generation asset type, which could be a coal, combined cycle gas turbine (CCGT), or variable renewable power plant.

The bidding data with available capacity values that are lower than $\eta_i Q_{i,0}^{\max}$ are removed, where $Q_{i,0}^{\max}$ is DU i's maximum available capacity over all time.

10.3.2 Sampling Bidding Curves

Since the bidding data are in the format of price-quantity pairs, which differs from the commonly analyzed data series format, it is difficult to process, compare and analyze them. To address this problem, a bidding curve sampling procedure is proposed.

First, the bidding data of DU i, which consist of price-quantity pairs, are transformed into a bidding curve. Then, the sampling procedure is applied to the bidding curve, with a sampling range of $\left[0, Q_{i,0}^{\max}\right]$ and a sampling resolution of $\Delta q_i = Q_{i,0}^{\max}/M$, where M is the number of samples in the bidding curve. To realize the sampling function, M should be set to a value that exceeds the number of bidding pairs, namely, N.

After sampling the bidding curve, all the DU i price-quantity pairs, namely, $\mathbf{B}_{i,t}^{\text{PR}}$, could be transformed into a series of bidding prices, namely, $\mathbf{P}_{i,t}^{\text{SP}}$, that have the same sampling resolution. The format is as follows:

$$\mathbf{P}_{i,t}^{\text{SP}} = \left[p_{i,t}^{\text{SP}}(1), \ldots, p_{i,t}^{\text{SP}}(m), \ldots, p_{i,t}^{\text{SP}}(M)\right] \tag{10.2}$$

where $p_{i,t}^{\text{SP}}(m)$ is the bidding price value at sampling point m of DU i at time period t after sampling. The corresponding quantity value for $p_{i,t}^{\text{SP}}(m)$ at sampling point m is denoted as $q_{i,t}^{\text{SP}}(m)$. For clarification, the generated series, namely, $\mathbf{P}_{i,t}^{\text{SP}}$, is plotted to generate the bidding curves.

10.3.3 Unifying Data Length

Due to the DUs' physical withholding behaviors, the available capacity values usually vary among time periods. In any bidding curve, the value of $p_{i,t}^{\text{SP}}(m)$ will be empty if its corresponding quantity value, namely, $q_{i,t}^{\text{SP}}(m)$, exceeds the available capacity value $Q_{i,t}^{\max}$. Overall, the length inconsistency among bidding curves would hinder clustering.

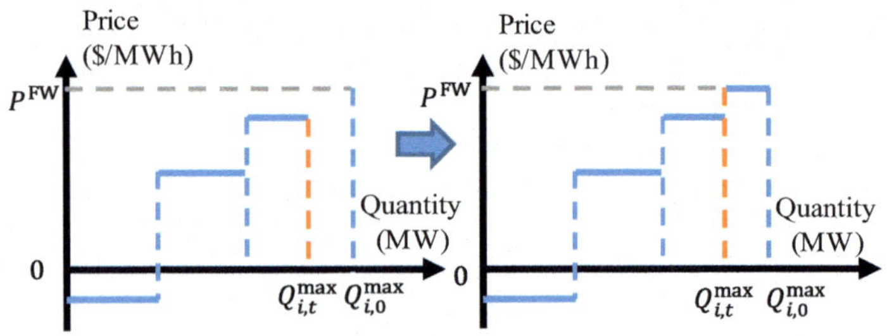

Fig. 10.2 Schematic diagram of physical withholding replacement

To address this problem, an equivalent conversion method is utilized on the bidding curves with physical withholding. The physical withholding behaviors are replaced by financial withholding behaviors with a sufficiently high bidding price, which is denoted as P_0^{FW}. In a bidding curve $\mathbf{P}_{i,t}^{SP}$, if $q_{i,t}^{SP}(m)$ is larger than $Q_{i,t}^{max}$, the empty value, namely, $p_{i,t}^{SP}(m)$, will be replaced by P_0^{FW}, which can be expressed as:

$$p_{i,t}^{UN}(m) = \begin{cases} p_{i,t}^{SP}(m) & , q_{i,t}^{SP}(m) \le Q_{i,t}^{max} \\ P_0^{FW} & , q_{i,t}^{SP}(m) > Q_{i,t}^{max} \end{cases} \tag{10.3}$$

A schematic diagram of this replacement is shown in Fig. 10.2. Although the financial withholding bid and the physical withholding bid seem like different bidding patterns, they are equivalent if the market clearing prices are lower than the chosen price P_0^{FW}, namely, if the financial withholding capacity has not been cleared.

In addition, to facilitate comparison among the bidding curves of all the DUs, all the bidding curves are normalized based on their individual maximum available capacities $Q_{i,0}^{max}$:

$$q_{i,t}(m) = \frac{q_{i,t}^{SP}(m)}{Q_{i,0}^{max}} \quad , \forall m \tag{10.4}$$

where $q_{i,t}(m)$ is the bidding quantity value at sampling point m of DU i at time period t after data normalization.

With the normalization of the bidding quantity, the bidding curves of various DUs with various available capacity values could be easily compared in the same data range.

10.3.4 Clipping Extreme Prices

The clearing price usually ranges from tens to hundreds of dollars per megawatt hour in most power markets. However, the ranges of the allowed upper and lower caps of the bidding prices in these markets are much larger, such as tens of thousands of dollars per megawatt hour as an upper cap and negative thousands of dollars per megawatt hour as a lower cap. Although most of the effective bidding prices lie in a narrow price range, many DUs submit bids at extremely high prices for a segment of capacity as a test to potential high market clearing prices, which could be verified by the DUs' actual bidding behaviors shown in Sect. 10.6. The extremely high bidding prices will not have any influence on the market clearing most of the time; however, they will generate a large amount of insignificant information and therefore, will interfere with the analysis. Thus, extremely high bidding prices should be treated specifically.

The electricity spot prices are approximately normally distributed [42], and the distribution can be denoted as $N\left(\mu, \sigma^2\right)$. According to the 3-sigma rule, approximately 99.7% of the prices are within three standard deviations σ of the mean price, μ. Thus, we believe that the price range, $[\mu - 3\sigma, \mu + 3\sigma]$, is sufficiently representative for the bidding behavior analysis. Then, the effective price range for the analysis can be expressed as $\left[P_0^{\min}, P_0^{\max}\right]$. The prices outside the range can be treated as extreme prices and clipped to the adjacent range boundary. The detailed expression is as follows:

$$p_{i,t}(m) = \begin{cases} P_0^{\max} & , p_{i,t}^{\mathrm{UN}}(m) \geq P_0^{\max} \\ p_{i,t}^{\mathrm{UN}}(m) & , P_0^{\min} \leq p_{i,t}^{\mathrm{UN}}(m) \leq P_0^{\max} \\ P_0^{\min} & , p_{i,t}^{\mathrm{UN}}(m) \leq P_0^{\min} \end{cases} \tag{10.5}$$

where $p_{i,t}(m)$ is the adjusted bidding price after clipping on the mth sampling point of DU i's bidding curve at time t.

Now, the bidding curves are ready for clustering with the following format:

$$\mathbf{P}_{i,t} = \left[p_{i,t}(1), \ldots, p_{i,t}(m), \ldots, p_{i,t}(M)\right] \tag{10.6}$$

10.4 Adaptive Clustering of Bidding Behaviors

To identify the typical bidding strategies of a DU, an adaptive clustering method is proposed. The adaptive clustering method uses the Wasserstein metric as the distance measure and the K-medoids clustering method to ensure that the extracted bidding patterns preserve the stepwise features. An iterative adaptive clustering method is proposed for selecting a suitable number of clusters.

10.4.1 Distance Measurement

There are many ways to measure the distance between two objects, such as the Manhattan distance, the Euclidean distance, and the Minkowski distance [43]. In this chapter, the Wasserstein distance method is used. The Wasserstein distance method is a well-developed and prevalent method for measuring the distance between data series based on their probability distributions [37]. The Wasserstein distance is the weight times the distance to move one probability distribution to the other, and its form is similar to the process of changing from one bidding strategy to another. The distance between DU i's bidding curves at times t_1 and t_2 is calculated as $s_i^{t_1,t_2}$:

$$s_i^{t_1,t_2} = \inf_{\gamma \in \Gamma(\Phi_{i,t_1}, \Phi_{i,t_2})} \mathbb{E}_{(p_1, p_2) \sim \gamma} [\|p_1 - p_2\|] \tag{10.7}$$

where Φ_{i,t_1} and Φ_{i,t_2} are the probability distributions of the bidding curves at times t_1 and t_2, respectively; set Γ denotes all the couplings of Φ_{i,t_1} and Φ_{i,t_2}; and $\gamma\ (p_1, p_2)$ represents the probability that price p_1 is in Φ_{i,t_1} and price p_2 is in Φ_{i,t_2}.

10.4.2 K-Medoids Clustering

After the clustering process, a set of clustering centers will be obtained based on the clustering rule. The clustering centers are defined as the extracted *bidding patterns* of the DUs. However, many prevalent clustering methods, such as K-means clustering [44] and hierarchical clustering [45], cannot preserve the stepwise format of the clustering centers, which is a critical feature that reflects DUs' individual bidding characteristics.

To retain the stepwise characteristics in the clustering center results, the K-medoids clustering method is used [46]. As a type of prototype-based clustering that is similar to the prevalent K-means clustering method, the K-medoids method also minimizes the distance between the data points that are labeled as belonging to a cluster in designating a data point as the center of that cluster. The difference is that the K-medoids clustering method generates a clustering center curve, namely, $\mathbf{P}_{i,k}^{\mathrm{C}}$, from the bidding curves in the corresponding cluster set, namely, $\Omega_{i,k}^{\mathrm{C}}$.

10.4.3 Adaptive Clustering Procedure

Since the exact number of bidding patterns of a DU is unknown prior to clustering, a predefined K value limits the adaptivity and effectiveness of the method. In addition, the variational levels of each DU's bidding style differ; hence, it is inappropriate to use a uniform predefined K value for all the DUs in the analysis.

Thus, the K-medoids clustering method is modified with an iterative procedure for the adaptive K value. First, the concept of a separation threshold $S_{i,k}^{\mathrm{TH}}$ is established, which is used to represent the allowed separation level in cluster k. The separation threshold $S_{i,k}^{\mathrm{TH}}$ can be calculated as:

$$S_{i,k}^{\mathrm{TH}} = \theta_0 S_{i,k}^{\mathrm{ref}} \tag{10.8}$$

where θ_0 is separation threshold ratio, which ranges from 0 to 1, and $S_{i,k}^{\mathrm{ref}}$ is the distance between DU i's k-th clustering center curve, $\mathbf{P}_{i,k}^{\mathrm{C}}$, and a reference bidding curve, P_0^{ref}. The distance is calculated via Eq. (10.7).

After clustering, the distances, $S_{i,k}^{\mathrm{C},t_1}$, between each bidding curve are calculated. We consider the bidding curve at time t_1 and its cluster center as an example. Then, the proportion, $\rho_{i,k}$, is calculated based on the number of bidding curves with distances that exceed the threshold, $S_{i,k}^{\mathrm{TH}}$. If $\rho_{i,k}$ exceeds the tolerance rate, ρ_0, for any cluster, then the distances in the clusters are too large and K must be increased.

$S_{i,k}^{\mathrm{TH}}$ should vary among DUs and clustering center curves due to the various bidding price levels.

If the external environment changes, e.g., if new elements are added to the market or the market policy changes, the data should be divided into separate datasets based on the external environment changes. Then, the clustering method could be applied on the dataset that we want to analyze, and the typical bidding patterns of the dataset could be identified.

10.4.4 Clustering Algorithm

The clustering algorithm is presented in detail as Algorithm 10.1. First, a K-Medoids clustering algorithm with an initial value of $K = K_0$ is utilized, which produces K_0 clusters and the corresponding clustering center curves $\mathbf{P}_{i,k}^{\mathrm{C}}$. Then, the proportion $\rho_{i,k}$ of each cluster is calculated and compared with the tolerance rate ρ_0. If any $\rho_{i,k}$ exceeds the tolerance rate, the K value will be incremented by one. Then, the K-Medoids algorithm will be run again until no $\rho_{i,k}$ is larger than ρ_0. Last, each DU's bidding data will generate K_T clustering center curves, which are known as bidding patterns.

Algorithm 10.1 Adaptive K-medoids clustering algorithm

procedure
Required: bidding curves $\mathbf{P}_{i,t}$ for DU i, initial and maximum numbers of clusters K_0 and $K_{\max}$, separation rate threshold θ_0 and tolerance rate ρ_0
  set $K = K_0$
  while $K \leq K_{\max}$ **do**
    run K-medoids with K on all bidding curves $\mathbf{P}_{i,t}$, and generate K clustering center curves from the bidding curves
    for all clusters **do**
      calculate $S_{i,k}^{\mathrm{TH}}$ via Eq. (10.8)
      for all $\mathbf{P}_{i,t} \in \Omega_{i,k}^{\mathrm{C}}$ **do**
        calculate all $s_{i,k}^{\mathrm{C},t}$ via Eq. (10.7)
        compare $s_{i,k}^{\mathrm{C},t}$ with $S_{i,k}^{\mathrm{TH}}$
      end for
      calculate the proportion $\rho_{i,k}$
      if $\rho_{i,k} > \rho_0$ **then**
        $K \leftarrow K + 1$
        break
      end if
    end for
  end while
  set $K_{\mathrm{T}} = K$
  return K_{T} bidding patterns $\mathbf{P}_{i,k}^{\mathrm{C}}$
end procedure

10.5 AEM Data Description

10.5.1 Description of Market Participants

There are approximately 120 GENCOs with more than 450 DUs in the AEM power market, according to [47]. There are 182 fossil-fired plants, 82 hydropower plants, 72 wind-power plants, 44 solar-energy plants, and 50 power plants of other types.

All DUs can choose a participant type from *scheduled*, *semi-scheduled* and *non-scheduled*. When we focus on a GENCO with scheduled power plants, there are only fewer than 40 GENCOs. Information on the "Big Five" ENCOs is presented in Table 10.1.

The largest GENCO in the AEM only accounts for a 13.5% market share. The total market share of the "Big Five" is 52.6%. The Herfindahl-Hirschman index (HHI) of the AEM is calculated to be 744.2 via the function in [48]. According to the standards of the U.S. Department of Justice [49], the AEM could be theoretically regarded as an unconcentrated competitive market. Thus, the GENCOs in the market should be basically regarded as price-takers with competitive bidding behaviors.

Table 10.1 Information on the "Big Five" GENCOs

GENCO	Source type	Capacity (MW)	Market share (%)
Origin Energy Electricity Limited	Fossil-Fired and Water	4955	13.5
AGL Macquarie Pty Limited	Fossil-Fired	4665	12.7
Stanwell Corporation Limited	Fossil-Fired and Water	3862	10.5
CS Energy Limited	Fossil-Fired and Water	3624	9.9
AGL Loy Yang Marketing Pty Ltd	Fossil-Fired	2210	6.0

10.5.2 Description of Bidding Data

Each DU is assigned a unique identification number (DUID). With the DUID tag, the bidding data on the AEMO website [32] could be linked to a DU and used for analysis. The time range of the collected bidding data for the analysis in this chapter is one year, namely, from October 18th, 2017, to October 17th, 2018.

Each DU could submit 10 price-quantity pairs every 30 min for the next day. The bids are in the stepwise format. The upper and lower caps of the bidding prices are 14,200 and −1,000$/MWh, respectively. The DUs have a high degree of freedom in manipulating their bidding prices and available capacities.

With the bidding data, each DU's bidding strategy at any time could be identified. For example, the bidding strategy of GSTONE1 on November 7th, 2017, is plotted in Fig. 10.3. The time coordinate axis represents the time over the 24 h from 04:30 to 04:00 on the next day. The capacity coordinate axis represents the capacity ratio from 0 to 100%. The price coordinate axis represents the bidding prices. Many bidding curves are similar among different periods, which supports the strategy of extracting typical patterns to depict the DUs' bidding behaviors.

In addition, the bidding dataset size is rather large. Each DU has 48 bids in one day. There are 186 DUs that can be analyzed. Thus, the total dataset contains 3.25 million bids, each of which contains 10 price-quantity pairs. A large amount of bidding data renders the bidding analysis even more difficult.

10.6 Bidding Pattern Analysis

To analyze the bidding patterns of various fuel-type generators, several representative DUs are selected for analysis and comparison. Then, DUs with similar attributes, such as the fuel type, capacity, and technology, are compared to identify the different individual subjective perspectives that influence the bidding behaviors.

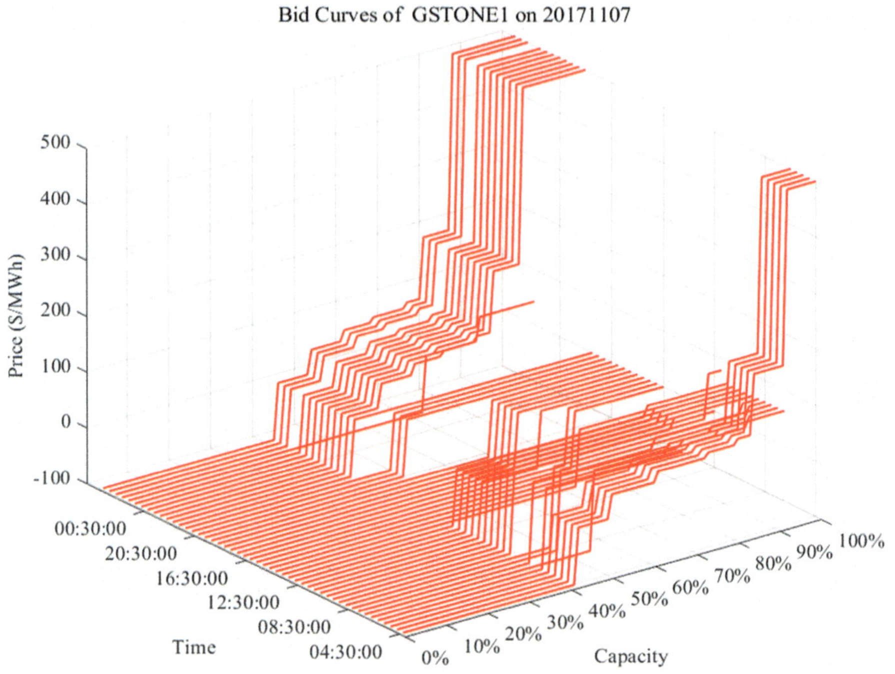

Fig. 10.3 Bidding data of GSTONE1 on November 7th, 2017

10.6.1 Parameter Setting

To avoid the influence of bids during abnormal time periods, such as maintenance, the threshold ratios for filtering the bidding data, η_i, for coal-fired, gas-fired, and renewable energy units are set as 0.5, 0.3, and 0.2, respectively. To balance the performance and the computational burden, the number of samples in a DUs' bidding series, namely, M, is set as 100. For simplicity, the high bidding price P_0^{FW} that is used to convert physical withholding to financial withholding is set as the bidding price upper cap P_0^{UC}, namely, 14,200 \$/MWh. Based on the AEM historical data, the spot market prices would lie in the range of –100 to 500 \$/MWh with over 99.9% probability. Thus, the upper and lower caps for the extreme price clipping procedure, namely, P_0^{max} and P_0^{min}, are set as 500 and –100 \$/MWh, respectively.

To initialize and complete the adaptive clustering algorithm, the lower and upper caps on K, namely, K_0 and K_{max}, must be set. Since the clustering algorithm should be applicable if K is larger than 2, the lower cap, namely, K_0, is set as 2. In this chapter, the upper cap, namely, K_{max}, is set as 30 for calculational efficiency. The separation threshold ratio, namely, θ_0, for the adaptive K-medoids clustering algorithm is set as 1%. The tolerance rate, namely, ρ_0, is set as 0.1%.

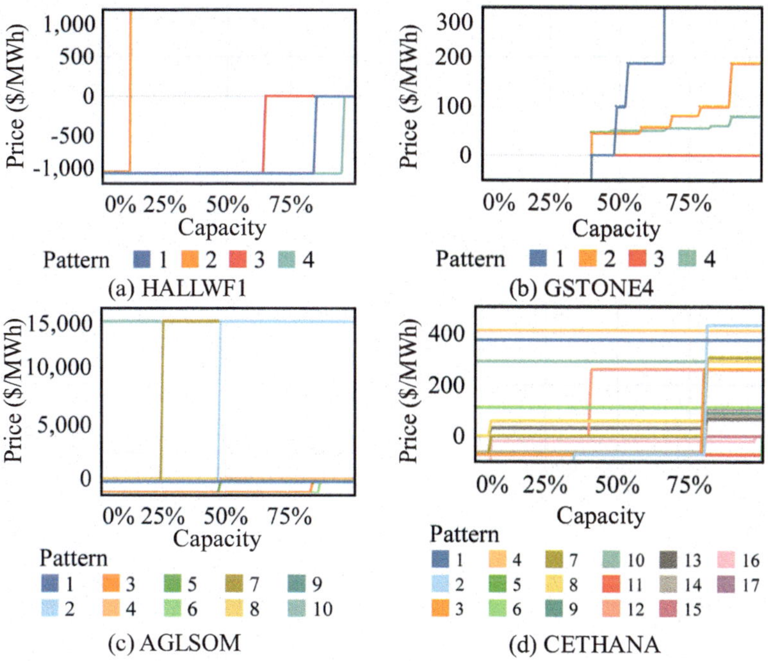

Fig. 10.4 Typical bidding patterns of 4 selected DUs

10.6.2 Bidding Patterns of DUs by Fuel Type

First, four DUs are selected to present the bidding patterns of four types of generators in terms of the fuel type, whose bidding patterns are plotted in Fig. 10.4. Each color represents a bidding pattern. The DUs with various fuel types have different bidding styles, as demonstrated by the two key features, namely, the number and shape, of the bidding patterns. Although the bidding price ranges are clipped during the data standardization processing procedure, the extracted bidding patterns have been restored to the original bidding curves for clearer observation.

HALLWF1 (wind power, 95 MW) is clustered into 4 typical bidding patterns, which can be categorized into two main bidding attitudes. The first attitude is to ensure that most of its capacity is cleared, which is referred to as quantity-preferred, and the second attitude is to withhold most of its capacity, which is referred to as capacity-withholding. In the first attitude, HALLWF1 bids most of its quantity at a negative price (almost –1000$/MWh) and bids the remainder at zero price, which is observed in patterns 1, 3, and 4. In the second attitude, HALLWF1 might not want to participate in the market. Thus, HALLWF1 bids a small amount at low prices and bids the remainder of the quantity at over 10,000$/MWh, as pattern 2 shows.

Pattern 2 accounts for only 0.15% of all bids. Thus, HALLWF1 bids as quantity-preferred in more than 99.8% of the time periods. Although HALLWF1 bids at negative prices, since the market prices are calculated based on the marginal prices, the cleared prices of HALLWF1 are determined by all the DUs' bids and are mostly positive. For example, the proportions of five regions when the market clearing prices are negative from 2017/10/18 to 2018/10/17 are NSW, 0%; QLD, 0.1%; SA, 1.3%; TAS, 2.1%; and VIC, 0.3%.

The special bidding attitudes of HALLWF1 are largely caused by the operational characteristics of wind power units: low-cost, intermittent, and nonstorable generation. Due to the near-zero operational cost, it is beneficial for wind power units to sell energy if the market clearing prices are positive. Due to the intermittent operation and inability to store power, wind power units cannot choose which time to sell their power. Thus, the optimal strategy is to sell as much energy as possible only if they can generate energy. Hence, HALLWF1 bids with the quantity-preferred attitude to ensure that all its energy can be cleared.

GSTONE4 (coal-fired, 285 MW) is also clustered into 4 typical bidding patterns. The typical bidding patterns can also be categorized into two main bidding attitudes. The first attitude is quantity-preferred, in which bids from a negative price to a zero price are submitted, which corresponds to pattern 3. The second attitude is to tentatively balance the market clearing price and its cleared quantity for a profitable result by submitting bids with a strong zigzag-increasing trend, which is referred to as *price-preferred* and corresponds to patterns 1, 2, and 4.

Compared to low-cost wind power units, the operational costs of coal-fired units are much higher; hence, GSTONE4 focuses on the prices of its cleared power. GSTONE4 tries to ensure that its power is cleared at a profitable price by submitting zigzag-increasing bids. In addition, the tentative degrees of typical patterns 1, 2, and 4 differ, where pattern 1 is the most aggressive, and pattern 4 is the most cautious. The proportions of the 4 typical patterns (11%, 14%, 9%, and 66%) demonstrate that GSTONE4 submits price-preferred bids most of the time (91%) and mainly follows the most cautious bidding pattern (pattern 4 with a proportion of 66%). Due to the high on/off switching costs, GSTONE4 always bids at negative prices under its 40% capacity to maintain operational status in all its typical bidding patterns.

AGLSOM (gas-fired, 170 MW) is clustered as 10 typical bidding patterns. However, pattern 10, whose bidding prices are all set as 13,944 $/MWh for its entire capacity, accounts for 89% of all bids, namely, AGLSOM conducts financial withholding most of the time. To more clearly present AGLSOM's bidding patterns, the vertical axis is shortened to –100 to 500 $/MWh to present its other bidding patterns in Fig. 10.5a. In addition, the time distributions of all its bidding patterns are shown in Fig. 10.5b. The vertical axis represents the 48 time periods in the considered day from 0:00 to 23:30, with a resolution of 30 min. The horizontal axis represents the date from October 18th, 2017, to October 17th, 2018.

According to Fig. 10.5a, the bidding attitudes are much more complex than those of the two DUs described above. Diverse in terms of both the price and quantity ranges, the various typical bidding patterns could be roughly categorized into 3 attitudes: (1) quantity-preferred, which is represented by patterns 1, 3, and 6 and accounts for 5.5% of all bids; (2) price-preferred, which is represented by patterns

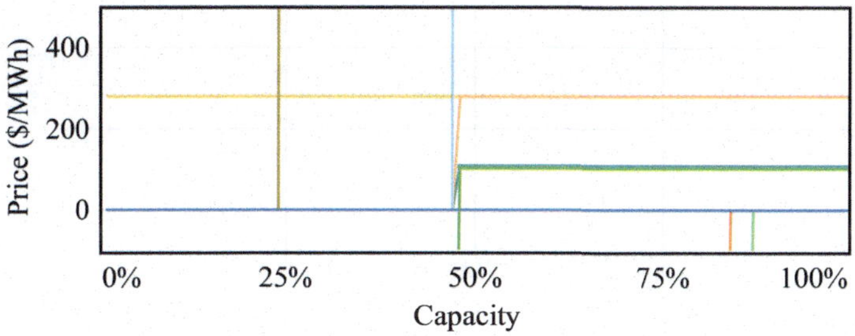

(a) Typical Bidding Patterns

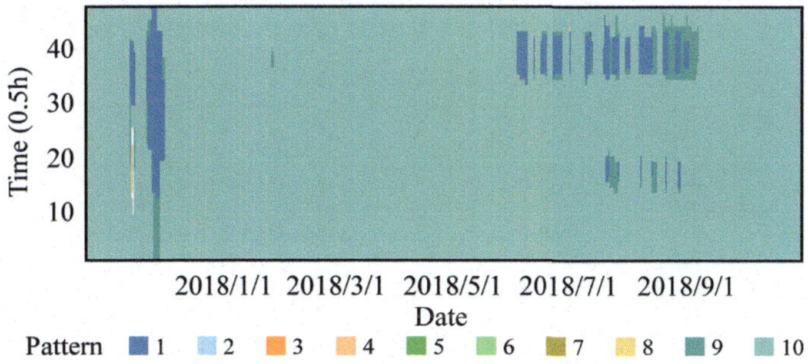

(b) Time Distribution

Fig. 10.5 AGLSOM's typical bidding patterns and the time distributions

4, 5, 8 and 9 and accounts for 5.3% of all bids; and (3) capacity-withholding, which is represented by patterns 2, 7 and 10 and accounts for 89.2% of all bids.

From the proportions of the three bidding attitudes, we deduce that AGLSOM stays off-status most of the time by submitting capacity-withholding bids. AGLSOM is only willing to be cleared in 11% of the time periods. Then, according to Fig. 10.5b, the capacity-withholding bids of AGLSOM are mainly submitted in valley and flat time periods, thereby preventing the high-cost gas-fired units to be cleared in low-price time periods. At peak time periods, AGLSOM also submits quantity-preferred bids, which are represented by pattern 1, and price-preferred bids, which are represented by pattern 9.

Compared with the coal-fired units, the gas-fired units have higher costs and more flexibility. The pressure to pursue payoffs is generated by higher variable costs and the ability to switch on/off states frequently due to higher flexibility, and both of these factors induce the gas-fired DUs to pay more attention to the price of its cleared power by attempting to be cleared at the time periods with the highest prices by following various bidding patterns.

CETHANA (hydropower, 100 MW) is clustered as 17 typical bidding patterns. Similar to gas-fired AGLSOM, the typical bidding patterns of CETHANA could also be categorized into 3 main bidding attitudes: (1) quantity-preferred, which is represented by patterns 5, 11, 15, and 16 and accounts for 22.9% of all bids; (2) price-

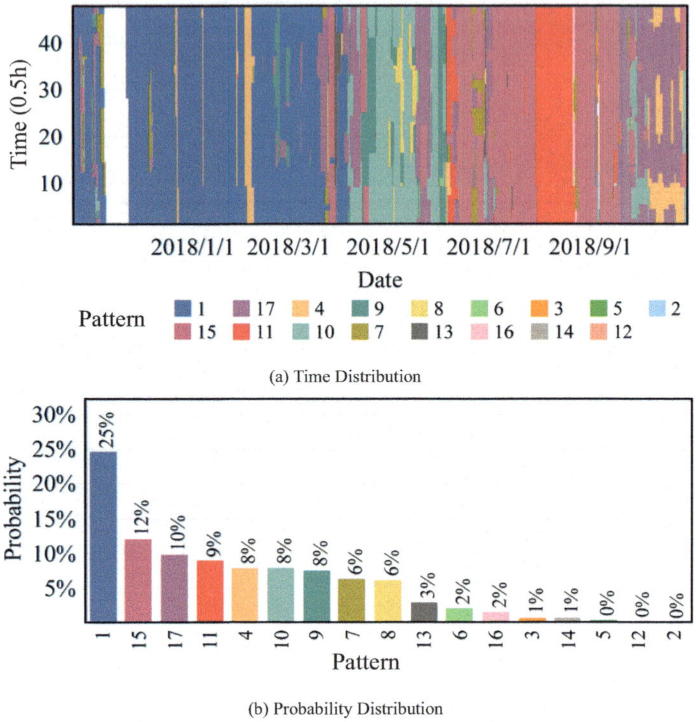

Fig. 10.6 Time and probability distributions of CETHANA's bidding patterns

preferred, which is represented by patterns 2, 3, 6, 7, 8, 9, 12, 13, 14 and 17 and accounts for 36.6% of all bids; and (3) capacity-withholding, which is represented by patterns 1, 4 and 10 and accounts for 40.5% of all bids.

To more clearly present CETHANA's bidding patterns, the time and probability distributions of its bids are presented in Fig. 10.6a and b. The bidding patterns frequently change with the date and time and 11 typical bidding patterns have probabilities that exceed 2%, thereby demonstrating hydropower CETHANA's active bidding characteristics.

Although they are both low-cost renewable energy units, hydropower CETHANA and wind-power HALLWF1 differ substantially in terms of their bidding characteristics. The main reason is that the energy of CETHANA is storable. With the ability to store energy, hydropower units are more motivated and able to strategically bid in the market. Thus, CETHANA would try to get cleared in the most profitable periods by frequently adjusting the bidding prices according to its predictions regarding the market scenario.

Overall, it is concluded that DUs differ in terms of their bidding characteristics according to the various fuel types. The variable cost level, startup/shutdown cost level, on/off switch flexibility, and energy storage ability might all influence a unit's

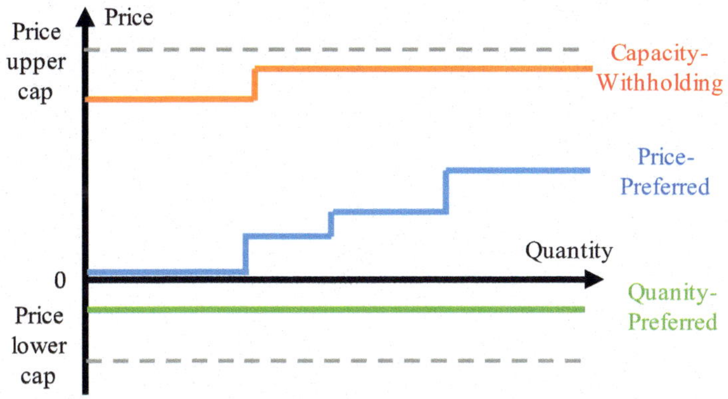

Fig. 10.7 Sketch maps of three main bidding characteristics

bidding characteristics. Although the typical bidding patterns vary among generators, three main bidding attitudes are identified: (1) quantity-preferred, when DUs submit bids at negative or low positive prices to ensure that most of their capacity could be cleared; (2) price-preferred, when DUs submit bids in zigzag-increasing shapes to tentatively balance their cleared quantity and the corresponding market clearing prices; and (3) capacity-withholding, when DUs intend to physically or financially withhold more than half of their capacities. Sketch maps of the three main bidding characteristics are presented in Fig. 10.7. Most of the bidding patterns could be derived from the three main bidding attitudes or the hybrids of them.

10.6.3 Comparison of Similar DUs

In this case, two DUs with similar attributes are selected for comparison. TARONG#3 and STAN-2 both have 385 MW of maximum available capacity and are fueled by black coal. The bidding patterns of TARONG#3 and STAN-2 are presented in Fig. 10.8. For TARONG#3, only 2 typical bidding patterns are clustered, whereas for STAN-2, 11 typical bidding patterns are clustered.

Although TARONG#3 has 2 distinct typical bidding patterns, they both belong to price-preferred bidding attitudes. The difference is that pattern 1 withholds approximately 25% of the capacity by bidding at 13,791 \$/MWh, while the highest price of pattern 2 is 292 \$/MWh.

However, the typical bidding patterns are much more diversified for STAN-2. STAN-2 has 2 main bidding attitudes: (1) price-preferred, which is represented by all typical patterns except pattern 9 and accounts for 99% of all bids, and (2) capacity-withholding, which is represented by pattern 9 and accounts for 1% of all bids.

To facilitate comparison of the two DUs' bidding characteristics, the time distributions of their patterns are plotted in Fig. 10.9.

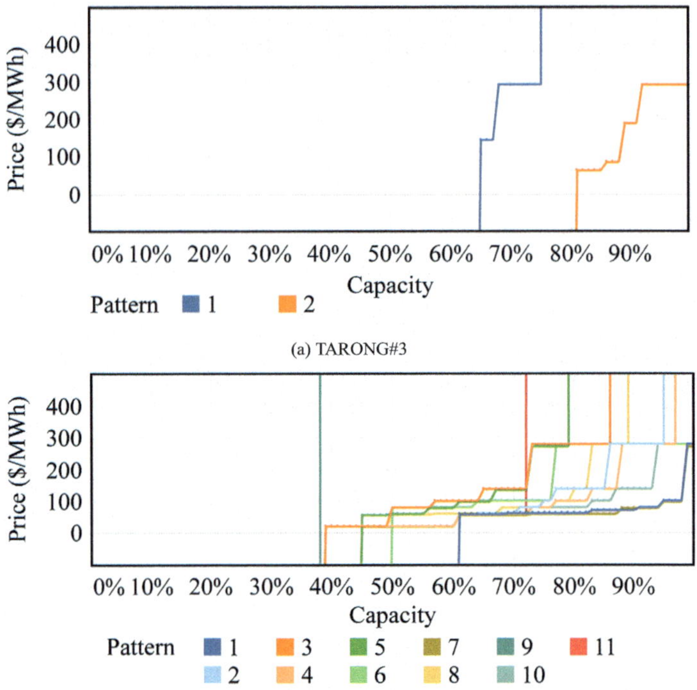

(a) TARONG#3

(b) STAN-2

Fig. 10.8 Bidding patterns of 2 similar DUs

For TARONG#3's bids, pattern 1 is mainly identified during the valley time periods and pattern 2 is mainly identified during the peak time periods. When the date and time change, TARONG#3 will switch from one bidding pattern to the other, with few changes according to the market scenario. In contrast, STAN-2 has more active bidding characteristics. STAN-2 frequently changes the bidding patterns in its strategy space to earn acceptable payoffs toward various market scenarios, as reflected by the mixed time distribution in Fig. 10.9b.

Since the parameters of the adaptive clustering algorithm on these two DUs are the same, the reason for the completely opposite bidding characteristics of the two DUs with similar attributes is due to the difference in the individual subjective perspectives. The above analysis demonstrates that it is inappropriate and perhaps even inaccurate to simulate all the DUs' behaviors based on a uniform model. The distinguished bidding characteristics that are extracted from historical data should be both considered and integrated into the model for more realistic and accurate simulation or forecasting.

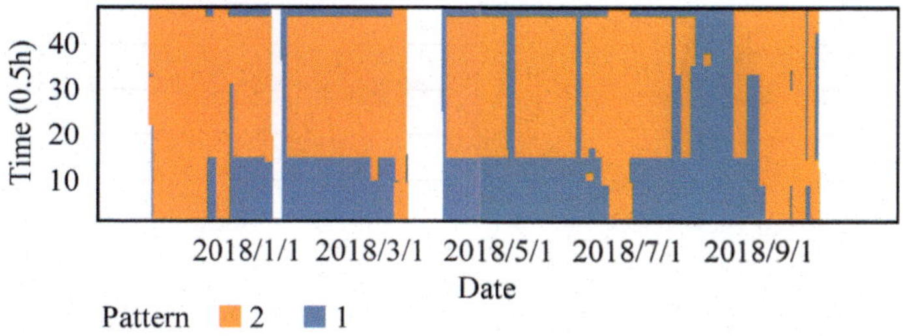

(a) TARONG#3

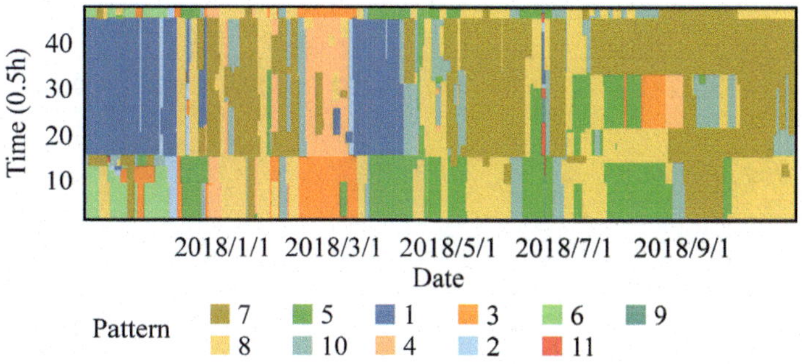

(b) STAN-2

Fig. 10.9 Time distributions of bidding patterns of TARONG#3 and STAN-2

10.6.4 Discussion

Since the Australian Electricity Market can be regarded as a fully competitive market (proven in Sect. 10.5.1), all market participants are assumed to participate in the market as price-takers according to economics theories. However, with our analysis on the bidding behaviors, we find that most generators exhibit strategic bidding behaviors, which are reflected in the deviations between their bids and their true costs and available capacities. Thus, we conclude that many price-takers also behave as strategic bidders in the market. There is no clear boundary between price-takers and strategic bidders.

10.7 Feature Analysis on Bids

In this section, we analyze three bidding features: discretization features, probability distribution features, and time distribution features.

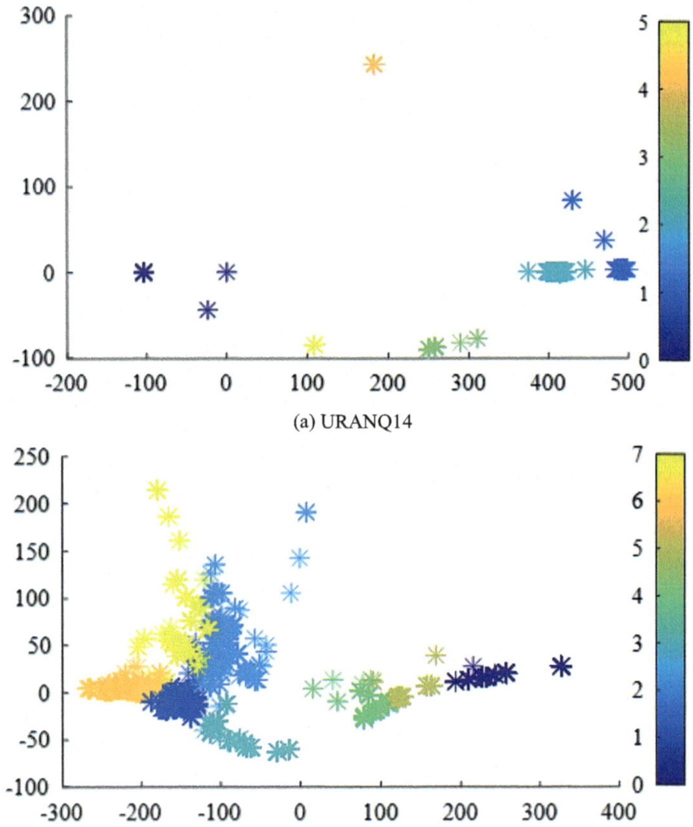

(a) URANQ14

(b) TRIBUTE

Fig. 10.10 Multidimensional scaling results of URANQ14 and TRIBUTE

10.7.1 Discrete Aggregation Feature

To present the distribution and clustering results of the high-dimensional bidding curves, we map them into a 2D plane via the multidimensional scaling (MDS) method. The MDS method is prevalently used to transform the pairwise distances among points in a dataset into a lower dimensional space to facilitate visualization [50].

Two DUs, namely, URANQ14 (gas-fired, 166 MW) and TRIBUTE (hydropower, 92 MW), are selected as examples, and their MDS results are presented in Fig. 10.10.

Each point in the plane represents a bid. Points of the same color are categorized into the same cluster. The numbers of valid bidding curves for URANQ14 and TRIBUTE are 9430 and 14512, respectively. However, there seem to be fewer mapped bidding-curve points. By calculating the distance matrices of the two DUs' bids, we find that the numbers of distinct bidding cures of URANQ14 and TRIBUTE are 221

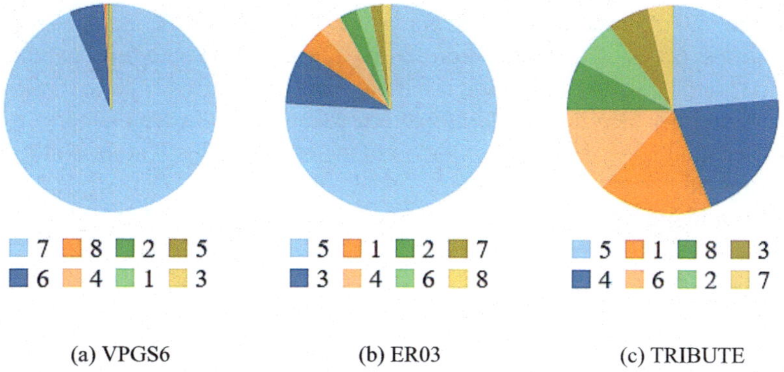

Fig. 10.11 Probability distributions of bidding patterns of VPGS6, ER03 and TRIBUTE

and 551, respectively, which are much smaller than the corresponding numbers of bids. On average, 43 bids share the exact same curve shapes for URANQ14, and 26 share the exact same curve shapes for TRIBUTE.

It could be concluded that the bidding curves have prominent discretely aggregated characteristics; hence, the DUs' bidding behaviors could be represented by several discrete patterns. Thus, analyzing the probabilistic and time distribution characteristics of DUs' bidding patterns and integrating them into the model building process would facilitate bidding behavior simulation.

In addition, the bids from different clusters are distributed distinctly. The data in a cluster are close to one another and are relatively far from the data in the other clusters, which supports the effectiveness and validity of the proposed framework.

10.7.2 Probability Distribution Feature

To demonstrate the probability distribution features of DUs' bids, three DUs, namely, VPGS6 (gas-fired, 60 MW), ER03 (coal-fired, 750 MW), and TRIBUTE (hydropower, 92 MW), which share the same number of typical bidding patterns, are analyzed. The probability distributions of their patterns are presented in Fig. 10.11.

For VPGS6, pattern 7 accounts for 94% of all bids. The top 2 patterns account for almost 100%. Thus, VPGS6's bidding strategy could be approximately by the top 2 bidding patterns with different probabilities. VPGS6 can be considered with low-activity bidding characteristics.

Comparatively, the probability distribution of ER03 becomes more dispersed. Although the top two patterns still account for 84% of all bids, the other patterns cannot be neglected due to their substantial contributions.

For TRIBUTE, the probability distribution of its bidding patterns becomes more balanced. The most frequently occurring pattern, namely, pattern 5, accounts for only

24% of all bids. Compared with the above two DUs, TRIBUTE changes its bidding patterns most frequently, thereby representing the most active bidding characteristics.

Since the probability distribution feature varies among DUs and is independent from the number of typical bidding patterns, it serves as crucial supporting information for analyzing the DUs' bidding characteristics. This feature could be used as a significant feature to describe, simulate and evaluate DUs' strategic bidding behaviors and to further integrate them into a data-driven market equilibrium market analysis.

10.7.3 Time Distribution Feature

To represent the time distribution features of DUs' typical bidding patterns, BW02 (coal-fired, 700 MW), BBTHREE2 (gas-fired, 49 MW), and KAREEYA1 (hydropower, 22 MW) are selected as examples. The detailed time distributions are plotted in Fig. 10.12. The format is the same as the format of Fig. 10.5b. The DUs' bidding patterns have strong temporal correlational characteristics, which could be categorized into time-varying and season-varying features.

As a coal-fired DU, BW02 has time-varying bidding features. During the valley time, from 0:00 to 10:30, pattern 3 tends to be used as the bidding strategy. During the peak time, from 11:00 to 24:00, BW02 is more likely to use pattern 5. AGLSOM and TARONG#3, which are analyzed in the previous section, also have time-varying features in their bids, as shown in Figs. 10.5b and 10.9a.

As a gas-fired DU, BBTHREE2 has season-varying bidding features. During spring and summer, BBTHREE2 is more willing to select patterns 1 and 6 as bidding strategies. During fall and winter, BBTHREE2 becomes willing to bid using patterns 2, 3, and 4. CETHANA, which was analyzed in the previous section, also has season-varying features in its bids, as shown in Fig. 10.6a.

In total, the time distribution feature of DUs' typical bidding patterns could be regarded as another piece of critical supportive information for describing the DUs' strategic bidding behaviors. By giving more consideration to the different bidding characteristics of market participants in various seasons and time periods, we could describe, analyze and simulate the time-varying market competition more elaborately.

10.8 Conclusions

Utilizing the large-scale public bidding data for the AEM, a data-driven analysis framework is proposed for more effectively analyzing the individual bidding behaviors in a power market. A data standardization processing method is proposed for addressing the data particularities. Using an adaptive clustering method that was

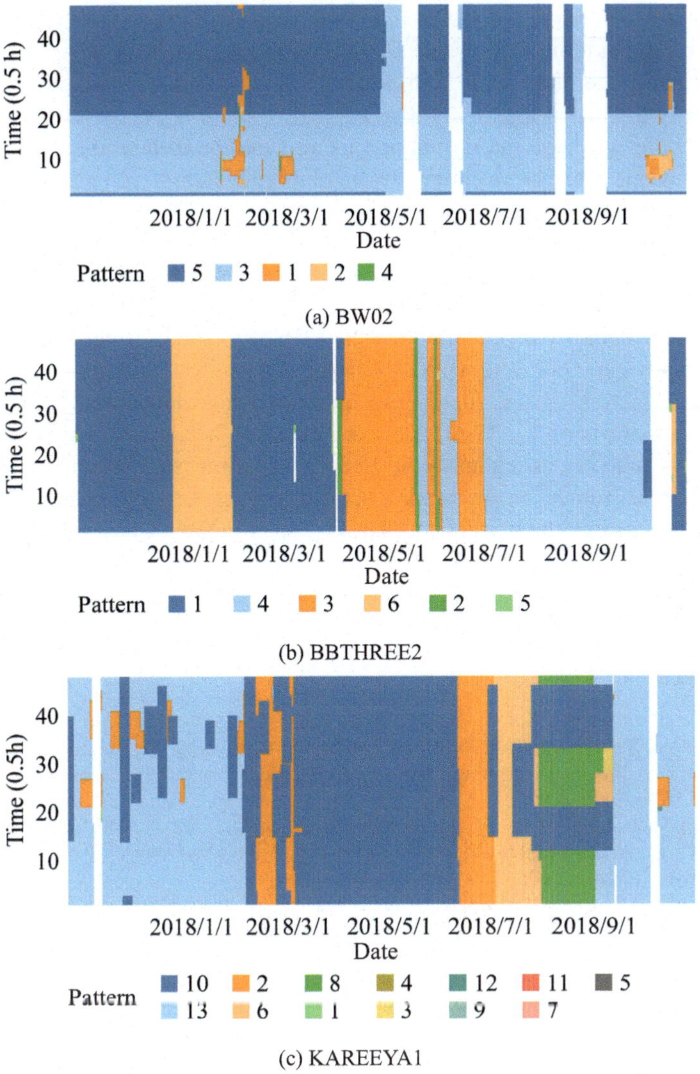

Fig. 10.12 Temporal pattern distributions of BW02, BBTHREE2 and KAREEYA1

developed from the K-medoids method and the Wasserstein distance measure, the DUs' bidding patterns are extracted from the AEM data.

The results demonstrate remarkable differences in the bidding behaviors among individual generators that differ in terms of fuel type. Three main bidding attitudes (price-preferred, quantity-preferred, and capacity-withholding) are summarized and could represent most of the bidding patterns. The individual bidding behaviors could be distinguished from the traditional idealized assumptions by three features, namely, discretization, diversity, and patterning, with time-varying probability distributions. In addition, the bidding behaviors vary widely among different types of generators. Even for two generators with similar attributes, the bidding behaviors can differ substantially due to various individual subjective bidding perspectives. These conclusions could not have been obtained using traditional theoretical and idealized assumptions but could be explained by an elaborate data-driven analysis on a massive bidding dataset.

In the future, additional efforts could be made to further increase the accuracy of the proposed framework, such as identifying the most suitable distance measurement, clustering algorithm, and adaptive clustering parameters.

In addition, data-driven machine-learning-based methods, such as artificial neural networks and deep neural networks, could be utilized to link the bidding behaviors with external factors (system supply, transmission outage, load forecasting, fuel price, and weather information). Then, if certain external factor is provided, the bidding behaviors of market participants could be further forecasted based on the learning model that was trained on the market data. If high accuracy is realized in bidding behavior forecasting, the market could be further simulated based on the forecasting results.

References

1. Borenstein, Severin, James B Bushnell, and Frank A Wolak. 2002. Measuring market inefficiencies in california's restructured wholesale electricity market. *American Economic Review* 92 (5): 1376–1405.
2. Harvey, Scott M, William W Hogan, and Todd Schatzki. 2004. A hazard rate analysis of Mirant's generating plant outages in California. *KSG Faculty Research Working Paper Series RWP05-027*, March 2005.
3. Bosong Li, Xu., Mohammad Shahidehpour Wang, Chuanwen Jiang, and Zhiyi Li. 2019. Der aggregator's data-driven bidding strategy using the information gap decision theory in a non-cooperative electricity market. *IEEE Transactions on Smart Grid* 10 (6): 6756–6767.
4. Xiaoyuan, Xu., Zheng Yan, Mohammad Shahidehpour, Zhiyi Li, Mingyu Yan, and Xiangrui Kong. 2019. Data-driven risk-averse two-stage optimal stochastic scheduling of energy and reserve with correlated wind power. *IEEE Transactions on Sustainable Energy* 11 (1): 436–447.
5. Dehghanpour, Kaveh, M Hashem Nehrir, John W Sheppard, and Nathan C Kelly. 2016. Agent-based modeling in electrical energy markets using dynamic bayesian networks. *IEEE Transactions on Power Systems* 31 (6): 4744–4754.
6. Guo, Hongye, Qixin Chen, Xichen Fang, Kai Liu, Qing Xia, and Chongqing Kang. 2020. Efficiency loss for variable renewable energy incurred by competition in electricity markets. *IEEE Transactions on Sustainable Energy* 11 (3): 1951–1964.
7. Guo, Hongye, Qixin Chen, Qing Xia, and Chongqing Kang. 2019. Market power mitigation clearing mechanism based on constrained bidding capacities. *IEEE Transactions on Power Systems* 34 (6): 4817–4827.
8. Zou, Peng, Qixin Chen, Yu. Yang, Qing Xia, and Chongqing Kang. 2017. Electricity markets evolution with the changing generation mix: An empirical analysis based on china 2050 high renewable energy penetration roadmap. *Applied Energy* 185: 56–67.
9. Ruiz, Carlos, Antonio J Conejo, and Yves Smeers. 2011. Equilibria in an oligopolistic electricity pool with stepwise offer curves. *IEEE Transactions on Power Systems* 27 (2): 752–761.

10. Li, Tao, and Mohammad Shahidehpour. 2005. Strategic bidding of transmission-constrained gencos with incomplete information. *IEEE Transactions on Power Systems* 20 (1): 437–447.
11. Baringo, Luis, and Antonio J Conejo. 2015. Offering strategy of wind-power producer: A multi-stage risk-constrained approach. *IEEE Transactions on Power Systems* 31 (2): 1420–1429.
12. Yazdani-Damavandi, Maziar, Nilufar Neyestani, Miadreza Shafie-khah, Javier Contreras, and Joao PS Catalao. 2017. Strategic behavior of multi-energy players in electricity markets as aggregators of demand side resources using a bi-level approach. *IEEE Transactions on Power Systems* 33 (1): 397–411.
13. Doraszelski, Ulrich, Gregory Lewis, and Ariel Pakes. 2018. Just starting out: Learning and equilibrium in a new market. *American Economic Review* 108 (3): 565–615.
14. Hortacsu, Ali, and Steven L Puller. Understanding strategic bidding in multi-unit auctions: a case study of the texas electricity spot market. *The RAND Journal of Economics* 39 (1): 86–114.
15. Kohansal, Mahdi, Ashkan Sadeghi-Mobarakeh, and Hamed Mohsenian-Rad. 2017. A data-driven analysis of supply bids in California iso market: Price elasticity and impact of renewables. In *2017 IEEE International Conference on Smart Grid Communications (SmartGridComm)*, 58–63. IEEE.
16. Kohansal, Mahdi, and Hamed Mohsenian-Rad. 2016. A closer look at demand bids in California ISO energy market. *IEEE Transactions on Power Systems* 31 (4): 3330–3331.
17. Wang, Long, Zijun Zhang, and Jieqiu Chen. 2016. Short-term electricity price forecasting with stacked denoising autoencoders. *IEEE Transactions on Power Systems* 32 (4): 2673–2681.
18. Wang, Yi., Dahua Gan, Mingyang Sun, Ning Zhang, Lu. Zongxiang, and Chongqing Kang. 2019. Probabilistic individual load forecasting using pinball loss guided LSTM. *Applied Energy* 235: 10–20.
19. Teeraratkul, Thanchanok, Daniel O'Neill, and Sanjay Lall. 2017. Shape-based approach to household electric load curve clustering and prediction. *IEEE Transactions on Smart Grid* 9 (5): 5196–5206.
20. Velloso, Alexandre, Alexandre Street, David Pozo, José M Arroyo, and Noemi G Cobos. Two-stage robust unit commitment for co-optimized electricity markets: An adaptive data-driven approach for scenario-based uncertainty sets. *IEEE Transactions on Sustainable Energy* 11 (2): 958–969.
21. Kwac, Jungsuk, June Flora, and Ram Rajagopal. 2014. Household energy consumption segmentation using hourly data. *IEEE Transactions on Smart Grid* 5 (1): 420–430.
22. Liang, Huishi, Jin Ma, Rongfu Sun, and Du. Yanling. 2019. A data-driven approach for targeting residential customers for energy efficiency programs. *IEEE Transactions on Smart Grid* 11 (2): 1229–1238.
23. Zheng, Kedi, Qixin Chen, Yi. Wang, Chongqing Kang, and Qing Xia. 2019. A novel combined data-driven approach for electricity theft detection. *IEEE Transactions on Industrial Informatics* 15 (3): 1809–1819.
24. Sun, Weigao, Mohsen Zamani, Mohammad Reza Hesamzadeh, and Hai-Tao Zhang. Data-driven probabilistic optimal power flow with nonparametric bayesian modeling and inference. *IEEE Transactions on Smart Grid* 11 (2): 1077–1090.
25. Wang, Ying, Can Wan, Zhi Zhou, Kaifeng Zhang, and Audun Botterud. 2017. Improving deployment availability of energy storage with data-driven AGC signal models. *IEEE Transactions on Power Systems* 33 (4): 4207–4217.
26. Bagheri, Ali, Jianhui Wang, and Chaoyue Zhao. 2016. Data-driven stochastic transmission expansion planning. *IEEE Transactions on Power Systems* 32 (5): 3461–3470.
27. Ruiz, Carlos, Antonio J Conejo, and Dimitris J Bertsimas. 2013. Revealing rival marginal offer prices via inverse optimization. *IEEE Transactions on Power Systems* 28 (3): 3056–3064.
28. Saez-Gallego, Javier, Juan M Morales, Marco Zugno, and Henrik Madsen. 2016. A data-driven bidding model for a cluster of price-responsive consumers of electricity. *IEEE Transactions on Power Systems* 31 (6): 5001–5011.
29. Tianguang, Lu., Zhaoyu Wang, Jianhui Wang, Qian Ai, and Chong Wang. 2018. A data-driven Stackelberg market strategy for demand response-enabled distribution systems. *IEEE Transactions on Smart Grid* 10 (3): 2345–2357.

30. Chen, Ruidi, Ioannis Ch Paschalidis, Michael C Caramanis, and Panagiotis Andrianesis. 2019. Learning from past bids to participate strategically in day-ahead electricity markets. *IEEE Transactions on Smart Grid* 10 (5): 5794–5806.
31. Mitridati, Lesia, and Pierre Pinson. 2017. A Bayesian inference approach to unveil supply curves in electricity markets. *IEEE Transactions on Power Systems* 33 (3): 2610–2620.
32. AEMO. 2019. Yesterdays bids reports. http://www.nemweb.com.au/REPORTS/ARCHIVE/Yesterdays_Bids_Reports/.
33. MISO. 2019. Market reports: Day-ahead cleared bids report. https://www.misoenergy.org/markets-and-operations/real-time--market-data/market-reports/#nt=/MarketReportType:Bids.
34. Sun, Mingyang, Yi Wang, Fei Teng, Yujian Ye, Goran Strbac, and Chongqing Kang. 2019. Clustering-based residential baseline estimation: A probabilistic perspective. *IEEE Transactions on Smart Grid* 10 (6): 6014–6028.
35. Jinxing, Hu, and Hongru Li. 2019. A new clustering approach for scenario reduction in multi-stochastic variable programming. *IEEE Transactions on Power Systems* 34 (5): 3813–3825.
36. Ping, Qing, Christopher C Yang, Sarah A Marshall, Nancy E Avis, and Edward H Ip. 2016. Breast cancer symptom clusters derived from social media and research study data using improved *k*-medoid clustering. *IEEE Transactions on Computational Social Systems* 3 (2): 63–74.
37. Arjovsky, Martin, Soumith Chintala, and Léon Bottou. 2017. Wasserstein GAN. arXiv preprint arXiv:1701.07875.
38. Shahmohammadi, Ali, Ramteen Sioshansi, Antonio J Conejo, and Saeed Afsharnia. 2018. The role of energy storage in mitigating ramping inefficiencies caused by variable renewable generation. *Energy Conversion and Management* 162: 307–320.
39. Xu, Bolun, Yuanyuan Shi, Daniel S Kirschen, and Baosen Zhang. 2018. Optimal battery participation in frequency regulation markets. *IEEE Transactions on Power Systems* 33 (6): 6715–6725.
40. He, Guannan, Qixin Chen, Panayiotis Moutis, Soummya Kar, and Jay F Whitacre. 2018. An intertemporal decision framework for electrochemical energy storage management. *Nature Energy* 3 (5): 404–412.
41. Hosna Khajeh and Asghar Akbari Foroud. 2017. Behavioural market power indices in a transmission-constrained electricity market. *IET Generation, Transmission and Distribution* 11 (18): 4608–4616.
42. Hajiabadi, Mohammad Ebrahim, and Habib Rajabi Mashhadi. 2013. Analysis of the probability distribution of LMP by central limit theorem. *IEEE Transactions on Power Systems* 28 (3): 2862–2871.
43. Aghabozorgi, Saeed, Ali Seyed Shirkhorshidi, and Teh Ying Wah. 2015. Time-series clustering-a decade review. *Information Systems* 53: 16–38.
44. Ershun, Du, Ning Zhang, Bri-Mathias Hodge, Qin Wang, Chongqing Kang, Benjamin Kroposki, and Qing Xia. 2018. The role of concentrating solar power toward high renewable energy penetrated power systems. *IEEE Transactions on Power Systems* 33 (6): 6630–6641.
45. Kwac, Jungsuk, June Flora, and Ram Rajagopal. 2016. Lifestyle segmentation based on energy consumption data. *IEEE Transactions on Smart Grid* 9 (4): 2409–2418.
46. Wang, Tiexing, Qunwei Li, Donald J Bucci, Yingbin Liang, Biao Chen, and Pramod K Varshney. 2019. K-medoids clustering of data sequences with composite distributions. *IEEE Transactions on Signal Processing* 67 (8): 2093–2106.
47. AEMO. 2019. NEM registration and exemption list. https://www.aemo.com.au/.
48. David, A Kumar, and Fushuan Wen. 2001. Market power in electricity supply. *IEEE Transactions on Energy Conversion* 16 (4): 352–360.
49. U.S. Department of Justice and the Federal Trade Commission. 2010. Horizontal merger guidelines. https://www.justice.gov/sites/default/files/atr/legacy/2010/08/19/hmg-2010.pdf.
50. Wang, Yi, Qixin Chen, Chongqing Kang, and Qing Xia. 2016. Clustering of electricity consumption behavior dynamics toward big data applications. *IEEE Transactions on Smart Grid* 7 (5): 2437–2447.

Chapter 11
Aggregated Supply Curves Forecasting

11.1 Introduction

Since the reconstruction of power markets, the decisions about bidding strategies have been one of the most significant concerns of generation companies (GENCOs). To address this problem, there have been many studies working on providing an optimal bidding strategy for GENCOs in recent years [1, 2].

In common, there are two pivotal steps to take for providing a feasible optimal bidding strategy. The first step is to estimate the future competition situations of power markets. The second step is to make optimal bidding strategies based on the estimated market information. Though the total number of studies on optimal bidding is huge, most of them mainly focus on the second step and pay little attention to the first step. For example, the optimal bidding strategies for hydro generators [3], wind farms [4, 5], storages [6, 7] and virtual power plants [8] are all provided with the premises on knowing exactly their individual rivals' bidding strategies beforehand. Thus, estimating the market competition situation is necessary and fundamental for most optimal bidding researches.

To do so, the most intuitive direction is to forecast rivals' individual bidding behaviors. Kazemi et al. proposed a robust bidding strategy for a hybrid energy GENCO, which has both energy generation and energy retailing business [9]. In this study, the rivals' bids are defined as equal to the predicted cleared prices and capacities on specific nodes. R. Chen et al. estimated rivals' variable cost functions using an inverse optimization approach based on historical data, such as market clearing prices and individual cleared capacities [10].

However, it is difficult to directly conduct individual bidding forecasting from a market participant's view. Here are several reasons. First, due to information privacy policies, the individual information of generators, especially the bids and the cleared capacities and prices for specific generators, are usually not released or are masked for publication. Thus, a market participant does not have the basic data foundation to forecast rivals' bidding strategies. Second, the bidding behavior of individual market participants has uncertain and stochastic features. Even if a market participant can get access to other rivals' historical bidding data, it is still complicated and difficult to forecast individual bidding behaviors accurately.

Q. Chen et al., *Data Analytics in Power Markets*,
https://doi.org/10.1007/978-981-16-4975-2_11

Though the hundreds or thousands of market participants in a power market mostly have uncertain bidding behaviors, the uncertainty of their aggregation is greatly decreased. As the aggregation of individual participants' bidding behaviors, aggregated supply function (ASC) information is always public in most power markets. Comparatively, it is more feasible to find the ASCs' patterns hidden in the historical data and conduct accurate forecasting. The forecasted ASC information could also provide reference for market participants to optimize their bidding strategies.

There have been several studies working on forecasting ASCs so far. Dehghanpour used a dynamic Bayesian network model to predict the residual demand curve (RDC), which is the difference value between ASC and system demand, in power markets [11]. However, the RDC formulation is simplified as a linear function, which is largely different from the reality. In [12], Mitridati and Pinson proposed a Bayesian inference approach to forecast the ASC in power markets based on Markov Chain Monte Carlo and sequential Monte Carlo methods. The multi-segmented ASCs were constituted by market participants' linearized bids, which gains improvement compared with the last one. The proposed method has relatively good performance. However, it still has some drawbacks in ASC format and data. First, the ASC used for forecasting has been simplified as 10 certain blocks, which cannot be applied in real power markets. Second, the bids submitted by participants are generated artificially during the research, rather than real data in power markets, which means the information behind the real market data cannot be considered. In [13], Kiannejad et al. proposed a feedforward neural network method to predict the ASCs. The proposed method was tested on a 12-participant market and a 6-participant market. The method was tested with better performance than a Bayesian inference method. However, the given dataset is too small and simplified compared with the real market data, which make it doubtful to be used in real power markets.

In conclusion, the research on forecasting ASC has become more practical and the applied algorithm become more advanced. Though the existing studies seem to bring many good-enough results right now, there have been several problems remained to be addressed. The first problem is *the excessively small dataset*. The proposed methods are usually tested on designed power markets with several artificial participants. Since the data processing technologies on small data and big data are widely divergent, it is doubtful whether the existing methods can be applied in real worlds. The second problem is *the oversimplification in ASC formulations*. For example, the ASCs are usually generated by gathering several participants' linear bids together. The piled-up ASC data have lost several significant features of the real ASC data, such as high-dimension and nonlinearity. This deviation further decreases the adaptivity of the existing methods. The third problem is *the missing of the real market data*. Most of the existing studies are conducted and tested on artificially generated datasets, which means much information hidden in the real data has not been fully identified and analyzed.

In fact, the boundedness of the existing ASC forecasting methods is mainly caused by the deficiency of real power market data and suitable analysis methods. However, power market data have become more public in recent years and the data-driven analysis framework have experienced a rapid development, which provides us an opportunity to propose ASC forecasting methods targeting on real market data. With overcoming the aforementioned problems, the proposed ASC forecasting method can be more practical and significant.

However, it is indeed quite difficult to directly utilize ASC data for data-driven analysis due to their particularities in four folds. First, in most power markets, the ASC data are not usually structured, participants can only have access to the masked individual bidding data. Due to the numerous numbers of participants and their bidding blocks, it is quite difficult to directly obtain structured ASC data. Second, since the ASC data are constituted by large number of participants' bids, they are in the format of high dimensional arrays (i.e., thousands of dimensions), which consist of multiple discrete price and capacity values. It is both inefficient and ineffective to directly utilize data-driven analysis methods on such high dimensional ASC data. Third, the raw ASC data have lots of redundant information, which is unimportant but usually deteriorates the performances of the normal forecasting methods. The reason is that the raw price ranges of ASC data are usually greatly larger than the normal-cleared-price ranges. However, only the information hidden in the latter one is the key target for forecasting. Forth, the key influencing factors related to ASC data are unknown, causing trouble to organize a data-driven forecasting model. For better performances, the most appropriate dataset should be identified and integrated in the proposed forecasting method.

As for forecasting methods, there have been numerous forecasting methods applied in power market areas [14–16]. In recent years, long-short term memory (LSTM) model has been widely adopted in forecasting tasks in power markets and proved with great performances [17, 18]. As an improved recurrent neural network (RNN) architecture, LSTM is good at dealing with data with lags of unknown duration between important events in a time series. It has been applied in areas such as renewable energy forecasting [19], load forecasting [20], and electricity price forecasting [21]. In [19], the application of different deep learning architectures is discussed. It is shown that LSTM and gated recurrent network exhibit promising performance for renewable energy forecasting. Tan et al. proposes an LSTM-based method to forecast industrial load in [20]. The experimental results show that LSTM outperforms classical benchmark methods including support vector regression (SVR), deep belief network (DBN), and Xgboost. Chang et al. uses LSTM with wavelet transform to perform electricity price prediction in [21]. The accuracy of the LSTM-based method is significantly higher than that of back propagation neural network and general regression neural network.

In total, the chapter proposes an ASC forecasting framework based on data-driven methods. First, facing the situation of no ready-made ASC data, a data integration method is proposed to formulate the ASC data from oceans of masked individual bidding data. Second, to deal with the particularities of the real ASC data, a feature extraction method is proposed based on nonlinear mapping and principal components analysis (PCA) [22, 23]. Thus, the redundant information of data can be filtered, and the data dimensions can be reduced, while preserving as much necessary information as possible. Third, an ASC forecasting method is developed based on the LSTM model. The proposed forecasting model can identify the relations between the ASC data and the potential influencing factors embedded in the abundant market public information. To test the performance of the proposed forecasting method, empirical studies based on real market data from MISO are provided.

The main contributions can be summarized as follows:

1. A data-driven forecasting framework for real ASC is proposed with full utilization of accessible market data. The real ASC data are collected, formulated, and forecasted for participants to estimate the market competition situations.
2. A data processing method suited for real ASC data is proposed, which solves the problem of obtaining data, extracting features, filtering unnecessary information, and reducing dimensionality.
3. A LSTM-based model is customized to capture the dynamic relations between ASC data and potential influencing factors, providing forecasting results for ASCs.
4. The real market data from the Midcontinent Independent System Operator (MISO) market have been used for empirical analysis, which indicates the adaptivity of the proposed method.

The remainder of the chapter is organized as follows. The market descriptions and the proposed framework are given in Sect. 11.2. In Sect. 11.3, the data processing method for ASCs is presented, including the data integration method and feature extraction method. In Sect. 11.4, the LSTM-based ASC forecasting model is developed considering multiple influencing factors. In Sect. 11.5, numerical cases based on the data from MISO are provided. Finally, conclusions are given in Sect. 11.6.

11.2 Market and Framework

11.2.1 Market Descriptions

The power market analyzed in this study mainly refers to the power markets with centralized clearing procedures, such as Pennsylvania-New Jersey-Maryland (PJM) power markets, MISO power markets, and Australian electricity market (AEM). An independent system operator, denoted as ISO, is assumed to run the market.

For simplicity, the proposed analysis framework is focused on the day-ahead (DA) power market, where the market participants need to submit their bids in a monotonically increasing format for the next day. The bids could be in a stepwise or piecewise format, which depends on the specific power market rules. The number of bidding curves of each market participant in a day depends on the granularity of the market clearing time intervals, such as 30 min or 1 h. After collecting the bids, the ISO will clear the power markets, providing the market clearing results such as locational market prices (LMPs) and generator dispatch schedules.

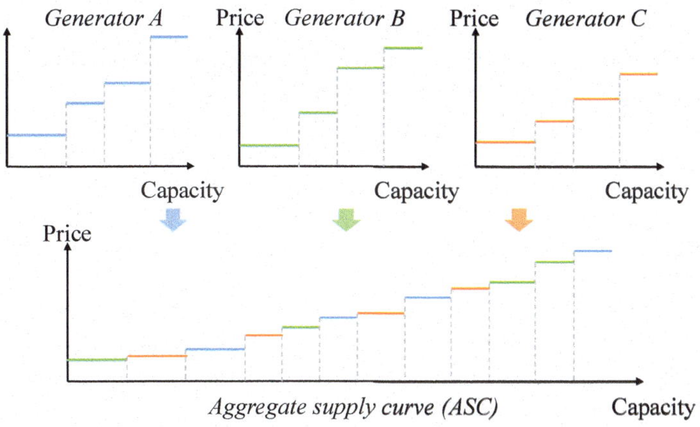

Fig. 11.1 Constitution process of ASC

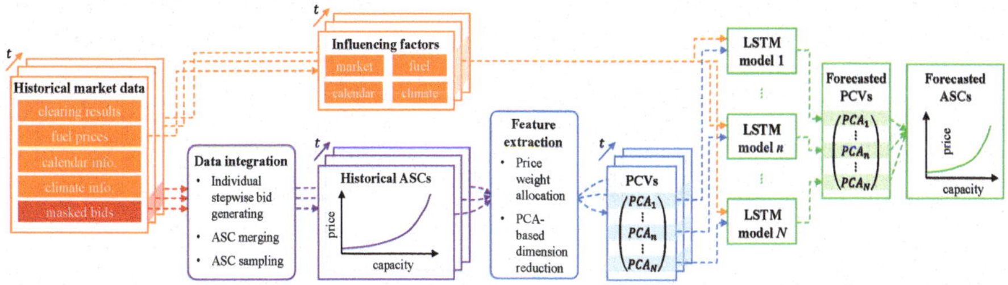

Fig. 11.2 Framework for ASC forecasting

The aggregation of all participants' bids constitutes an ASC. The detailed progress is shown in Fig. 11.1 as an example based on three generators with four bidding blocks.

As for a generator, it must know or infer the next-day information of power markets to make optimal bidding decisions. Since the topologies of power systems and the individual information of rivals are masked and difficult to obtain in most power markets, the scale of the reference information for generators can be reduced to the system load and ASCs. The former information is usually forecasted by ISOs and widely published for market participants. However, the forecasting information of the latter one (ASC) is barely exposed to the public and needs to be inferred by market participants themselves.

11.2.2 Forecasting Framework

The details of the proposed ASC forecasting framework are presented in Fig. 11.2.

First, all accessible market data are collected as historical market data. Although not all ISOs publish their ready-made ASC data, most of them publish their masked individual bidding data. For generality, a data integration method is proposed to generate ASC data from dispersive and numerous masked individual bidding data.

Second, different from the other data types as scalars, the ASC data have excessively high dimensions at one time period. In addition, there are other particularities in ASC data, such as large but nonsignificant price ranges, making it more difficult to conduct data-driven analysis. Thus, they cannot be treated similarly as the other data types with the commonly utilized methods. As a result, a feature extraction method is proposed to adaptively allocate weights among different prices and reduce the dimensionality of the ASC data. For a clear representation, the principal component vector of ASCs is denoted as PCV, and the variable on the nth dimension of a PCV is denoted as the PCV's nth component.

Third, to capture the dynamic relations among ASC data and influencing factors, the LSTM method is deployed and customized. Since the components of different dimensions in PCVs are linearly independent of each other, LSTM models are built and trained respectively for different components. In detail, the time series of the nth component of PCVs on all time periods is served as an input of the LSTM-n, as well as the time series of influencing factors. After the training of all LSTM models, we can forecast future PCVs with the input data. At last, the forecasted PCVs can be transformed back to forecasted ASCs.

11.3 Data Integration and Feature Extraction

11.3.1 Data Integration

When collecting all individual bidding data, we need to standardize, merge and sample them to generate ASCs in a uniform format for data-driven analysis. Since the treatment process is complex, the detailed process of data integration is presented in Fig. 11.3.

11.3.1.1 Individual Stepwise Bid Generating

There are mainly two different bidding formats allowed in power markets. The first is the stepwise format, $B_{i,d}^{\mathrm{S1}}$, whose bids consist of N_0 pairs of price and capacity blocks, $(p_{d,n}^{\mathrm{S1}}, q_{d,n}^{\mathrm{S1}})$. It should be noted that the indexes i and d represent the generator serial number and bidding data serial number, respectively. The accumulation of these price-capacity blocks constitutes a bidding curve. The second is the piecewise format, $B_{i,d}^{\mathrm{S0}}$, whose bids consist of N_0 pairs of price and capacity coordinate points, $(p_{(d,n)}^{\mathrm{S0}}, q_{(d,n)}^{\mathrm{S0}})$. The link lines of these price-capacity points constitute bidding curves.

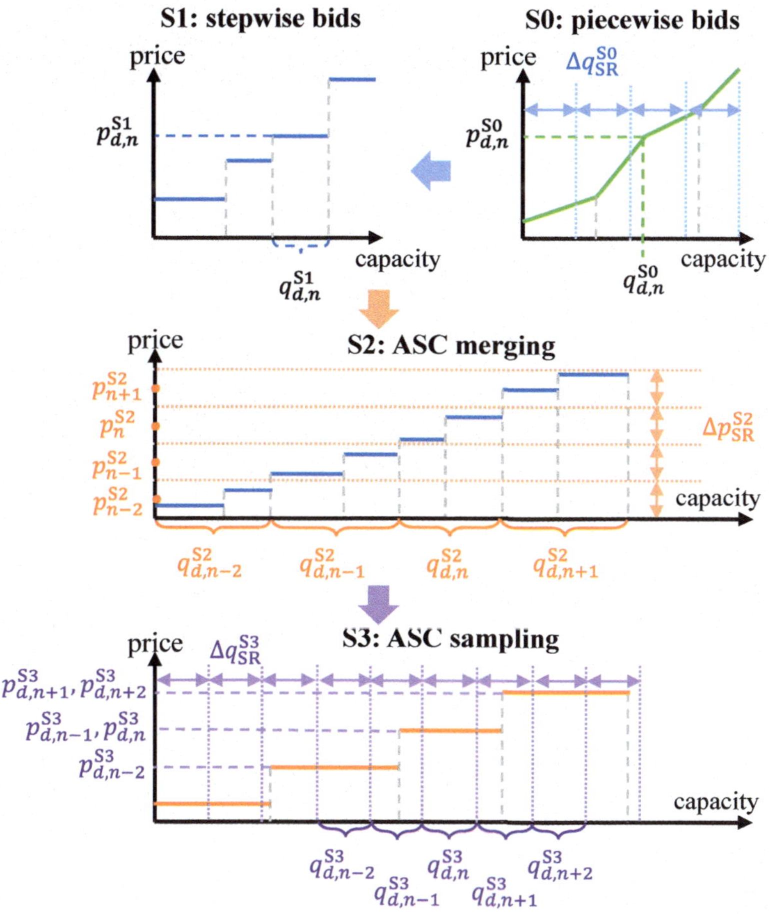

Fig. 11.3 Flowchart of data integration, S0 to S3

For better adaptivity, both formats are considered in the study. To accumulate all the bids for an ASC, the piecewise bids are transformed into stepwise formats through sampling. In detail, the piecewise bids are sampled with a uniform sampling resolution of the capacity, $\Delta q_{\mathrm{SR}}^{\mathrm{S0}}$. Then, a piecewise bid $\mathbf{B}_{i,d}^{\mathrm{S0}}$ will be transformed into a stepwise bid $\mathbf{B}_{i,d}^{\mathrm{S1}}$ with multiple price-capacity blocks.

11.3.1.2 ASC Merging

Since the bidding prices are complicatedly distributed in the range from lower to upper price caps as any possible floating-point numbers, it is ineffective to merge all the bids while maintaining the exact price values. Thus, we propose a series

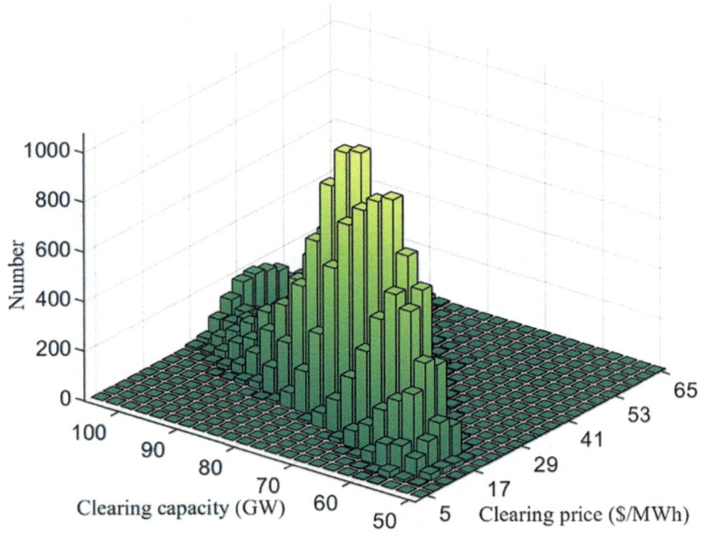

Fig. 11.4 Distributions of clearing prices and capacities in MISO market

of reference prices, $[p_1^{S2}, p_2^{S2}, \ldots, p_{N_0}^{S2}]$, with a preset resolution, Δp_{SR}^{S2}, and merge all the bids into their nearest price intervals. For example, the price interval for reference price p_n^{S2} is $(p_n^{S2} - 0.5\Delta p_{SR}^{S2},\ p_n^{S2} + 0.5\Delta p_{SR}^{S2})$. Then, we can obtain the merged stepwise ASC with multiple price-capacity blocks, as $(p_n^{S2}, q_{d,n}^{S2})$.

However, the boundaries of the reference price range remain unknown. In practice, the prices of ASC data range from negative hundreds of $/MWh to positive thousands of $/MWh based on the bidding price caps. However, the market clearing prices usually fluctuate around dozens of $/MWh. As a proof, the distribution of DA clearing prices and capacities in the MISO market [24] from January 1st, 2015 to June 30th, 2019 are presented in Fig. 11.4. The vertical axis represents the appearance number of a scene with certain clearing price and capacity in 4.5 years.

The clearing prices are mainly concentrated around 15 to 35 $/MWh. In detail, 99.5% of the clearing prices exist in the range from 10.14 to 77.49 $/MWh, which is much smaller than the allowed bidding price ranges. If we choose the whole available bidding price range as the reference price range, it will cause a greater calculation burden and generate a large amount of insignificant information.

To address this problem, the reference prices are fixed in a range $[P_{\min}^{S2}, P_{\max}^{S2}]$, which is determined by the distribution of the DA clearing prices to maintain an acceptable confidence level, CL_0, such as 99.5%. The bidding prices outside the reference price range will be clipped to its adjacent range boundary.

11.3.1.3 ASC Sampling

Even with the merged stepwise ASC data, it is still difficult to compare two ASCs without a uniform sampling resolution. Thus, we utilize a sampling method on the ASC data with a uniform sampling resolution, $\Delta q_{\mathrm{SR}}^{\mathrm{S3}}$. After sampling, all ASC data have been transformed into data series of bidding prices, $\mathbf{P}_d^{\mathrm{S3}}$, with N_1 price-capacity points. The detailed formats are written as:

$$\mathbf{P}_d^{\mathrm{S3}} = \left(p_{d,1}^{\mathrm{S3}}, \ldots, p_{d,n}^{\mathrm{S3}}, \ldots, p_{d,N_1}^{\mathrm{S3}}\right) \tag{11.1}$$

$$N_1 = \mathrm{int}\left(\sum_n q_{d,n}^{\mathrm{S3}} / \Delta q_{\mathrm{SR}}^{\mathrm{S3}}\right) \tag{11.2}$$

11.3.2 Feature Extraction

Though the ASC data have been integrated, they cannot be directly and effectively used for data-driven learning because of the uneven bidding price range distribution and high dimensional data. Thus, we propose a feature extraction method to allocate suitable weights among various prices and reduce the dimension of ASC data based on PCA methods.

11.3.2.1 Price Weight Allocation

In fact, even in the range of $\left[P_{\min}^{\mathrm{S2}}, P_{\max}^{\mathrm{S2}}\right]$, the distribution of the bidding prices are not even: most of the market clearing prices are distributed in a smaller range. For example, the range from 18.11 to 37.16 \$/MWh accounts for 80% probability with only 28.3% price span in the MISO market.

The probability distribution of the market clearing prices can be approximately considered as a t-distribution. In a market participant's view, the forecasting of the ASC should be more accurate around the prices with higher probabilities than that of the other prices. Thus, we should allocate different weights on various prices, mainly according to their appearance probabilities.

To do so, a logistic function is introduced to adaptively allocate different weights on various bidding prices, shown as:

$$p_{d,n}^{\mathrm{S4}} = \frac{1}{1 + e^{-r\left(p_{d,n}^{\mathrm{S3}} - P_{\mathrm{mid}}^{\mathrm{S4}}\right)}} \tag{11.3}$$

$$r = \begin{cases} r_0 / \left(P_{\max}^{\mathrm{S4}} - P_{\mathrm{mid}}^{\mathrm{S4}}\right), & p_{d,n}^{\mathrm{S3}} \geq P_{\mathrm{mid}}^{\mathrm{S4}} \\ r_0 / \left(P_{\mathrm{mid}}^{\mathrm{S4}} - P_{\min}^{\mathrm{S4}}\right), & p_{d,n}^{\mathrm{s3}} < P_{\mathrm{mid}}^{\mathrm{S4}} \end{cases} \tag{11.4}$$

where $P_{\max}^{S4}$ is the sigmoid midpoint of a logistic function, which can be chosen as the price at the 50% quantile point. Due to the DA prices' unbalanced probability distribution on the two sides of the midpoint, a different logistic growth rate r is proposed for the left and right sides based on a reference price span. In detail, $P_{\min}^{S4}$ and $P_{\max}^{S4}$ can be chosen as the prices at the marginal quantile points, such as the 10 and 90% quantile points.

11.3.2.2 PCA-Based Dimension Reduction

Due to the sampling procedure in Sect. 11.3.1.3, each ASC is a bidding price vector with extremely high dimensions. It is difficult and inefficient to directly forecast the sampled ASC. In addition, the high dimensional ASC series are redundant because of the high correlation that exists among the data. Thus, we introduce the PCA method to extract features from the data and reduce the dimensions of the sampled ASCs.

PCA is a linear dimension reduction method, which uses an orthogonal transformation to convert a set of possibly correlated variables into a set of linearly uncorrelated variables. It can retain most of the covariance of the data features with few artificial variables [25], and has been used for data dimension reduction in the areas of load consumption prediction [26], renewable energy forecasting [27] and early event detection [28].

First, the ASC series data, P_d^{S4}, need to be standardized on each of their dimension as defined by equation (11.5), where P_{mean}^{S4} and P_{std}^{S4} are the average value of standard deviation of $P_{d,n}^{S4}$.

$$p_{d,n}^{S5} = \left(p_{d,n}^{S4} - p_{\text{mean}}^{S4}\right) / p_{\text{std}}^{S4},\ n \in N_1 \tag{11.5}$$

$$\mathbf{P}_d^{S4} = \left[p_{d,1}^{S4}, \ldots, p_{d,n}^{S4}, \ldots, p_{d,N_1}\right] \tag{11.6}$$

Second, the covariance matrix $\mathbf{C}$ of the standardized ASC data, $\mathbf{X} = [P_1^{S5}, \ldots, P_d^{S5}, \ldots, P_{D_0}^{S5}]^\top$, can be calculated as:

$$\mathbf{C} = \mathbf{X}\mathbf{X}^\top \tag{11.7}$$

where D_0 is the number of bidding data.

Third, we conduct an eigenvalue decomposition on the covariance matrix $\mathbf{C}$ and acquire the eigenvalue matrix and eigenvector matrix, $\mathbf{W}$. Then, we rank the eigenvalues and combine the eigenvectors corresponding to the top N_2 largest eigenvalues. The principal component projection matrix $\mathbf{W}^*$ can be calculated with the dimension $N_1 \times N_2$. By projecting the ASC matrix $\mathbf{X}$ on the direction $\mathbf{W}^*$, the dimension-reduced principal component matrix $\mathbf{Y}$ can be derived:

$$\mathbf{Y} = \mathbf{X} \cdot \mathbf{W}^* \tag{11.8}$$

$$\mathbf{Y} = \left[\mathbf{P}_1^{S6}, \ldots, \mathbf{P}_d^{S6}, \ldots, \mathbf{P}_{D_0}^{S6}\right]^\top \tag{11.9}$$

$$\mathbf{P}_d^{S6} = \left[p_{d,1}^{S6}, \ldots, p_{d,n}^{S6}, \ldots, p_{d,N_0^{S6}}^{S6} \right] \quad (11.10)$$

The dimension of matrix Y is $D_0 \times N_2$, which means that the column number has been reduced from N_1 to N_2.

11.4 ASC Forecasting

11.4.1 LSTM Model

As we all know, the bidding behaviors of individuals are influenced by many factors, such as market clearing prices on the yesterday, the forecasting load on the operation day, etc. It is intuitive to use data-driven methods to find out the relationships between the individual bidding behaviors and the influencing factors. Since the ASCs are generated based on the combination of all participants' bids, we can use the similar idea to identify the relationship between ASC and "influencing factors".

In detail, the LSTM method [29] is employed in the chapter to capture the dynamic relationships among influencing factors and the PCV components of ASCs. Taking PCA-n as an example, the detailed structure of the proposed LSTM forecasting model is presented in Fig. 11.5.

First, the historical PCA-n data with t_I time steps are combined with influencing factors, such as market clearing prices and loads, serving as the inputs of the model. Second, the input data at each time step are connected to LSTM units at the same time step. The LSTM units at all time steps constitute an LSTM layer. Third, the outputs of the LSTM units at the last time step are connected to dense layers, which consist of fully connected units. At last, the PCA-n to be forecasted are outputted by the output layer.

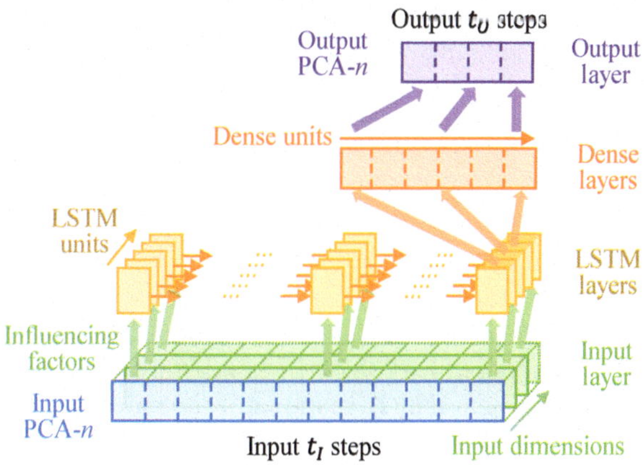

Fig. 11.5 Structure of the proposed LSTM forecasting model

Table 11.1 List of potential influencing factors of ASCs

No.	Category	Data type	Format
1	Market	Price	Float
2	Market	Load	Float
3	Market	Capacity	Float
4	Market	Generation mix	Float
5	Economy	Fuel price	Float
6	Calendar	Season	One-hot
7	Calendar	Weekday	Boolean
8	Calendar	Holiday	Boolean
9	Calendar	Hour	Float
10	Calendar	Time period	Integer
11	Climate	Temperature	Float

Since the specific structure of LSTM units and the training method of the LSTM model have been explained in many works [17, 30], they are not elaborated detailedly in this chapter.

11.4.2 *Influencing Factors*

Since ASCs are constituted by all market participants' bids, all potential factors which have influences on the individual bidding strategies should be considered as the influencing factors for ASCs. Thus, a list of potential influencing factors is proposed in Table 11.1.

In detail, the various data types can be roughly divided into 4 categories, which are market, economy, calendar, and climate. The market category contains four data types, which are market clearing prices (values on the previous days or forecasting values on the operation day), the forecasting system load on the operation day, the estimated system capacity with the consideration of scheduled outage, the generation mix of the system, which is the combination of the various fuels used to generate electricity in the power market. The economy category contains fuel price data. The calendar category contains season, weekday, holiday, hour and time period of the bids. The climate category contains temperature, irradiation and wind speed data types. It should be noted that the listed data types are only the potential influencing factors of ASC assumed by the authors, other data types can also be added in to improve the performance.

The proposed LSTM-based forecasting method is compatible with different data types, as long as they have the same time resolution, which is one-hour or half-hour, with the forecasting target, ASC data. As for some data with a larger time resolution, such as fuel price with daily time resolution, it should be extended to be consistent with the ASC data. In addition, the "hour" data has been processed before training. In detail, the hours from 1 to 24 have been mapped into 0 to 1 by sine functions, making data type "hour" with the float format.

11.4.3 Training and Forecasting

The training data should be divided into three datasets, which are the training, validation, and testing datasets. The training and validation datasets are used for hyperparameter tuning, where the former one is utilized to train the model and the latter is used to evaluate the performance and prevent overfitting. Then, the LSTM model trained with the best-performed hyperparameter set is tested on the testing dataset. The forecasting results can be used for model performance evaluation.

Since the bids are submitted on a daily frequency, the ASCs at different time periods on the next day should be forecasted simultaneously on this day. Thus, the allocation of the three datasets from the original dataset should be based on the daily division. For generality, the three datasets should be divided in date order.

11.4.4 Evaluation Criteria

In fact, the model training and testing are not based on direct ASC data, but based on the PCV of ASC data. Thus, evaluation criteria should be both given to PCV and ASC data. However, since the two kinds of data have different data characteristics, two different evaluation criteria are introduced for the two kinds of data respectively.

Since the values of PCVs have been normalized, their values fluctuate around zero. The mean squared error (MSE) is used to evaluate the performance. In detail, the MSE of principal component n in PCV on day d is written as:

$$MSE_d^{(n)} = \frac{1}{T} \sum_t \left[v_{d,t}(n) - \overline{v}_{d,t}(n) \right]^2 \tag{11.11}$$

where t is the index of time period; T is the total time period number in a day; $v_{d,t}(n)$ is the forecasted principal component n of PCV on day d time period t; and $\overline{v}_{d,t}(n)$ is the real value at the same time.

Different with PCV data, the ASC data are in the format of stepwise monotonously increasing curves. Thus, the mean absolute percentage error (MAPE) is used to evaluate the performance. In detail, the MAPE of ASC on day d is written as:

$$MAPE_d = \frac{1}{T} \sum_t \frac{1}{S} \sum_s \frac{\left| p_{d,t,s} - \bar{p}_{d,t,s} \right|}{\bar{p}_{d,t,s}} \tag{11.12}$$

where s is the index of the sampling points on ASC; S is the total number of the sampling points; $p_{d,t,s}$ is the forecasted price on sampling point s of ASC on day d time period t; and $\overline{p}_{d,t,s}$ is the real price on sampling point s at the same time.

11.5 Case Study

11.5.1 Dataset Description

To test the effectiveness and adaptivity of the proposed framework, a 4.5-year hourly dataset is collected from the real market data published by MISO, National Centers for Environmental Information (NOAA), U.S. Energy Information Administration (EIA), etc. In detail, the market data span from January 1st, 2015 to June 30th, 2019. The total dataset has 39408 h data.

The dataset includes tens of data types: market clearing price and its forecasted value for the next day [24], system load and its forecasted value for the next day [31], temperature and its forecasted value for the next day [32], the capacity of 6 fuel types (coal, gas, wind, nuclear, hydro and other) and its forecasted value for the next day [33], generation mix and its forecasted value for the next day, oil price and gas price [34], weekday, hour, season, time period. An illustration of several data has been shown in Fig. 11.6. Since the whole data set is too long to present, only the data of several data types in the first two weeks in 2015 have been presented. It should be noted that there are a few data published in the daily resolution, such as (e) temperatures and (f) oil prices in Fig. 11.6. Thus, they are filled with the same value on one day to transform into hourly resolution. If there are data with smaller resolution, they are also able to be utilized in the method.

The 39408-h original dataset has been divided into three sub-datasets for model training and evaluation, which are training, validation, and testing datasets. The proportions of the three sub-datasets are 0.8, 0.1, and 0.1. First, the proposed model is trained based on the training and validation datasets to tune hyperparameters, including the past time length, the size of LSTM, the structure of dense layers, the selected features for inputs. Then, the trained forecasting model of the hyperparameters with the best performances is tested based on the testing dataset, providing an objective performance evaluation.

11.5.2 Feature Extraction

The feature extraction procedure is demonstrated in this section, working as a data processing before learning.

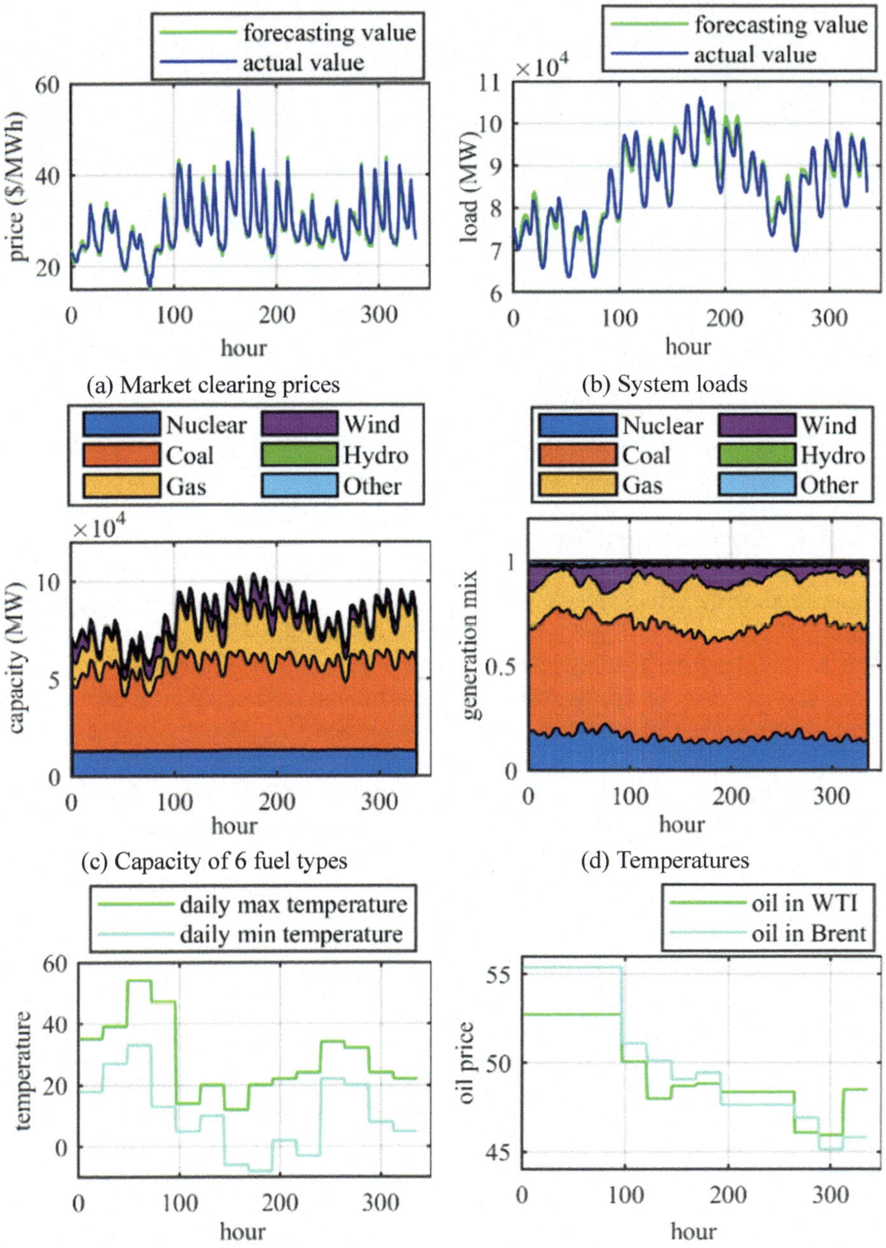

(a) Market clearing prices (b) System loads

(c) Capacity of 6 fuel types (d) Temperatures

(e) Capacity of 6 fuel types (f) Oil prices

Fig. 11.6 Market clearing prices

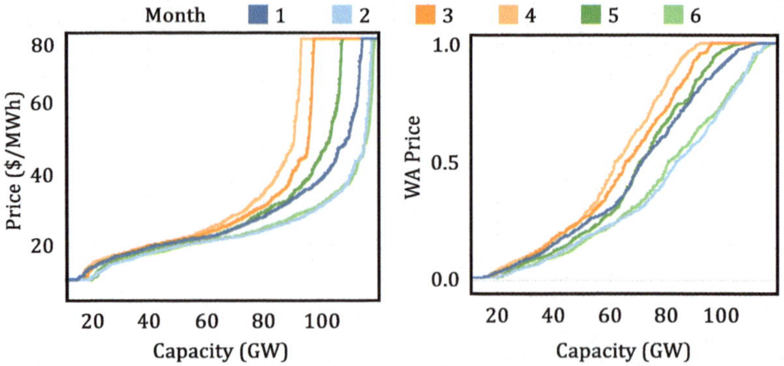

(a) Original ASCs (b) Weight-allocated ASCs

Fig. 11.7 Original ASCs and weight-allocated ASCs at 16 h on the 21st from January to June 2017

11.5.2.1 Price Weight Allocation

For presenting the effectiveness, the original ASC curves and weight-allocated (WA) ASCs are presented in Fig. 11.7. Since there are too many ASCs, only 6 ASCs at 16 h on the 21st from January to June 2017 are chosen to be presented.

As shown in Fig. 11.7a, the ASCs in various months have slight differences during the frequently-clearing price ranges but have tremendous differences when market clearing prices are higher. The unbalanced differences in different price ranges will inevitably harm the effectiveness of further forecasting.

From Fig. 11.7b, we can see that the main differences among ASCs are moved from the high clearing price ranges to the frequently-clearing price ranges, where WA price = 0.5 means the most frequently-clearing prices. The weigh-allocated ASCs are more suitable for machine learning.

11.5.2.2 PCA-Based Dimension Reduction

To demonstrate the effectiveness of PCA-based dimension reduction, the top 6 principal components' eigenvalues and their explained percentages of the total information are listed in Table 11.2.

Table 11.2 Top 6 principal component variances and the explained percentages

Component rank	1	2	3	4	5	6
Eigenvalue	163.7	32.4	4.3	2.9	1.1	0.8
Explained pct.	78.9	15.6	2.1	1.4	0.5	0.4
Total pct.	78.9	94.6	96.6	98.0	98.6	98.9

In detail, the explained percentage represents the percentage of the total variance of the raw data explained by a specific component, and the total percentage in rank n represents the total percentage explained from component 1 to n. As shown in Table 11.2, the top two principal components account for the most information of the total variance (i.e., 94.6%). With another 2 principal components, the extracted 4-dimension PCV could account for more than 98% information of the total variance.

Thus, we choose to use PCVs with the first 4 principal components to represent ASCs in the later ASC forecasting procedures.

11.5.3 ASC Forecasting

To better demonstrate the performance of the proposed method, there are two parts in this section for ASC forecasting. First, we use training and validation datasets to tune hyperparameters of the proposed LSTM-based model. Then, the best hyperparameter set is chosen to build an LSTM-based ASC forecasting model, which is later implemented on the testing dataset to evaluate the performances.

It should be noted that all data used for model training need to be normalized beforehand. To make sure the performance evaluation can be compared among different datasets, the MSE-based loss of PCV components is calculated based on their normalized values. Since the ASC forecasting results will only be analyzed on the testing dataset, we can calculate their MAPE-based loss based on the origin values.

11.5.3.1 LSTM-Based Model Hyperparameter Tuning

The hyperparameters to be tuned can be roughly divided into three categories: (1) past period, which is the historical time periods used as inputs to forecast the future 24 time periods' ASC's PCV components. There are 3 kinds of past periods used for hyperparameter tuning, which are 24 (one day), 48 (two days), and 72 (three days); (2) model size, which is the detailed configure of the LSTM-based model. The size of LSTM's units and dense layers and units can be modified. In this case, the number of LSTM's units is set as 25, 50, or 75; and the dense layer is set as 24 units in one layer, [24, 24] units in two layers, or [24] units in two layers. It should be noted the above hyperparameters of the model size are assumed based on the experience. In fact, the hyperparameters of model size can be any other values, which won't influence the main framework of the proposed method. Since there are hyperparameter tuning process, the assumed model size hyperparameters mainly work like an example; (3) feature group, which contains the data types serving as inputs. Since the types of the data inputs are numerous, several feature groups are proposed to contain specific data type combinations. The feature group detailed information can be seen in Table 11.3, where "1" means the group contains the corresponding data types and the empty means "not containing".

Table 11.3 Feature group information

Data type	Feature group (G)								
	1	2	3	4	5	6	7	8	9
Historical PCV components	1	1	1	1	1	1	1	1	1
Historical market prices	1	1	1	1	1	1	1	1	1
Forecasting market prices		1	1	1	1	1	1	1	1
Historical system loads	1	1	1	1	1	1	1	1	1
Forecasting system loads	1	1	1	1	1	1	1	1	1
Weekday	1	1	1	1	1	1	1	1	1
Hour	1	1	1	1	1	1	1	1	1
Historical max/min temperatures		1							
Oil price				1					
Gas price				1					
Forecasting capacity of fuel types					1				
Historical generation mix						1			
Forecasting temperatures							1		
Forecasting renewable capacity								1	
Forecasting generation mix									1

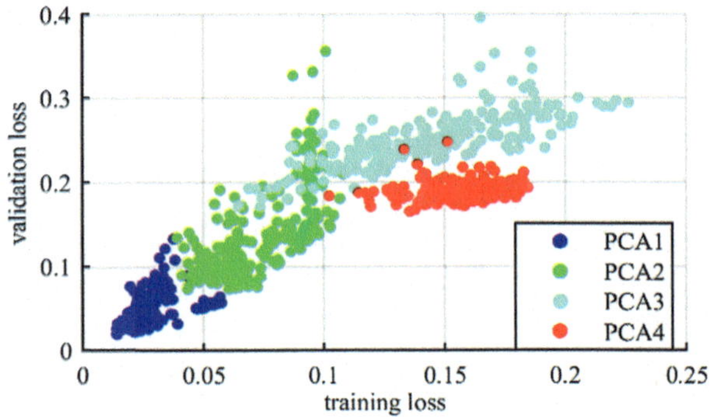

Fig. 11.8 Loss information of 972 hyperparameter sets

In total, the number of hyperparameter sets for each PCV component is 243. Since there are 4 PCV components, the total number of hyperparameter sets for tuning is 972.

Then, we use grid search method for hyperparameter optimization. The grid search method is simply an exhaustive searching through all the hyperparameter sets, whose number is 972 in this chapter. Since the models are trained based on the training dataset, the MAPE-based loss on the validation is considered as the performance metric for hyperparameter tuning. After the grid search, the MAPE-based loss values of 972 hyperparameter sets are produced, which can further be compared and analyzed (Fig. 11.8).

Table 11.4 Hyperparameter sets for each PCV component in grid search

PCV	Index	Loss (MSE)		Past period	Model size		Group
		Train	Val.		LSTM	Dense	
PCA1	183	0.0143	0.0199	72	25	[24]	G8
	102	0.0141	0.0206	48	25	[24]	G8
	84	0.0216	0.0224	48	25	[24]	G2
PCA2	52	0.0435	0.0729	24	50	[24, 24]	G9
	101	0.0605	0.0733	48	25	[48, 24]	G8
	4	0.0611	0.0741	24	25	[24, 24]	G2
PCA3	210	0.0735	0.1713	72	50	[24]	G8
	156	0.0644	0.1722	48	75	[24]	G8
	214	0.0648	0.1728	72	50	[24, 24]	G9
PCA4	177	0.1358	0.1650	72	25	[24]	G6
	186	0.1380	0.1673	72	25	[24]	G9
	165	0.1197	0.1706	72	25	[24]	G2

It is obvious that the forecasting performances of the proposed LSTM-based model on various PCV components are different. In detail, the performance on PCA1 is the best, reflected in the lowest training loss (mostly lower than 0.05) and validation loss (mostly lower than 0.1). The performance on PCA2 is not as good as PCA1, whose training loss mainly ranges from 0.05 to 0.1, whereas the validation loss mainly ranges from 0.1 to 0.2. In addition, there are considerable models having results with validation loss higher than 0.2. The performances on PCA3 and PCA4 get worse, shown in the increasing training and validation loss.

It should be noted that decreasing performances with the increase of PCV components is reasonable. With the increase of PCV components, the meaningful information quantity is decreasing, which means more random fluctuations are brought into the training data. Thus, the forecasting performance on the higher PCV components decreases. However, since the explained percentages decrease rapidly with the increase of PCV components, the effects of the large forecasting errors are limited in the final ASC forecasting.

To present a detailed result of the grid search, the top 3 hyperparameter sets for each PCV component with the highest validation loss are presented in Table 11.4, where "index" is the index of hyperparameter sets in the hyperparameter tuning process, from 1 to 243 in this chapter.

It can be roughly found that the LSTM-based forecasting model on various PCV components has different preferences on hyperparameters to accomplish the best performances. As for PCA1, the simplest model size suits the forecasting task better. In detail, the LSTM with the smallest size and one-layer dense network provide the top 3 best performances. The past period should be set as 72 or 48, indicating that the

Table 11.5 Feature group information

Group	Loss (MSE)		
	Training	Validation	Average
G8	0.0140	0.0206	0.0173
G2	0.0216	0.0224	0.0257
G5	0.0187	0.0264	0.0220
G4	0.0240	0.0274	0.0255
G7	0.0217	0.0293	0.0274
G3	0.0216	0.0333	0.0312
G9	0.0238	0.0386	0.0226
G6	0.0279	0.0534	0.0406
G1	0.0514	0.0539	0.0526

past period hyperparameters have little influences on the forecasting performance as long as it has more that two-day data length.

For better analysis on the selection of data types, we fix the settings of past period, LSTM and dense as 48, 25 and 24 as one of the best hyperparameter settings, and compare the loss of G1 to G9 in Table 11.5.

It can be concluded that the best feature group is G8, which has great performances on both training and validation datasets. Comparing G8 and G2 in Table 11.3, we can see that the only difference is "forecasting renewable capacity" data type. Thus, it can be concluded that the addition of "forecasting renewable capacity" data type will increase the forecasting performance on PCA1. As for G5, the addition of "forecasting capacity of fuel types" increases the accuracy in training process but decrease the accuracy in validation process, which means it might cause overfitting. As for G3, G4, G6, G7 and G9, the addition of new data types does not show promotion in performances but cause disturbance and decrease the accuracies. Comparing G1 and G2, we can see it is particularly important to include "forecasting market prices" data type in the training.

As for PCA2, it seems a more complex setting in dense layer might provide better performances. The forecasting generation mix and renewable capacity data types are both helpful to increase the forecasting performances. Due to space limitation, the hyperparameter preferences of the other 2 PCV components with low significances are not detailed here.

11.5.3.2 ASC Forecasting Test

Thus, we choose the hyperparameter sets with the lowest validation loss for each PCV component to build the LSTM-based model to conduct ASC forecasting on the testing dataset.

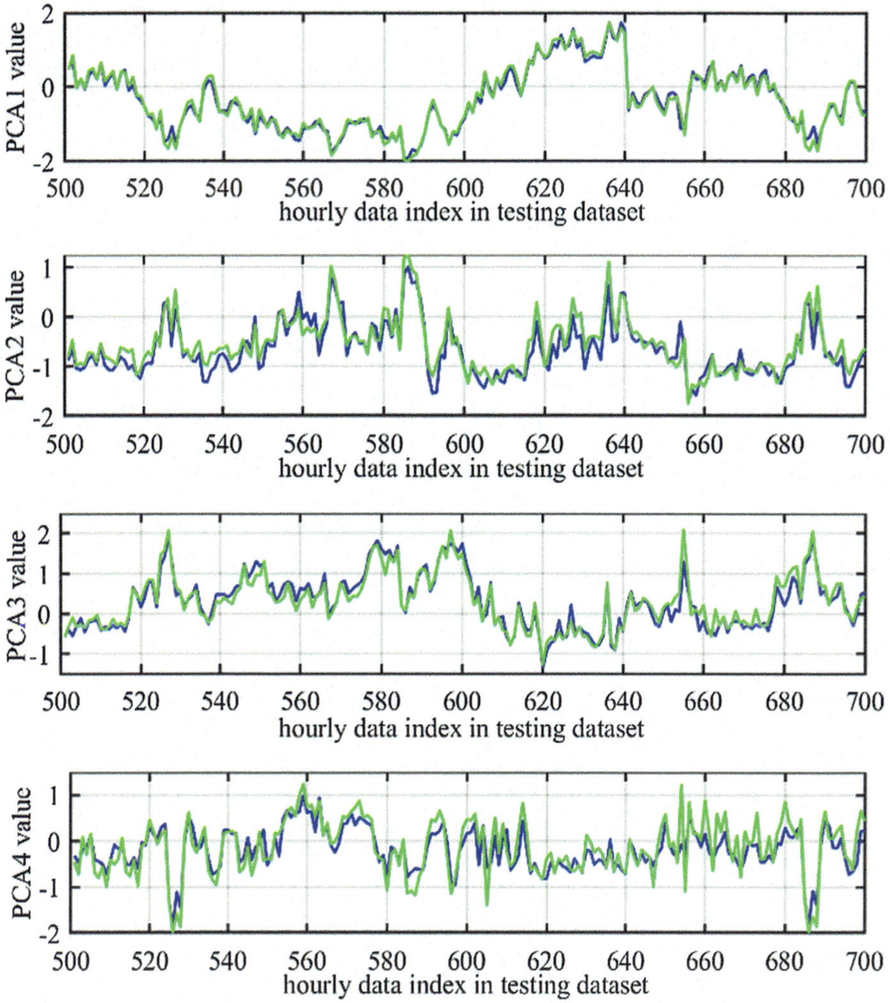

Fig. 11.9 Forecasting and real hourly PCV components from 501-h to 700-h in testing dataset. (green: real value, blue: forecasting value)

To provide a visual presentation of the PCV components forecasting results, the detailed forecasting and real hourly normalized PCV components of ASCs are drawn in Fig. 11.9. Since the data length of the testing dataset is too long (3840 h) to be shown, a 200-h result from 501st h to 700th h is used for representation. The green curves represent the real values, and the blue curves represent the forecasting values.

It can be observed that the forecasting PCV components have very similar trends with the real values and usually stick with them, representing the good performance of the proposed LSTM-based method. In addition, the forecasting performances of various PCV components seem different. It can be roughly concluded that the

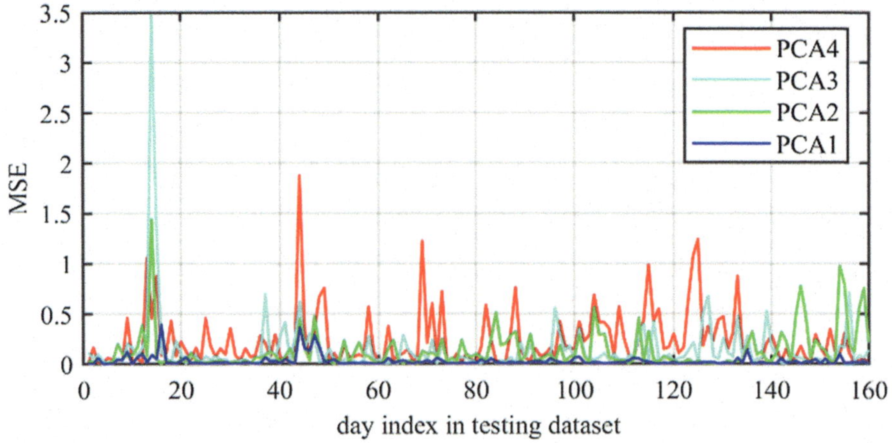

Fig. 11.10 Daily MSEs of forecasting results of PCV components in testing dataset

forecasting of PCA1 has the best performance, and the deviations between forecasting and real values of PCA2 to PCA4 seem larger.

For a better view of various PCV components' performance, the daily MSEs of PCV components forecasting results are presented in Fig. 11.10.

The daily MSEs of PCA1 forecasting are almost the lowest in the testing dataset. The daily MSEs of PCA2 get higher in several days, such as day-14, day-84, day-104, day-146, day-154 and day-159, which are Jan. 31st, Apr. 11th, May 1st, Jun. 12th, Jun. 20th and Jun. 25th. Further, the daily MSEs of PCA3 and PCA4 are getting much higher. It is reasonable that with the increase of PCV index, the invalid interference information becomes larger, and the ASC results are more difficult to be forecasted accurately. The average MSEs of 4 PCV components in the testing dataset are calculated in the last row of Table 11.6, whose results are consistent with the results found during the hyperparameter tuning process.

Then, the 4 forecasting PCV components are combined together, and transformed back to ASC forecasting results. Since the hourly forecasting results are in the format of high-dimensional curves, it is much more difficult to present all of them in a figure directly. Thus, we choose to present several hourly forecasting ASCs and their real values to give a direct picture of the forecasting performances. To prevent losing generalizability, we collect four days of data from day 40 at equal intervals (20 days). Then, the forecasting and real ASCs at hour-14 of the 4 selected days are presented in Fig. 11.11. It can be observed that the forecasting ASCs have similar shapes to the real ASCs.

Then, the daily MAPEs of ASC forecasting results are presented in Fig. 11.12. It can be observed that the daily MAPEs of forecasting ASCs fluctuate among different days, while most of the daily MAPEs are lower than 0.04. The information of the days with the highest MAPEs have been listed in Table 11.6. The average MAPE has also been listed in the last row of Table 11.6.

Table 11.6 Information of the days with the highest MAPEs

Day		ASC	PCV loss (MSE)			
Index	Date	MAPE (%)	PCA1	PCA2	PCA3	PCA4
45	2019/3/3	9.05	**0.153**	**0.157**	**0.139**	**0.332**
14	2019/1/31	8.50	0.087	**1.438**	**3.477**	**0.456**
46	2019/3/4	8.25	**0.126**	0.034	**0.304**	0.080
16	2019/2/2	7.64	**0.390**	0.009	0.113	0.085
48	2019/3/6	7.26	**0.166**	**0.151**	0.021	**0.669**
97	2019/4/24	7.23	0.013	**0.182**	**0.297**	**0.425**
Average		2.74%	0.025	0.119	0.127	0.195

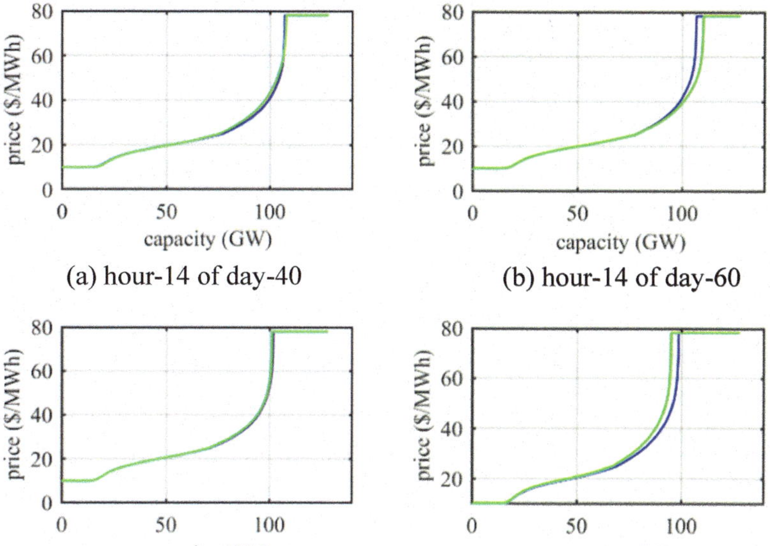

(a) hour-14 of day-40 (b) hour-14 of day-60

(c) hour-14 of day-80 (d) hour-14 of day-100

Fig. 11.11 ASC forecasting results of hour-14 in 4 days in testing dataset. (green: real value, blue: forecasting value)

Fig. 11.12 Daily MAPEs of ASCs in testing dataset

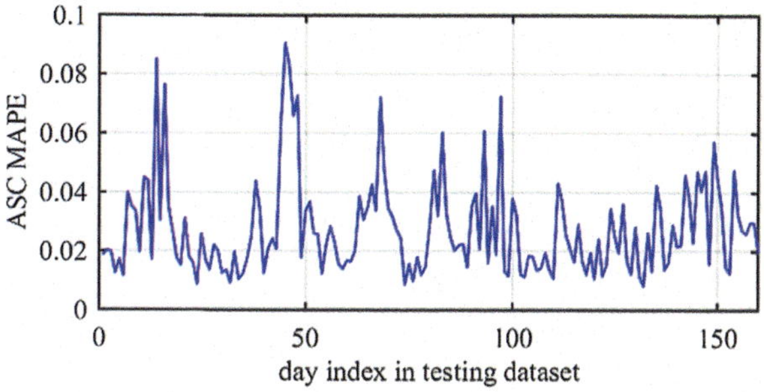

It can be observed that the high MAPEs of forecasting ASCs are mainly caused by the large deviations of forecasting results on various PCVs. To be clear, the data with large deviations have been marked as bold. For example, as for the forecasting results on day 45, the MSEs of PCA1 to PCA4 are all greatly larger than their average MSEs, which results in the highest MAPE in ASC forecasting. As for the other days, the MAPEs of ASC decrease with the decreases of forecasting deviations on various PCVs.

By comparing the results of Tables 11.6 and 11.4, we can find that the errors of most PCV components on the testing dataset are slightly larger than the results on the training and validation datasets. However, the error increases are slight and acceptable, demonstrating the adaptivity of the proposed model.

11.5.4 Calculation Information

The proposed LSTM-based models and the methods used for comparison are implemented using Scikit-Learn [35] and TensorFlow [36]. The model training is supported by CUDA 8.0 and an Nvidia GPU, 2080Ti.

The average training time of a hyperparameter set is 1 min 55 s. When the model has finished training, the calculation time of forecasting process only takes several seconds, which can be roughly neglected. In addition, since the grid search process contains multiple rounds of training, whose number is 972 in this study, the calculation time would inevitably increase. In this study, it takes about 30 h to finish the hyperparameter tuning of 4 PCVs. Since the forecasting model is trained based on years of data, it only needs to be updated on a relatively longer period, such as one month or one week. In addition, the hyperparameter tuning process can be run in a distributed mode. These facts make it practical to apply the proposed method in real power markets.

11.5.5 Methods Comparison

To demonstrate the advantages of the proposed method, the reference methods need to be tested and compared. Considering the existing studies on forecasting ASC are not designed for real ASC data, they cannot be directly applied for comparison. Thus, we choose several advanced forecasting algorithms and different data processing methods for comparison.

In detail, the compared forecasting algorithms include multilayer perceptron (MLP), support vector regression (SVR) and random forecast (RF). The compared data processing method is to naively sample ASC at several preset capacity data points. The names of each compared case are listed in Table 11.7, where CDP means *compared data processing* and PDP means *proposed data processing*. Different with PDP using PCA to extract features, CDP reduces the dimensionality of ASC data by

Table 11.7 Compared case information

Settings	MLP	SVM	RF	LSTM
CDP	C1	C2	C3	C4
PDP	C5	C6	C7	C8

Table 11.8 Performances of compared cases

Case	Method	PCV or data points		ASC
		Validation	Testing	Testing
C1	MLP+CDP	4.67%	4.95%	3.41%
C2	SVR+CDP	5.61%	6.22%	5.15%
C3	RF+CDP	4.08%	4.31%	3.21%
C4	LSTM+CDP	3.85%	3.72%	3.28%
C5	MLP+PDP	0.0512	0.0556	2.91%
C6	SVR+PDP	0.1378	0.1512	6.02%
C7	RF+PDP	0.0518	0.0551	3.32%
C8	**LSTM+PDP**	**0.0337**	**0.0448**	**2.74%**

directly sampling several data points with fixed resolution. C8 is the case with the proposed ASC forecasting method.

Then, C1 to C7 are trained with the same setting as the proposed method. The sampling points in CDP is selected as 12. The performances of each cases are shown in Table 11.8.

Comparing C1 to C4, we can find that the RF and LSTM algorithms have the best ASC forecasting performances, whose lowest MAPE is 3.21%. Comparing C5 to C8, our proposed method (C8) has the best performance on ASC forecasting, whose MAPE is 2.74%. Since the LSTM algorithm has great performances in both CDP and PDP circumstances, it can be regarded with great performance to deal with time-series ASC data.

Comparing C4 and C8, we can observe that the MAPE of C8 on ASC forecasting is 0.54% lower than the MAPE of C4, which demonstrates that the introduction of *data integration* and *feature extraction* procedures proposed in this chapter has positive effects on the LSTM-based ASC forecasting method.

11.6 Conclusion

This chapter proposes a data-driven framework to forecast ASC based on real market data. First, the ASC data are collected, formatted, merged, sampled, and transformed into a uniform format for analysis. Second, the prices in the original ASC data are assigned with different weights based on their appearance probability, and the dimension of the data is reduced based on a PCA method. Third, an LSTM-based learning model is proposed to identify the dynamic relations between the PCVs and potential influencing factors. After appropriate training, the LSTM-based model can forecast the future ASCs.

The real data from the MISO market from 2015 to 2019 have been used for numerical examples. The results demonstrate that the proposed method can effectively forecast real ASC. In detail, the proposed feature extraction method can effectively extract key information and reduce data dimensionality. The proposed LSTM-based forecasting framework is indicated with good performance on ASC forecasting. The various components in PCVs have different preferences on input data types and model structures. The "forecasting renewable capacity" data type and "forecasting market price" data type are proved to be positive for ASC forecasting.

Based on a comparison example, the proposed ASC forecasting method is compared with other 3 algorithms with two data pretreatment methods. The results demonstrate that the LSTM algorithm is good at forecasting ASC data, and the proposed data pretreatment method is helpful to increase the accuracy of ASC forecasting.

The forecasting ASC data can be further used as inputs to decision-making problems, such as strategic bidding and system planning, in power markets. Our future work is to study the probabilistic forecasting method for real ASC data.

References

1. Li, Gong, Jing Shi, and Qu Xiuli. 2011. Modeling methods for genco bidding strategy optimization in the liberalized electricity spot market'a state-of-the-art review. *Energy* 36 (8): 4686–4700.
2. Ye, Yujian, Dawei Qiu, Mingyang Sun, Dimitrios Papadaskalopoulos, and Goran Strbac. 2020. Deep reinforcement learning for strategic bidding in electricity markets. *IEEE Transactions on Smart Grid* 11 (2): 1343–1355.
3. Moiseeva, Ekaterina, and Mohammad Reza Hesamzadeh. 2018. Bayesian and robust nash equilibria in hydrodominated systems under uncertainty. *IEEE Transactions on Sustainable Energy* 9 (2): 818–830.
4. Baringo, Luis, and Antonio J. Conejo. 2011. Strategic offering for a wind power producer. *IEEE Transactions on Power Systems* 28 (4): 4645–4654.
5. AlAshery, Mohamed Kareem, Dongliang Xiao, and Wei Qiao. 2020. Second-order stochastic dominance constraints for risk management of a wind power producer's optimal bidding strategy. *IEEE Transactions on Sustainable Energy* 11 (3): 1404–1413.
6. He, Guannan, Qixin Chen, Chongqing Kang, Pierre Pinson, and Qing Xia. 2016. Optimal bidding strategy of battery storage in power markets considering performance-based regulation and battery cycle life. *IEEE Transactions on Smart Grid* 7 (5): 2359–2367.
7. Bolun, Xu, Yuanyuan Shi, Daniel S. Kirschen, and Baosen Zhang. 2018. Optimal battery participation in frequency regulation markets. *IEEE Transactions on Power Systems* 33 (6): 6715–6725.
8. Kardakos, Evaggelos G., Christos K. Simoglou, and Anastasios G. Bakirtzis. 2016. Optimal offering strategy of a virtual power plant: A stochastic bi-level approach. *IEEE Transactions on Smart Grid* 7 (2): 794–806.

9. Kazemi, Mostafa, Hamidreza Zareipour, Mehdi Ehsan, and William D. Rosehart. 2017. A robust linear approach for offering strategy of a hybrid electric energy company. *IEEE Transactions on Power Systems* 32 (3): 1949–1959.
10. Chen, Ruidi, Ioannis Ch Paschalidis, Michael C. Caramanis, and Panagiotis Andrianesis. 2019. Learning from past bids to participate strategically in day-ahead electricity markets. *IEEE Transactions on Smart Grid* 10 (5): 5794–5806.
11. Kaveh Dehghanpour, M., Hashem Nehrir, John W. Sheppard, and Nathan C. Kelly. 2016. Agent-based modeling in electrical energy markets using dynamic bayesian networks. *IEEE Transactions on Power Systems* 31 (6): 4744–4754.
12. Mitridati, Lesia, and Pierre Pinson. 2018. A bayesian inference approach to unveil supply curves in electricity markets. *IEEE Transactions on Power Systems* 33 (3): 2610–2620.
13. Kiannejad, Mohammad, Mohammad Reza Salehizadeh, Majid Oloomi-Buygi, and Miadreza Shafie-khah. 2020. Artificial neural network approach for revealing market competitors' behaviour. *IET Generation, Transmission and Distribution* 14 (7): 1292–1297.
14. Guo, Hongye, Qixin Chen, Qing Xia, Chongqing Kang, and Xian Zhang. 2018. A monthly electricity consumption forecasting method based on vector error correction model and self-adaptive screening method. *International Journal of Electrical Power and Energy Systems* 95: 427–439.
15. Zheng, Kedi, Yi. Wang, Kai Liu, and Qixin Chen. 2020. Locational marginal price forecasting: A componential and ensemble approach. *IEEE Transactions on Smart Grid* 11 (5): 4555–4564.
16. Álvarez, Veróonica, Santiago Mazuelas, and José A. Lozano. 2021. Probabilistic load forecasting based on adaptive online learning. *IEEE Transactions on Power Systems* 36 (4): 3668–3680.
17. Wang, Yi., Dahua Gan, Mingyang Sun, Ning Zhang, Lu. Zongxiang, and Chongqing Kang. 2019. Probabilistic individual load forecasting using pinball loss guided LSTM. *Applied Energy* 235: 10–20.
18. Kong, Weicong, Zhao Yang Dong, and David J. Hill, Fengji Luo, and Yan Xu. 2018. Short-term residential load forecasting based on resident behaviour learning. *IEEE Transactions on Power Systems* 33 (1): 1087–1088.
19. Wang, Huaizhi, Zhenxing Lei, Xian Zhang, Bin Zhou, and Jianchun Peng. 2019. A review of deep learning for renewable energy forecasting. *Energy Conversion and Management* 198: 111799.
20. Tan, Mao, Siping Yuan, Shuaihu Li, Su. Yongxin, Hui Li, and Feng He. 2019. Ultra-short-term industrial power demand forecasting using LSTM based hybrid ensemble learning. *IEEE Transactions on Power Systems* 35 (4): 2937–2948.
21. Chang, Zihan, Yang Zhang, and Wenbo Chen. 2019. Electricity price prediction based on hybrid model of adam optimized LSTM neural network and wavelet transform. *Energy* 187: 115804.
22. Yi Wang, Qixin Chen, Dahua Gan, Jingwei Yang, Daniel S Kirschen, and Chongqing Kang. 2019. Deep learning based socio demographic information identification from smart meter data. *IEEE Transactions on Smart Grid* 10 (3): 2593–2602.
23. Pelagatti, Matteo. 2013. Complex Models and Computational Methods in Statistics, 203–213. Milano: Springer Milan.
24. MISO. 2019. Market reports: Day-ahead pricing. https://www.misoenergy.org/markets-and-operations/real-time--market-data/market-reports/.
25. Wang, Yi., Qixin Chen, Chongqing Kang, and Qing Xia. 2016. Clustering of electricity consumption behavior dynamics toward big data applications. *IEEE Transactions on Smart Grid* 7 (5): 2437–2447.
26. Li, Pan, Baosen Zhang, Yang Weng, and Ram Rajagopal. 2017. A sparse linear model and significance test for individual consumption prediction. *IEEE Transactions on Power Systems* 32 (6): 4489–4500.
27. Burke, Daniel J., and Mark J. O'Malley. 2011. A study of principal component analysis applied to spatially distributed wind power. *IEEE Transactions on Power Systems* 26 (4): 2084–2092.
28. Xie, Le., Yang Chen, and P.R. Kumar. 2014. Dimensionality reduction of synchrophasor data for early event detection: Linearized analysis. *IEEE Transactions on Power Systems* 29 (6): 2784–2794.

29. Shi, Xingjian, Zhourong Chen, Hao Wang, Dit-Yan Yeung, Wai-kin Wong, and Wang-chun Woo. 2015. Convolutional LSTM network: A machine learning approach for precipitation nowcasting. *Advances in Neural Information Processing Systems* 28: 802–810.
30. Weicong Kong, Zhao Yang Dong, Youwei Jia, David J Hill, Yan Xu, and Yuan Zhang. 2019. Short-term residential load forecasting based on LSTM recurrent neural network. *IEEE Transactions on Smart Grid* 10 (1): 841–851.
31. MISO. 2019. Market reports: Daily forecast and actual load by local resource zone. https://www.misoenergy.org/markets-and-operations/real-time--market-data/market-reports/.
32. NOAA. Climate data online. https://www.ncdc.noaa.gov/cdo-web/.
33. MISO. Market reports-generation fuel mix. https://www.misoenergy.org/markets-and-operations/real-time--market-data/market-reports/#nt=%2FMarketReportType%3ASummary%2FMarketReportName%3AGeneration%20Fuel%20Mix%20(xls)&t=10&p=0&s=MarketReportPublished&sd=desc.
34. EIA. Data of petroleum & other liquids. https://www.eia.gov/petroleum/data.php#prices.
35. Pedregosa, Fabian, Gaël. Varoquaux, Alexandre Gramfort, Vincent Michel, Bertrand Thirion, Olivier Grisel, Mathieu Blondel, Peter Prettenhofer, Ron Weiss, Vincent Dubourg, et al. 2011. Scikit-learn: Machine learning in python. *Journal of Machine Learning Research* 12: 2825–2830.
36. Abadi, Martín, Paul Barham, Jianmin Chen, et al. 2016. In *Tensorflow: A system for large-scale machine learning*, vol. 16), 265–283. Savannah, GA: USENIX Association.

Chapter 12
Learning Individual Offering Strategy

12.1 Introduction

Since the early 2000s, the restructuring process of power markets worldwide has developed rapidly. Although market deregulation provides larger optimization space and more accurate price signals for power system operation, it also creates more opportunities for participants to offer or bid strategically and even manipulate prices [1].

To maintain the order of power markets, effective power market simulation methods are necessary for market regulators, such as independent system operators (ISOs), governments, and independent third party organizations [2]. A qualified market simulation method should be able to capture the participants' offering strategies, describe the game relationships in the market, quantify the influences of potential factors on market operation, and further accurately predict the potential market equilibrium state in the future. An accurate market simulation result can help to prevent market power abuse, provide supportive information for market mechanism design, etc.

To accomplish this target, many studies on market simulation have been proposed in recent years. In general, they can be approximately divided into two categories: market equilibrium analysis and agent-based market simulation. Though they have been widely applied in many scenarios in the last decades, they mostly have inevitable limitations in practice due to their model characteristics.

The studies in the first category mainly simulate market operation by constructing a traditional game-theoretical mathematical programming framework, such as Cournot equilibrium [3, 4], Stackelberg equilibrium [5, 6], and Bayesian equilibrium [7, 8]. The equilibrium models usually consist of two or more levels of problems that separately simulate strategic bidding behaviors of participants and market operations expected by the strategic participants [9, 10]. Thus, the equilibrium models cannot be directly solved. Based on the solving methods, the equilibrium programming models can be divided into two types: the first is equilibrium problem with equilibrium constraints (EPEC) [11–13] and the second is iterative diagonalization approach [14, 15].

Since the equilibrium models usually have explicit and rigorous mathematical structures and the market results are based on the concept of Nash equilibrium, such models are usually highly explanatory, which is an advantage in simulating market

Q. Chen et al., *Data Analytics in Power Markets*,
https://doi.org/10.1007/978-981-16-4975-2_12

operations. However, their limitations are also obvious. First, to ensure that simulation models can meet the requirements of specific game theories and have sufficient solvability, many simplifications and assumptions should be made during power system modeling or offering strategy developing. For example, the participants are usually assumed to have full knowledge of market operation information, including the strategies of its competitors, the operative parameters of power systems, and the detailed clearing process of power markets. The oversimplifications and unrealistic assumptions often result in a considerable gap between the simulation and the real situation, greatly weakening the adaptivity of the methods. Second, the market participants in these models are usually modeled as rational men [16], which means they make use of all information they can get and only pursue the highest profits without consideration on other aspects. However, this is also an over-idealized hypothesis which is revealed by several studies [16, 17].

The studies in the second category usually simulate the market operation by modeling agents who can make individual strategies, learn from past experience and adapt to environmental changes [18, 19]. Mathematical tools from reinforcement learning [20, 21], probabilistic graphical models [22], and statistical inference [23] have been applied widely for agent learning processes. There have been several well-known power market simulation platforms established based on agent-based simulation [24–26].

The agent-based market simulation methods can model individuals variously, and their iterative-based solution methods allow modeling to be more complex compared to the first category. However, the modeling of individual participants still requires detailed information of their bidding characteristics, such as operation costs, startup and shutdown costs, and risk preferences. In fact, the detailed information is usually kept as secrets of market participants and unable to be acquired by the researchers or operators who run market simulations. Thus, the bidding characteristics are usually defined by theoretical prior assumptions or market statistical data. The mismatch between the prior assumptions and the real market reduces authenticity and effectiveness. In addition, due to iterative solving methods [27], oscillation and divergence often occur during the calculation process in some studies, and global optimality cannot be guaranteed. Furthermore, to accelerate the learning process of agent models, offering strategies or market clearing procedures are also simplified compared to the real world in some studies, such as using linear functions as supply functions [22] or simplifying the market clearing procedure from an optimization problem to a linear equation set calculation [28].

In essence, a qualified power market simulation method should have the ability to accurately infer individual strategic offering behaviors while maintaining as many power system operation features as possible. To accomplish this target, two elements are indispensable. The first is an abundant real market dataset containing individual offering strategies and potential influencing factors. With the appropriate market datasets, we can identify individuals' real strategic characteristics and forecast their behaviors. The second is a bottom-up market simulation framework that can decouple the individual behavior simulation and market clearing process. Thus, power market modeling can be complex enough to be realistic.

It should be noted that the individual offering data were usually masked and unavailable previously. Therefore, the existing studies failing to consider real offering data is both an inherent defect and a compelling situation. However, in recent years, information disclosure in power markets has become more transparent. Increasing market data have been published on the ISO websites and are available to the public, which provides research institutes with opportunities to use real market data to simulate power markets. For example, the explicit offering data of each generator in the last 13 months in the Australian energy market (AEM) have been published and updated on its website, open to the public [29].

There have been some improvements to the data-driven individual behavior simulation. For example, Chen et al. used an inverse optimization method to infer the rivals' true cost functions from market data [28]. However, in this study, the market clearing procedures are oversimplified, and the market data used for examples are artificially generated. Based on the hidden Markov model, Mitridati and Pinson proposed a Bayesian inference approach in a day-ahead (DA) electricity market to reveal the aggregate supply curve, which can be seen as an integration of all individuals' behaviors [23]. However, the study also addresses artificial predefined data rather than real historical market data. Without the utilization of real market data, it is hard to capture the real information hidden in the data and to test whether the methods perform well in real applications.

In fact, it is quite difficult to directly use individual offering data for market simulation. The main obstructive factor is their unique data forms. In detail, the form of offering data is usually a step function with multiple quantity-price steps, which is totally different from the normally data-driven-analyzed market data, such as prices, loads, and power flows. Thus, to use data-driven methods to infer individual offering strategies, the original high-dimensional and unstructured data need to be transformed and mapped into a concise and structured data form. A mapping method is inspired by reference [16]. The chapter has proposed a pattern extraction method, which demonstrates that the individual offering behaviors have discretization, diversity, and patterning features, and that they can be represented by some typical offering patterns. Thus, the individual typical offering patterns can be seen as a concise version of individual offering behaviors and used for individual offering data-driven simulation.

In total, to make full use of the information hidden inside real market data and maintain real power system operation features while simulating power markets, this chapter proposes a data-driven power market simulation framework based on simulating individual strategic offering behaviors in a bottom-up framework. Four major procedures are included in this framework. First, the individual strategic offers are clustered into different patterns and indexed utilizing the pattern extraction method proposed in [16] based on real market data. Second, the relationships between the potential influencing factors and the individual offering patterns are captured with the utilization of deep neural networks (DNNs). Third, the trained DNN-based model is further used to simulate the detailed individual offering strategies based on future influencing factors. Finally, with the forecasted individual offers as bidding inputs, a market clearing model is proposed to simulate future market operations. It should be noted that since the individual behavior forecasting and power market simulation are decoupled, the market clearing process can be modeled as complexly as the real world incorporating exact market rules and power system operation features.

The main contributions can be summarized as follows:

1. A bottom-up data-driven power market simulation method is proposed based on forecasting individual offering strategies and realistic power market clearing models.
2. A DNN-based model is proposed to forecast the individual offering strategies based on learning the relationships between the individual offering patterns and the potential influencing factors from the historical market data.
3. A realistic power market clearing model is proposed to simulate market operation. The market fine-grained operation information, including individual cleared capacities, market prices and power flows, can all be provided during simulation.
4. An empirical example is proposed based on real market data from the Australian electricity market from March to May 2018. The detailed simulation results, including individual offering strategies, market clearing prices, dispatch schedules, and power flows, are presented.

The proposed data-driven simulation framework offers a novel method for market operators and researchers who have access to detailed market data, to simulate individual offering behaviors and whole power market operations based on real market data. The simulation results can be further used for power market operation monitoring, operation accident early warning, third-party testing, market power abuse detection, tacit collusion analysis, etc.

The remainder of the chapter is organized as follows. The market assumptions and the method to cluster and index offering data are given in Sect. 12.2. In Sect. 12.3, a DNN-based learning model for simulating individual offering strategies is developed. In Sect. 12.4, the market clearing model is presented. In Sect. 12.5, a case study based on the data from the Australian electricity market is provided. The forecasting results of individual offering strategies and power system operation are both presented. Finally, conclusions are given in Sect. 12.6.

12.2 Data-Driven Market Simulation Framework

12.2.1 Market Assumptions

The DA market, which is a fundamental component in power market operation, is simulated in this chapter. A market operator, denoted as ISO, is assumed to run the market. There are I_0 suppliers in the market, which specifically, are generators. Each generator submits stepwise offering curves with multiple blocks in a monotonically increasing form for the next day, indicating its intention on price and quantity. The offer curve quantity, T_0, of each generator in a day depends on the granularity of the market clearing time intervals, which can be 30 min or 1 h. On the demand side, the consumers are assumed to participate in the markets represented by load service entities (LSEs). LSEs are permitted to submit their monotonically decreasing stepwise bidding curves with multiple blocks.

The operation procedure of the DA market is listed as follows. First, the ISO releases the forecasting information of the next day market, such as demand and the outage information. Second, the suppliers and LSEs make their bidding decisions based on public information and then submit their bids to the ISO. Third, the ISO clears the market based on their bids considering the power system constraints, such as power output constraints, transmission constraints, power balance constraints, and ramping constraints. Finally, the ISO releases the market clearing results to individuals.

12.2.2 Offering Data Clustering and Indexing

In most power markets, generators submit offers in the format of a step function with multiple price-quantity pairs. Due to the offering data particularities, it is quite difficult to directly compare and analyze individual offering behaviors based on raw market data. To describe and demonstrate offering behaviors, the raw offering data need to be transformed into stepwise curves. However, the curves with various resolutions are still hard to compare. Thus, the curves need further to be sampled with the same resolution (i.e. a capacity interval) into a uniform format, which dramatically increases the dimension of offering data. The above procedure can be seen in Fig. 12.1. In detail, i represents the index of generators, t represents time tag of an offer, $P_{i,t,b}^{G,SW}$ and $Q_{i,t,b}^{G,SW}$ are the price and capacity in block b of generator i's stepwise offer at time t, s represents the index of sampling points on the offering curves, S represents the total number of sampling points, and $P_{i,t,s}^{G,UR}$ is the price value of s sampling point in generator i's offer with unified resolution at time t.

The high-dimensional format increases the difficulty in identifying the relationships between the potential influencing factors and offering behaviors and further deteriorates the forecasting performance of individual offering strategies. Thus, to accomplish individual offering strategy forecasting, an efficient method to deal with raw offering data is essential. An intuitive idea is to map high-dimensional data into a concise data form with low-dimensions. According to the research results in [16], the offering behaviors of generators usually have discrete aggregation features, which means that they are discretely distributed in the offering space and can be assembled into several groups.

For example, the offering behaviors of a generator in a year in AEM, whose one-day data are shown in Fig. 12.2a as an example, can be extracted as 4 typical offering patterns, shown in Fig. 12.2b. It is demonstrated and proved that the extracted typical offering patterns can represent most offering behaviors of the generator.

Then, each offer can be tagged with different typical offering pattern indexes (i.e. 1 to 4 in Fig. 12.2b), according to their distances to each typical offering pat-

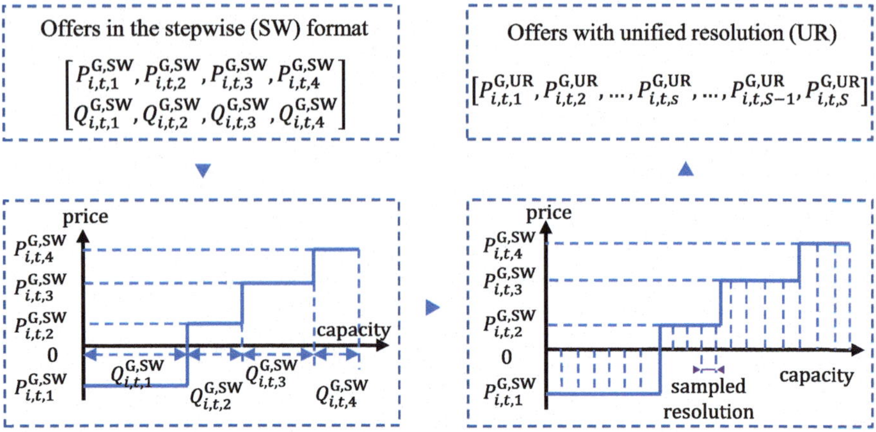

Fig. 12.1 Transformation procedure of offers as steps functions

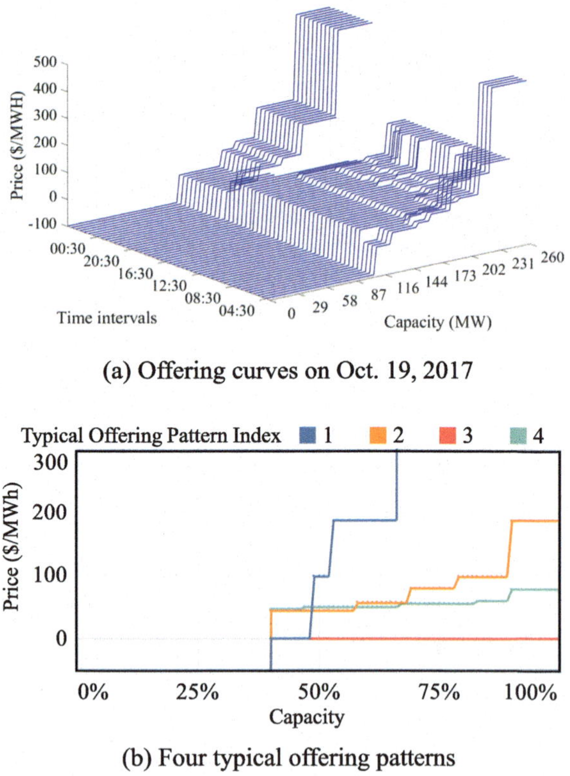

Fig. 12.2 Offering curves and typical offering patterns of generator GSTONE4 in AEM

tern. According to the discretization features of generators' offering behaviors, it is feasible to use their typical offering pattern set to represent their individual offering behaviors. Then, the offering pattern index can be transformed into the one-hot format, which can be directly used in classification models. The one-hot index of generator i's at time t is denoted as $O_{i,t}^{idx}$. It should be noted that the quantity of the typical offering pattern, C_O, can be defined during the extraction procedure. A larger quantity could provide higher accuracy. Since the detailed procedure of typical offering pattern extraction has been described and tested in detail in the study [16], it will not be expressed in this chapter.

After offering pattern extraction and indexing, each high-dimensional offering data can be replaced by a typical offering pattern index, which can be processed by many classification models in machine learning areas. After this step, the information density improves, and the connections between the potential influencing factors and offering behaviors are easier to identify.

12.3 Individual Offering Strategy Learning

Considering the fact that individuals usually consider many factors when they make offering decisions, it is an effective way to find the connection relationships between individual offering behaviors and the factors they take into account, which are later called "influencing factors". To do so, a DNN-based learning model is proposed. The DNN-based model includes several multilayer feedforward neural network (MFNN) models [30], which are used to learn and intimate the strategic offering decision-making process of one generator.

The learning process includes two separate procedures.

First comes the training procedure. The proposed MFNN model is trained for one generator at once. The model is trained based on the real historical market data, to acquire the connection relationships between the inputs and outputs. The inputs and outputs are the influencing factors and the offering behaviors of a generator at one time. In detail, the inputs are the market data that generators have access to, such as date, forecasting demand and historical price, etc. The outputs are the generator's offering indexes, defined in the last procedure, in the one-hot format to represent their belongings.

Then comes the forecasting procedure. The proposed MFNN model is used to intimate the generator's strategic offering decision-making process. When given the future market situation, which corresponds to a set of potential influencing factors, the MFNN model can forecast the most likely typical offering pattern chosen by the generator, whose data is used for this MFNN model training.

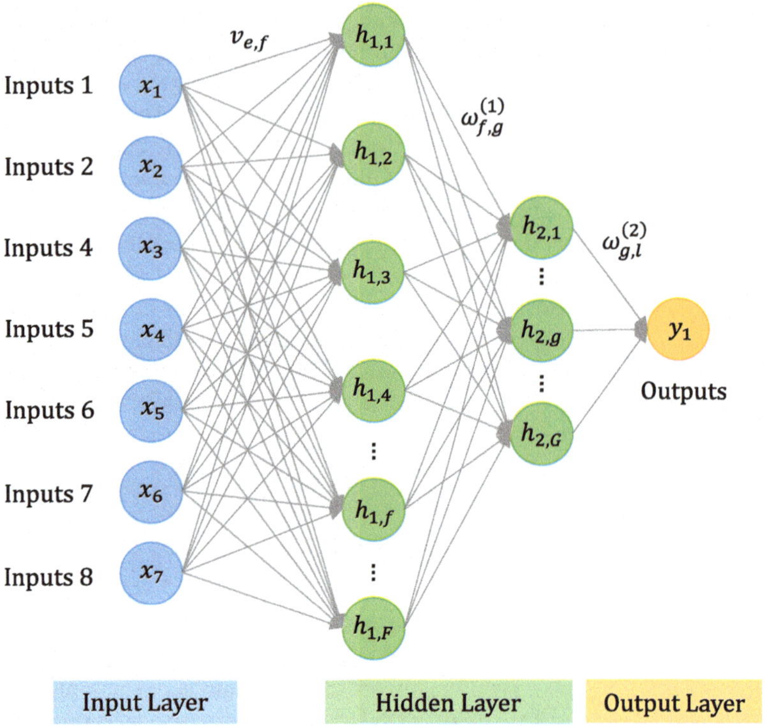

Fig. 12.3 Structure of the MFNN model

12.3.1 MFNN Model Structure

Considering the good performances and simple principle of MFNN in the forecasting fields of electricity load [30] or price [31], we use the MFNN model to learn the strategic decision-making process of each generator. It should be noted that MFNN is not the only way to learn the individual strategic offering behaviors, many other machine learning models made for classification problems can also be applied in this framework.

In detail, each MFNN model contains several sequential layers, including one input layer, multiple hidden layers, and one output layer [20]. In the section, we use an MFNN with two hidden layers as an example, shown in Fig. 12.3.

The model inputs include the historical data that can be accessed by generators and have potential influences on their offering behaviors, at time t. The model output is the index of generator i's offer at time t, $O^{idx}_{i,t}$, in the one-hot format.

The inputs of unit h in layer n are defined as $x^{(n)}_h$ and calculated as:

$$x^{(n)}_h = \sum_g \mu^{(n-1)}_{g,h} y^{(n-1)}_g \tag{12.1}$$

where $y_g^{(n-1)}$ is the output of unit g in layer $n-1$, and $\mu_{(g,h)}^{(n-1)}$ is the weight of the connection between unit g in layer $n-1$and unit h in layer n. The sigmoid function is adopted as the activation function of each unit in the hidden layers and output layer. Thus, the output of unit h in layer n can be calculated as:

$$y_h^{(n)} = \frac{1}{1 + e^{-x_h^{(n)}}} \tag{12.2}$$

12.3.2 MFNN Model Inputs

An adequate selection of inputs is the foundation to effectively capturing the relationships between the potential influencing factors and the individual offering strategies.

The input factors can be divided into three main categories: (1) *calendar information*, which contains season, month, time in a day, day type in a week, holiday, weather, and temperature at specific time periods; (2) *external factors*, such as the forecasted demand published by the ISO at specific time periods. The electricity prices, fuel prices, and temperatures can also be categorized into this type. The external factors represent the environmental information, usually working as the main driving factor for a generator to make offering strategies, and have been used in many power market studies, such as forecasting on consumer behaviors [32], electricity prices [33] or loads [34]; (3) *relevant offers*: since obvious relevance can be observed in offers in adjacent time periods, the historical offering patterns have important reference values for future generator decision-making.

The detailed input types and their formats for the MFNN model are listed in Table 12.1. Before training the MFNN model, all inputs should be concatenated together as a vector and used as the inputs of the MFNN model.

Table 12.1 List of input types of the MFNN model

No. category	Type	Format	Dimension	
1	Calendar	Season	One-hot	4
2	Calendar	Month	One-hot	12
3	Calendar	Time	One-hot	24 or 48
4	Calendar	Day in a week	One-hot	7
5	Calendar	Holiday	Boolean	1
6	External	Demand	Float	1
7	External	Electricity price	Float	1
8	External	Fuel price	Float	1
9	External	Temperature	Float	1
10	Relevant offer	Historical offer index	One-hot	C_0

12.3.3 MFNN Model Training

To train the model, the available historical data can be divided into three datasets: training, validation, and testing. The first two datasets are used for model iterative training, and the third dataset is used for model performance validation.

In fact, the MFNN model is used to capture the dynamic decision-making process of generators when they face various market situations. Thus, the historical data used for training should be closer to the offering strategies to be forecasted. In the common sense of electricity price or load forecasting, the historical data will present a significant decay of reference value over time [35], which makes researchers pay more attention to the recent data. Thus, when choosing the data used for training MFNN models, the data with appropriate length will be satisfied, which is able to capture the dynamic features of strategic offering and also not bring too much computation burden for the total framework.

Since the DA power market is run on a daily frequency, the power generators' offering strategies from different time periods in a day should be simulated together. Thus, the allocation of the three datasets from the original dataset should be based on a daily division.

Since forecasting the index of offering behavior is a classification problem, we use the cross-entropy loss function as the optimization objective [36].

$$L = -\sum_{s} \gamma_s \log\left(p_s\right) \tag{12.3}$$

where γ_s is the indicator variable for class s. If the output has the same class as the sample, γ_s equals 1 and vice versa. p_s is the probability of class s calculated by the MFNN model.

The scaled conjugate gradient backpropagation algorithm is used as the training algorithm [37].

There are two kinds of performance evaluation methods of the learning model MFNN-i on day u. The first is the classification accuracy, detailed as:

$$\text{acc}(i, u) = \frac{1}{T} \sum_{t \in \Gamma_u} if\left(O_{i,t}^{idt} = \dot{O}_{i,t}^{idx}\right) \tag{12.4}$$

where Γ_u represents the time set on day u and $\dot{O}_{i,t}^{idx}$ is the simulated class of typical offering patterns of generator i at time t.

The second is the mean absolute percentage error (MAPE) of the forecasted offer curve and the real offer curve, detailed as:

$$\text{MAPE}_{\text{offer}}\left(i, u\right) = \frac{1}{T} \sum_{t \in \Gamma_u} \frac{1}{S} \sum_{s} \frac{\dot{P}_{i,t,s}^{\text{G,UR}} - P_{i,t,s}^{\text{G,UR}}}{P_{i,t,s}^{\text{G,UR}}} \tag{12.5}$$

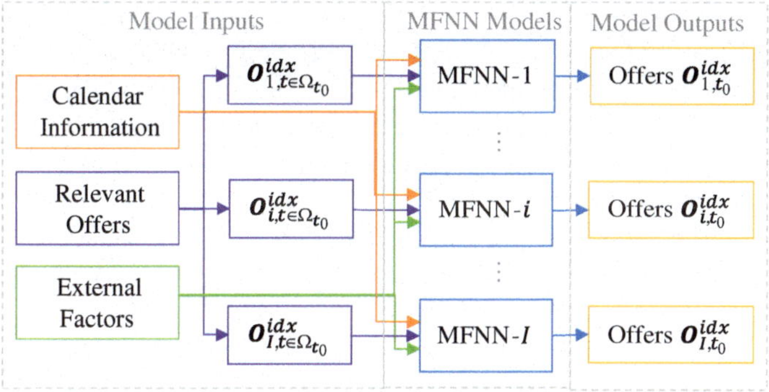

Fig. 12.4 DNN-based learning model for multiple generators offering strategies

where $P^{\mathrm{G,UR}}_{i,t,s}$ is the price at sampling point s on the real offering curve at time period t, and $\dot{P}^{\mathrm{G,UR}}_{i,t,s}$ is the forecasting result of $P^{\mathrm{G,UR}}_{i,t,s}$.

12.3.4 DNN-Based Model Structure

Considering that the strategic offering decision-making process of generators might be entirely different from each other, each generator is assigned with an independent MFNN model to learn their individual offering styles.

Thus, the total structure of the DNN-based learning model for all generators is shown in Fig. 12.4. MFNN-i represents the MFNN model for generator i. It can be observed that all generators' MFNN models share the same public information set, which consists of calendar information and external factors, as inputs. Since the relevant offer information is seen as private information in most power markets, each generator is assumed to only know their own historical offering data. In detail, Ω_{t_0} is used to represent the historical $O^{idx}_{i,t}$ datasets used for MFNN training for the offer at time period t_0.

With the trained MFNN models, the DNN-based individual offering strategy learning model is established. Given a future time node with external factors provided by ISOs, the proposed model can forecast the offering behaviors of all generators.

12.4 Market Clearing Simulation

After collecting the offers and bids, the ISO will clear the DA market with the objective function to maximize social welfare. It should be noted that the forecasted typical offering pattern indexes should be transformed into the original format of the multiple price-quantity pairs before market clearing simulation.

In fact, there are usually two main procedures in DA market clearing. First, a unit commitment (UC) should be conducted, providing a UC schedule for the operation day. Second, an economic dispatch (ED) should be conducted based on the UC schedule, providing a detailed power output schedule for generators as well as locational marginal prices (LMPs). Considering the two procedures are in the similar optimization form and the results will not change the core findings of the chapter, only the ED procedure is simulated here for simplicity and readability. In fact, the proposed framework is compatible to integrate with another UC procedure.

The detailed market clearing problem based on ED is stated as:

$$\min_{q^{\mathrm{O}}_{t,t,b}, q^{\mathrm{D}}_{d,t,k}} \sum_{t} \left[\sum_{i,b} \dot{P}^{\mathrm{G,SW}}_{i,t,b} q^{\mathrm{G}}_{i,t,b} - \sum_{d,k} P^{\mathrm{D,SW}}_{d,t,k} q^{\mathrm{D}}_{d,t,k} \right] \Delta t \tag{12.6}$$

subject to:

$$\sum_{i\in\Psi^{\mathrm{S}}_{n},b} q^{\mathrm{G}}_{i,t,b} - \sum_{d\in\Psi^{\mathrm{D}}_{n},k} q^{\mathrm{D}}_{d,t,k} - \sum_{m\in\Psi^{\mathrm{N}}_{n}} B_{n,m}\left(\delta_{n,t} - \delta_{m,t}\right) = 0 : \lambda_{n,t}, \forall n, \forall t \tag{12.7}$$

$$0 \le q^{\mathrm{G}}_{i,t,b} \le \dot{Q}^{\mathrm{G,SW}}_{i,t,b}, \forall i, \forall b, \forall t \tag{12.8}$$

$$Q^{\mathrm{G,dn}}_{i} \le \sum_{b} q^{\mathrm{G}}_{i,t,b} - \sum_{b} q^{\mathrm{G}}_{i,t-1,b} \le Q^{\mathrm{G,up}}_{i}, \forall i, \forall b, \forall t > 1 \tag{12.9}$$

$$0 \le q^{\mathrm{D}}_{d,t,k} \le Q^{\mathrm{D,SW}}_{d,t,k}, \forall d, \forall k, \forall t \tag{12.10}$$

$$B_{n,m}\left(\delta_{n,t} - \delta_{m,t}\right) \le Q^{\mathrm{Lmax}}_{n,m}, \forall n, \forall m \in \Psi^{\mathrm{N}}_{n}, \forall t \tag{12.11}$$

$$-\pi \le \delta_{n,t} \le \pi, \forall n, \forall t \tag{12.12}$$

$$\delta_{1,t} = 0 \tag{12.13}$$

where i and d are the generator and consumer indexes, b and k are the indexes of offering or bidding blocks of generators and consumers, t is the time index, n and m are the node indexes.

In detail, $\dot{P}^{\mathrm{G,SW}}_{i,t,b}$ is the forecasted price of generator i's offering block b at time t; $q^{\mathrm{G}}_{i,t,b}$ is the power produced through offering block b of generator i's offers at time t; $P^{\mathrm{D,SW}}_{d,t,k}$ is the consumption utility of block k of consumer d at time t; $q^{\mathrm{D}}_{d,t,k}$ is the power consumed by consumer d through block k at time t; Δt is the time interval; Ψ^{S}_{n}, Ψ^{D}_{n} and Ψ^{N}_{n} are the sets of generators and consumers nodes at node n, and nodes connected to node n; $B_{m,n}$ is the admittance of the line between node m and node n; $\delta_{n,t}$ is the voltage angle of bus n at time t; $\dot{Q}^{\mathrm{G,SW}}_{i,t,b}$ is the forecasted available capacity upper caps of offering block b in generator i's offer at time t; $Q^{\mathrm{G,up}}_{i}$ and $Q^{\mathrm{G,dn}}_{i}$ are the ramping upper and lower constraints of generator i; $Q^{\mathrm{D,SW}}_{d,t,k}$ is the demand quantity upper caps of bidding block k in consumer d's bid at time t; and $Q_{n,m}^{\mathrm{Lmax}}$ is the transmission power capacity of the line between nodes n and m.

The objective function (12.6) represents the maximization of social welfare. The constraint (12.7) represents the energy balance at node n at time t. $\lambda_{n,t}$ is the dual variable of equation (12.7), working as the LMP at node n at time t. The constraints (12.8)–(12.12) represent the power output constraints, generation ramping constraints, demand consumption constraints, transmission line capacity constraints and power angle constraints. The constraint (12.13) denotes that the reference node of the power angle is node 1 and set as 0.

Based on the market clearing model, we can simulate the market operation with the forecasted offers. Then, the fine-grained market clearing results, such as market clearing prices, power dispatch schedules, and power flows, can be obtained.

The forecasting performance of this model on market results is evaluated by MAPE between the forecasted and actual values, taking LMP at node n on day u as an example:

$$\mathrm{MAPE}_{\mathrm{market}}(u, n) = \frac{1}{T}\sum_{t\in\Gamma_u}\frac{\left|\lambda_{n,t} - \dot{\lambda}_{n,t}\right|}{\lambda_{n,t}} \tag{12.14}$$

where $\dot{\lambda}_{n,t}$ is the forecasted LMP at node n at time period t on day u.

12.5 Case Study

12.5.1 Basic Data

The public data in AEM from March to May 2018 are used for the empirical example. According to the generation fuel mix of AEM [38], the proportions of coal, gas, wind, water, and others are 43.4%, 22.8%, 11.6%, 15.1% and 7.1%, respectively. For better presentation, a reduced example of a similar generation mix with the original market is proposed. In detail, 57 generators with different fuel types are selected. The detailed information of the selected 57 generators is shown in Table 12.2.

Table 12.2 Generator information

Fuel type	Generator quantity	Capacity (MW)	Proportion (%)
Coal	7	3990	45.3
Gas	21	2019	22.9
Wind	7	940	10.7
Water	13	1230	14.0
Others	8	629	7.1

The main purpose of the empirical example is to test the performance of the proposed framework with real market data. First, the tests on the proposed individual offering behavior forecasting procedure are conducted in Sect. 12.5.2, to demonstrate its accuracies. Second, the market simulation procedure is also tested in Sect. 12.5.3 to identify its effectiveness. Then, a comparison with the current price forecasting method is proposed. At last, the calculation efficiency of the proposed method is stated and the adaptivity is clarified.

(1) Individual Offering Behavior Forecasting Settings
The parameters in the individual offering behavior forecasting procedure are set as follows. The quantity constraint of typical offering patterns is set from 2 to 20, which means each generator's offering behaviors will be clustered into 2 to 20 categories according to their sparse degree adaptively. According to the data availability, the input data types for the learning model contain calendar information, external factor data, and historical individual offering data. The number of time periods in day T_0 is 48. The quantity of hidden layers is set to 2. The unit quantities in the first hidden layer are calculated based on experience as the square of the product of the input unit number and the output unit number plus a constant, which is 5 in the study. The second hidden layer is set based on experience to 15 units. The proportions of training, testing, and validation datasets are 80%, 10%, and 10%, respectively.

(2) Market Simulation Settings
With the individual offering behavior forecasting results, the market simulation procedure can be conducted. Since the detailed topology of the Australian electricity market is not open to the public, the IEEE 118-bus system is used as an illustrative example to run the market simulation.

In detail, the 57 generators are distributed to the 54 generation nodes in the original 118-bus system topology, 51 with one plant and 3 with 2 plants. Since the supply is reduced from the real power market, the practical system demand data are also adjusted according to the reduction ratio and distributed to the IEEE 118-bus system's 91 demand nodes.

Admittedly, the market simulation results using the IEEE 118-bus system will deviate somewhat from the real operation of the market. However, the deviation degrees can be accepted. In fact, since the individual offering behavior forecasting procedure and market simulation procedure are decoupled, the applied testing system here mainly works as a clearing and demonstration platform. Though there might be deviations between the proposed results and the results using the real system topology in the detailed magnitudes and times of price, power outputs and power flows, the main conclusions to demonstrate the effectiveness of market simulation will not be harmed.

(3) Data Complexity
The model is trained based on the data from March 1, 2018, to May 20, 2018. The model is tested based on the data from the last week of the datasets, from May 21 to 27, 2018. The original dataset for the empirical example includes 21,120 calendar data, 232,320 external factor data, including demand, temperature, and historical price, and 240,768 individual offering data.

Table 12.3 Generator information

Accuracy	100%	(80%, 100%)	(60%, 80%]	(50%, 60%]
Number of generators	13	26	15	3
Proportion	22.8%	45.6%	26.3%	5.3%

12.5.2 Individual Offering Behavior Forecasting

Using the proposed method, the offering strategies of generators from May 21 to 27, 2018 are forecasted. The performances are demonstrated by two indexes as stated in equations (12.4) and (12.5).

(1) Performance based on classification accuracy
The forecasting accuracy was calculated based on Eq. (12.4). The summary is presented in Table 12.3.

Nearly 70% of generators are forecasted with accuracies higher than 80%, and only 5.3% of generators are forecasted with accuracies lower than 60%. Thirteen generators, which can be regarded as suppliers with obvious and stable patterns in their offering strategies, are forecasted with 100% accuracy.

In total, the average forecasting accuracy of all generators based on the capacity weights is 91.1%.

To present a more specific view of generator offering strategy forecasting, a detailed result of forecasting accuracy is presented in Fig. 12.5, where each star represents a generator; various colors represent different fuel types. It can be observed that the accuracies of individual forecasting offering behaviors are different. There are several factors influencing forecasting accuracies.

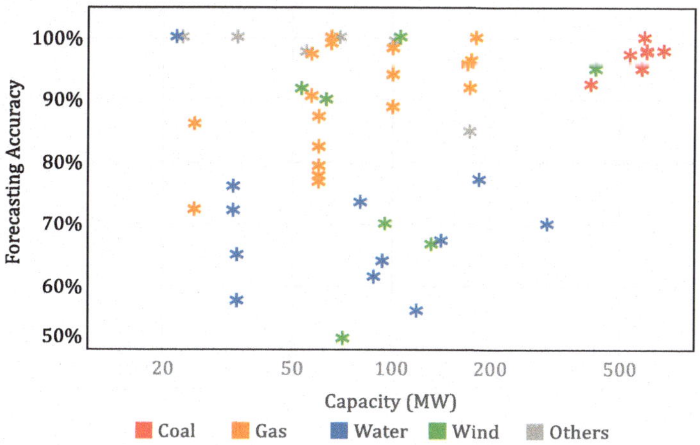

Fig. 12.5 Forecasting accuracy of all simulated generators

The first factor includes the different offering styles of various fuel-types. According to the results in study [16], generators with various fuel types have different offering styles. In detail, the coal-fired and wind-power generators are more likely to submit steady and patterned offers, which accounts for the high accuracy in forecasting coal-fired generators' offering behaviors. In addition, gas-fired generators and hydro-generators are more active in strategic offering, reflected in more offering patterns and more various time distribution. Thus, the gas-fired generators can be observed with more diversity in forecasting accuracies than coal-fired generators.

The second factor influencing the forecasting accuracy is that there are still several external factors not considered in the study, such as wind speed and reservoir water. Without the information of wind speed, even though the offering behaviors of wind-power generators are steady and patterned, the forecasting accuracies are largely impacted. Without the detailed information of each hydro-power generator's reservoir water, their offering behaviors become harder to be forecasted accurately.

(2) Performance based on MAPE
Then, the forecasted individual offering strategy indexes are transformed back to offering curve forms. To better present the forecasting performances, the detailed forecasting results of generator TARONG#4 on May 23, 2018 are shown in Fig. 12.6.

The forecasting offering curves of TARONG#4 are depicted in Fig. 12.6a, and the deviations between the forecasting and the real offering curves are depicted in Fig. 12.6b. It can be observed that forecasting errors usually occur at the rising position of offering curves. Based on Eq. (12.5), the MAPE of TARONG#4's forecasting offering curve on May 23, 2018 is 2.3%. Due to the chapter limit, other generators' forecasting results are not presented.

The daily average forecasting errors calculated by MAPE of all generators based on the capacity weights are presented in Table 12.4. It can be observed that the average MAPEs range from 2 to 7%, demonstrating the high forecasting accuracy of the proposed method. The weekly MAPE is calculated as 4.2%.

12.5.3 Market Simulation

The simulation results of market clearing prices, dispatch schedules, and power flows are presented to demonstrate the effectiveness of the proposed method.

(1) Market prices
For better presentation, the forecasting performances on market prices are tested based on system marginal price (SMP) and LMP.

First, the forecasted SMP results on two days (May 22 and 25) are shown in Fig. 12.7. There are two types of price curves on each day: (1) the forecasted price, green line, generated by the proposed method; (2) the real price, blue line.

On May 22 and 25, the forecasted price curves both have similar shapes as the real price curves. Specifically, the forecasted price curves are closer to the real price

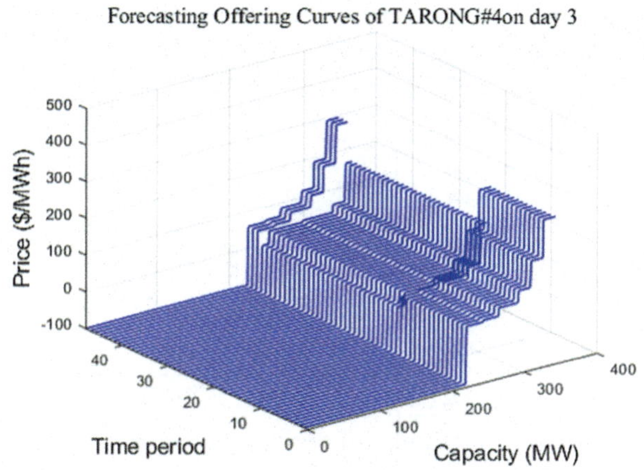

(a) Forecasting offering curves

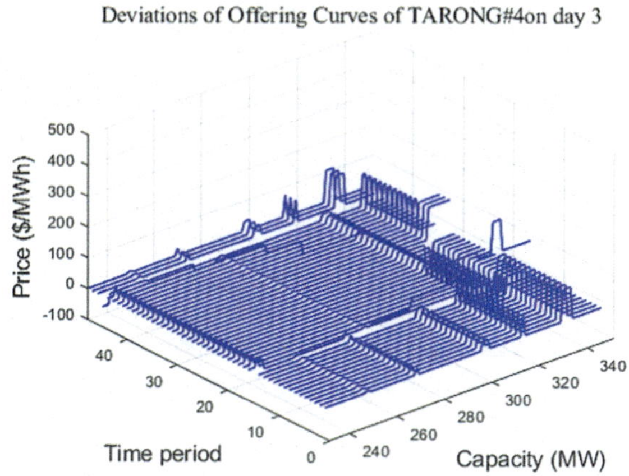

(b) Deviations between forecasting and real curves

Fig. 12.6 Forecasting results of TARONG#4 on May 23, 2018

Table 12.4 Forecasting errors of individual offering strategies

Date	5/21	5/22	5/23	5/24	5/25	5/26	5/27
MAPE (%)	5.3	4.6	4.3	6.2	3.7	3.0	2.3

curves from time period 1 to 22 (i.e., 0:30 to 11:00). The largest deviation period occurs during time period 32 to 38 (i.e., 16:00 to 19:00).

Fig. 12.7 Market price forecasting results on **a** May 22, **b** May 25

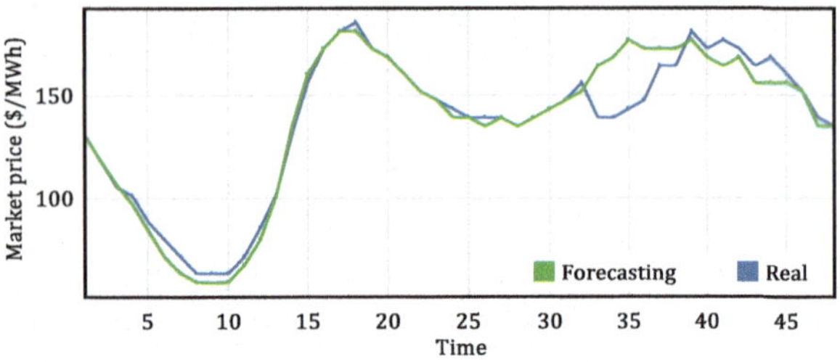

(a) May 22, Tuesday

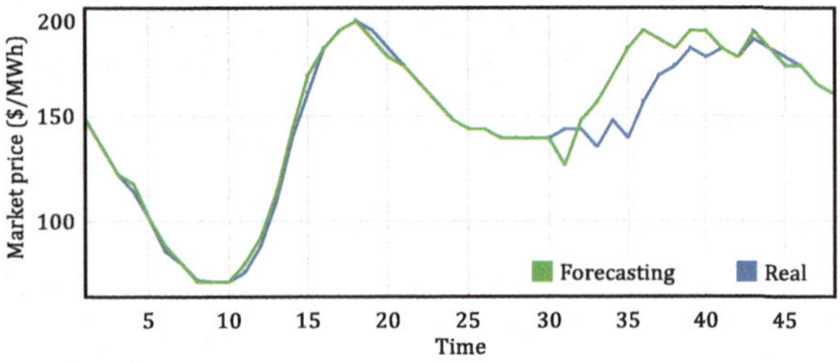

(b) May 25, Friday

To present the effectiveness of the proposed method, the MAPEs of SMP-based market prices in the simulated week are listed in Table 12.5.

It can be observed that the daily forecasting errors vary in a week, mainly from 1 to 6%, which verifies that the performance of the proposed method is stable. In total, the weekly MAPE of forecasting market prices is 3.3%.

In addition, the forecasted LMP results on three nodes (#5, #55, and #105) are also presented in Table 12.6.

It can be observed that the MAPEs of LMPs are different according to their locations. The MAPEs of LMPs on these three nodes mainly range from 0.5 to 6.8%,

Table 12.5 Forecasting errors of market prices

Date	5/21	5/22	5/23	5/24	5/25	5/26	5/27
MAPE (%)	2.8	4.0	5.5	4.1	3.0	2.7	0.9

Table 12.6 Forecasting errors of LMPs on three nodes

Node	5/21	5/22	5/23	5/24	5/25	5/26	5/27
#5	4.3	5.1	4.7	3.2	2.4	4.5	2.1
#55	4.2	6.1	6.8	6.1	3.7	2.5	0.5
#105	3.8	5.4	6.3	5.3	3.4	2.7	0.7

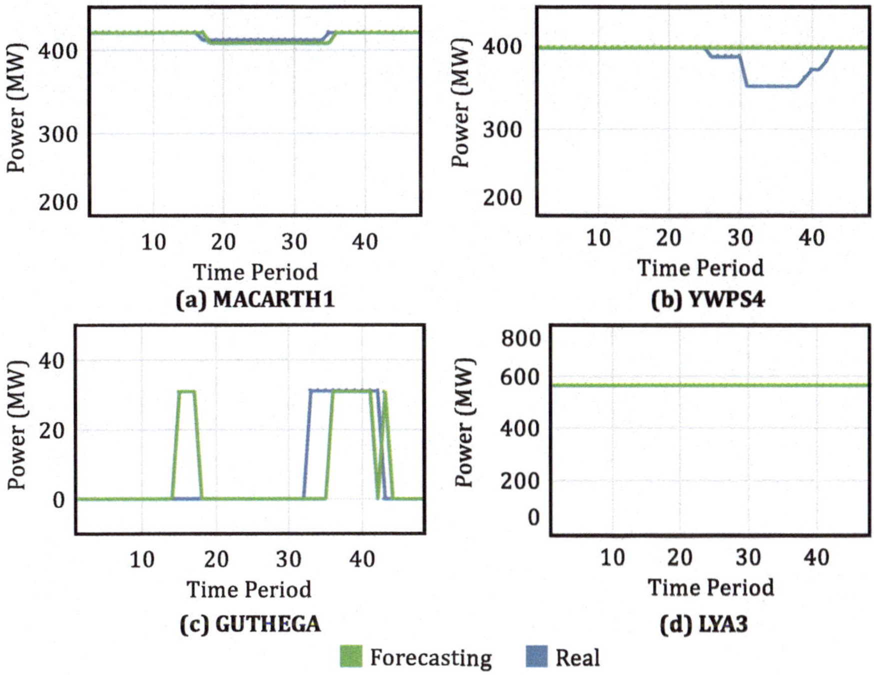

Fig. 12.8 Power outputs of four generators on May 22

which is wider than the SMP forecasting errors but still represents a high forecasting accuracy (Table 12.6).

(2) Dispatch schedules

To present the dispatch schedule simulation performance, the power outputs of several generators on May 22 are shown in Fig. 12.8.

For MACARTH1, deviations between the real and simulation results occur during the valley time between the two peaks in one day. The deviations of YWPS4 occurred during the second peak of that day. The deviations of GUTHEGA occur in both peak time periods. There is no deviation in the simulation of LYA3.

The total dispatch schedule forecasting performances are evaluated by power-weighted summing all generators' daily MAPEs and are shown in Table 12.7. The forecasting MAPEs range from 0.5 to 1.6%. The weekly MAPE can be calculated as 1.2%.

Table 12.7 Forecasting errors of dispatch schedules

Date	5/21	5/22	5/23	5/24	5/25	5/26	5/27
MAPE (%)	1.2	1.4	1.4	1.6	1.0	1.1	0.5

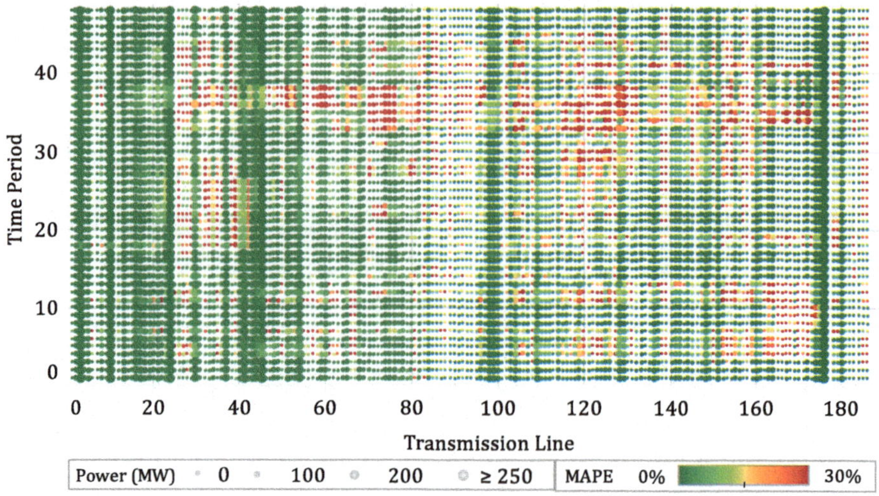

Fig. 12.9 Simulation errors of power flows in each transmission line on May 22

Table 12.8 Forecasting errors of power flows

Date	5/21	5/22	5/23	5/24	5/25	5/26	5/27
MAPE (%)	5.3	5.6	8.1	8.2	4.9	4.1	1.5

(3) Power flows
The power flow information can also be obtained from the market simulation. The detailed MAPE-based simulation errors of power flows of each transmission line on May 22 are presented in Fig. 12.9.

In the IEEE 118-bus system, there are 186 transmission lines, presented as the horizontal axis in the figure. The vertical axis is the time period in a day. The power flow on a transmission line and its simulated accuracy are represented by a specific point. The colors represent the simulation errors from 0 to 30%. The sizes represent the power transmitted on the line.

It can be observed that the power flows in most transmission lines are forecasted with MAPEs lower than 15%. Higher simulation errors mainly occur in the lines with lower flow. In total, the average power-weighted simulation error of all transmission lines on May 22 is 5.6%. The simulation performance of the whole test week is shown in Table 12.8. The forecasting MAPEs range from 1.5 to 8.2%. The weekly MAPE is 5.4%, demonstrating the high forecasting accuracy.

From the results of individual offering behavior forecasting and the market simulation, we can find that the errors of the two procedures have positive correlation to some extent. In detail, if the error of individual offering behavior forecasting is larger, the errors of the corresponding market simulation are likely to be larger. However, since the market simulation is a complex optimization problem, the extent of influences cannot be concluded as a universal quantified correlation.

12.5.4 Comparison with Current Price Forecasting Methods

To test the effectiveness of the proposed method, several mainly used forecasting methods are tested and compared with the proposed method. According to the current literature, since forecasting on power flows and generation schedules are too fine-grained for most of the works, this chapter compares price forecasting to verify the effectiveness of the proposed framework. In detail, the compared methods include naïve forecasting, multilayer perceptron (MLP), support vector machine (SVM), and extremely randomized trees (ERT). The SMP forecasting errors based on the MAPE of the compared methods are presented in Table 12.9.

The naïve forecasting method uses yesterday's prices as the foresting prices today, which can be used as a reference case. Though it has good performances on May 23 and 24, its performances on the other days are the worst, and their average MAPE is the highest. The forecasting performances of MLP, SVR, and ERT are better, both reflected in the forecasting accuracy and stability. It can be observed that the proposed method has better performances than the compared methods both in accuracy and stability. In detail, the weekly price forecasting MAPE of the proposed method is 3.3%, less than half of the MAPEs of the other methods. In addition, its forecasting error range is from 0.9% to 5.5%, much narrower than the compared methods, which are 5.0% to 13.2% of ERT.

It should be noted that the chosen compared methods here are the commonly used basic forecasting methods, which might not be as good as the most advanced and customized forecasting method. However, the comparison results still demonstrate the high forecasting performances of the proposed method.

Table 12.9 Forecasting errors of market prices of compared methods

MAPE %	5/21	5/22	5/23	5/24	5/25	5/26	5/27	Avg
Naïve	17.0	9.4	4.9	5.4	8.8	12.3	14.3	10.3
MLP	15.9	7.8	5.6	5.8	7.7	7.7	13.0	9.1
SVR	15.1	4.7	5.1	8.2	5.1	9.7	10.9	8.4
ERT	12.3	5.0	6.5	8.7	5.6	7.3	13.2	8.4
Proposed	2.8	4.0	5.5	4.1	3.0	2.7	0.9	3.3

12.5.5 *Calculation Efficiency*

With the training dataset, the MFNN models for each generator are trained one by one. The example is conducted using Deep Learning Toolbox in MATLAB on a PC with Intel 6 dual-core processors running at 4.3 GHz and 48 GB of RAM.

The training time for each generator's MFNN model is about 2 to 3 s. Since the forecasting performances have randomness in different training, we choose to train 6 MFNN models for one generator at once, and then choose the one with the best performance based on testing results on the validation datasets to forecast the future individual offering behaviors. Thus, the total training time for a generator is 15 to 20 s. With training all 57 generators, the total training time is about 16 min.

The forecasting time for each generator is less than 1 s, which can be neglected.

12.6 Conclusions

To address the problems of the existing market simulation methods, a data-driven method is proposed to simulate power markets based on learning on individual offering strategies. After clustering and indexing the individual offering behaviors, a DNN-based learning model is developed to capture the relationships between the individual offering strategies and potential influencing factors. Then, the power market is simulated by forecasting individual offering strategies, providing market clearing results, including market prices, dispatch schedules, and power flows.

An empirical example is proposed to demonstrate the effectivity, accuracy, adaptivity, and stability of the proposed method. Real AEM data from March to May 2018 are utilized, and 57 generators are used. The data in a week are used to test the performance of the proposed method.

The individual offering strategy forecasting method reaches 91.1% accuracy in terms of classification accuracy and 95.8% accuracy in terms of MAPE. Then, the market forecasting results, including market prices, dispatch schedules, and power flows, are presented. The weekly MAPEs of the three key factors are 3.3, 1.2 and 5.4%, demonstrating the accuracy and effectiveness of the proposed market simulation method. The daily MAPEs of the three key factors vary in a small range, demonstrating the adaptivity and stability of the proposed method.

Finally, the price forecasting performance of the proposed method is compared with four other common methods. The results show that the proposed method has better accuracy and stability than commonly used methods. There are still further studies on several different aspects that can be done in the future. It is worthwhile to analyze if other DNN models, such as RNN (recurrent neural network) or LSTM (long short-term memory), have better performances on individual offering behavior forecasting. Another topic is to integrate more external factors, which have noteworthy impacts on the individual offering decision-making process, such as wind speed and reservoir water. It will be also valuable to test the performance of the proposed method in a real power market topology, if one can have the entire data of a power system.

References

1. Karthikeyan, S Prabhakar, I Jacob Raglend, and Dwarkadas P Kothari. 2013. A review on market power in deregulated electricity market. *International Journal of Electrical Power & Energy Systems* 48: 139–147.
2. Guo, Hongye, Qixin Chen, Qing Xia, and Chongqing Kang. 2019. Market power mitigation clearing mechanism based on constrained bidding capacities. *IEEE Transactions on Power Systems* 34 (6): 4817–4827.
3. Guo, Hongye, Qixin Chen, Qing Xia, and Chongqing Kang. 2019. Electricity wholesale market equilibrium analysis integrating individual risk-averse features of generation companies. *Applied Energy* 252: 113443.
4. Virasjoki, Vilma, Afzal S. Siddiqui, Behnam Zakeri, and Ahti Salo. 2018. Market power with combined heat and power production in the nordic energy system. *IEEE Transactions on Power Systems* 33 (5): 5263–5275.
5. Nasrolahpour, Ehsan, Jalal Kazempour, Hamidreza Zareipour, and William D Rosehart. 2018. A bilevel model for participation of a storage system in energy and reserve markets. *IEEE Transactions on Sustainable Energy* 9 (2) :582–598.
6. Rahi, Georges El, S. Rasoul Etesami, Walid Saad, Narayan B. Mandayam, and H. Vincent Poor. 2019. Managing price uncertainty in prosumer-centric energy trading: A prospect-theoretic stackelberg game approach. *IEEE Transactions on Smart Grid* 10 (1) :702–713.
7. Moiseeva, Ekaterina, and Mohammad Reza Hesamzadeh. 2018. Bayesian and robust nash equilibria in hydrodominated systems under uncertainty. *IEEE Transactions on Sustainable Energy* 9 (2): 818–830.
8. Yang, Jie, Zhenhua Tian, Kai Ma, Congshan Wang, and Zhixin Liu. 2018. A bayesian game approach for noncooperative pricing among multiple utility companies in smart grid. *IEEE Access* 6: 68576–68585.
9. Pozo, David, Enzo Sauma, and Javier Contreras. 2017. Basic theoretical foundations and insights on bilevel models and their applications to power systems. *Annals of Operations Research* 254 (1–2): 303–334.
10. Moiseeva, Ekaterina, Mohammad Reza Hesamzadeh, and Darryl R Biggar. 2015. Exercise of market power on ramp rate in wind-integrated power systems. *IEEE Transactions on Power Systems* 30 (3) :1614–1623.
11. Ruiz, Carlos, Antonio J Conejo, and Yves Smeers. 2012. Equilibria in an oligopolistic electricity pool with stepwise offer curves. *IEEE Transactions on Power Systems* 27 (2) :752–761.
12. Zou, Peng, Qixin Chen, Qing Xia, Guannan He, Chongqing Kang, and Antonio J. Conejo. 2016. Pool equilibria including strategic storage. *Applied Energy* 177: 260–270.
13. Dai, Ting, and Wei Qiao. 2017. Finding equilibria in the pool-based electricity market with strategic wind power producers and network constraints. *IEEE Transactions on Power Systems* 32 (1): 389–399.
14. Ye, Yujian, Dimitrios Papadaskalopoulos, and Goran Strbac. 2018. Investigating the ability of demand shifting to mitigate electricity producers' market power. *IEEE Transactions on Power Systems* 33 (4): 3800–3811.
15. Kardakos, Evaggelos G., Christos K. Simoglou, and Anastasios G. Bakirtzis. 2013. Short-term electricity market simulation for pool-based multi-period auctions. *IEEE Transactions on Power Systems* 28 (3): 2526–2535.

16. Guo, Hongye, Qixin Chen, Gu. Yuxuan, Mohammad Shahidehpour, Qing Xia, and Chongqing Kang. 2020. A data-driven pattern extraction method for analyzing bidding behaviors in power markets. *IEEE Transactions on Smart Grid* 11 (4): 3509–3521.
17. Vahid-Pakdel, MJ, Sina Ghaemi, Behnam Mohammadi-Ivatloo, Javad Salehi, and Pierluigi Siano. 2019. Modeling noncooperative game of GENCOs' participation in electricity markets with prospect theory. *IEEE Transactions on Industrial Informatics* 15 (10): 5489–5496
18. Frank, Sensfuß, Mario Ragwitz, M. Massimo Genoese, and Möst Dominik. 2007. Agent-based simulation of electricity markets: a literature review. *Energy Studies Review* 15 (2): 19–47.
19. Ringler, Philipp, Dogan Keles, and Wolf Fichtner. 2016. Agent-based modelling and simulation of smart electricity grids and markets - a literature review. *Renewable & Sustainable Energy Reviews* 57: 205–215.
20. Lu, Renzhi, Seung Ho Hong, and Mengmeng Yu. 2019. Demand response for home energy management using reinforcement learning and artificial neural network. *IEEE Transactions on Smart Grid* 10 (6): 6629–6639.
21. Ye, Yujian, Dawei Qiu, Jing Li, and Goran Strbac. 2019. Multi-period and multi-spatial equilibrium analysis in imperfect electricity markets: A novel multi-agent deep reinforcement learning approach. *IEEE Access* 7: 130515–130529.
22. Dehghanpour, Kaveh, M Hashem Nehrir, John W Sheppard, and Nathan C Kelly. 2016. Agent-based modeling in electrical energy markets using dynamic bayesian networks. *IEEE Transactions on Power Systems* 31 (6): 4744–4754.
23. Mitridati, Lesia, and Pierre Pinson. 2018. A Bayesian inference approach to unveil supply curves in electricity markets. *IEEE Transactions on Power Systems* 33 (3): 2610–2620.
24. Li, Hongyan, and Leigh Tesfatsion. 2009. Development of open source software for power market research: The AMES test bed. *Journal of Energy Markets* 2 (2): 111–128.
25. Battula, Swathi, Leigh Tesfatsion, and Thomas E. McDermott. 2020. An ERCOT test system for market design studies. *Applied Energy* 275: 115182.
26. Vale, Zita, Tiago Pinto, Isabel Praca, and Hugo Morais. 2011. MASCEM: electricity markets simulation with strategic agents. *IEEE Intelligent Systems* 26 (2): 9–17.
27. Shafie-khah, Miadreza, and Joao P. S. Catalao. 2015. A stochastic multi-layer agent-based model to study electricity market participants behavior. *IEEE Transactions on Power Systems* 30 (2): 867–881.
28. Chen, Ruidi, Ioannis Ch Paschalidis, Michael C Caramanis, and Panagiotis Andrianesis. 2019. Learning from past bids to participate strategically in day-ahead electricity markets. *IEEE Transactions on Smart Grid* 10 (5): 5794–5806.
29. AEMO. Yesterdays bids reports, 2020. http://www.nemweb.com.au/REPORTS/ARCHIVE/Yesterdays_Bids_Reports/.
30. Wang, Yi., Ning Zhang, Qixin Chen, Daniel S. Kirschen, Pan Li, and Qing Xia. 2018. Data-driven probabilistic net load forecasting with high penetration of behind-the-meter PV. *IEEE Transactions on Power Systems* 33 (3): 3255–3264.
31. Wang, Long, Zijun Zhang, and Jieqiu Chen. 2017. Short-term electricity price forecasting with stacked denoising autoencoders. *IEEE Transactions on Power Systems* 32 (4): 2673–2681.
32. Saez-Gallego, Javier, Juan M Morales, Marco Zugno, and Henrik Madsen. 2016. A data-driven bidding model for a cluster of price-responsive consumers of electricity. *IEEE Transactions on Power Systems* 31 (6): 5001–5011.
33. Ziel, Florian, and Rick Steinert. 2016. Electricity price forecasting using sale and purchase curves: The X-model. *Energy Economics* 59: 435–454.
34. Hong, Tao, and Shu Fan. 2016. Probabilistic electric load forecasting: A tutorial review. *International Journal of Forecasting* 32 (3): 914–938.
35. Kong, Weicong, Zhao Yang Dong, Youwei Jia, David J Hill, Yan Xu, and Yuan Zhang. 2019. Short-term residential load forecasting based on LSTM recurrent neural network. *IEEE Transactions on Smart Grid* 10 (1): 841–851.
36. Zhang, Zhilu, and Mert Sabuncu. 2018. Generalized cross entropy loss for training deep neural networks with noisy labels. In S. Bengio, H. Wallach, H. Larochelle, K. Grauman, N. Cesa-Bianchi, and R. Garnett, editors, *Advances in Neural Information Processing Systems*, vol. 31, 8778–8788. Curran Associates, Inc.

37. Martin Fodslette Møller. 1993. A scaled conjugate gradient algorithm for fast supervised learning. *Neural Networks* 6 (4): 525–533.
38. AEMO. Exsiting and new development by fuel-technology category, 2020. https://www.aemo.com.au/Electricity/National-Electricity-Market-NEM/Planning-and-forecasting/Generation-information.

Chapter 13
Reward Function Identification of GENCOs

13.1 Introduction

Due to the deregulation of power markets worldwide, research on the strategic bidding behaviors of market participants has attracted a substantial amount of attention. Many methods have been developed to address this problem. State-of-art frameworks for modeling strategic bidding behaviors include the equilibrium model [1], bilevel optimization [2], reinforcement learning [3], etc. These frameworks have been extensively utilized to model the bidding decision-making process of thermal generators [4], renewables [5], energy storages [6], prosumers [7], load service entities [8] and their various combinations [9].

Regardless of the model, the reward function (or objective function) is considered the main driver of participants' bidding behaviors, which fundamentally determines the effectiveness of the simulation methods. However, most of the commonly employed reward function modeling approaches are stereotyped. In these models, participants are usually considered rational men, and their objective functions are usually defined as individual profit maximization problems. For example, the objective function of a generation company (GENCO) that participates in the day-ahead (DA) wholesale market is usually defined as the individual profit, which is the difference value between the operation costs and the market revenues [10]. The same modeling of the reward function of GENCOs is also presented in reinforcement-learning-based market studies [11, 12].

Although this objective function format has been applied in most studies, it has two serious but easily missed problems. Due to the information barriers, the operation cost parameters, which comprise personal information in most markets, cannot be obtained in normal ways, even in this era with gradually transparent data disclosure policies. Thus, the parameters of reward functions are usually designed according to baseless or subjective assumptions, which inevitably cause deviations during the market simulation process. On the other hand, the normal reward functions are designed assuming all participants are "completely rational", which is another factor that devastates the performances of existing methods. Considering the bounded rationality of generators [13], profit is not the only factor of concern for GENCOs when they are making decisions. Many GENCOs might also consider market risk, cleared capacity,

Q. Chen et al., *Data Analytics in Power Markets*,
https://doi.org/10.1007/978-981-16-4975-2_13

or social recognition, which are difficult to formulate in common algebraic methods for reward functions.

Conclusively, the conventional reward function formulation encounters the challenges incurred from data acquisition and decision modeling. It is urgent and expected to propose a novel reward function modeling method that can more realistically represent the individual driven core during the bidding decision-making process.

Fortunately, the big-data era gives us an opportunity to solve this problem in a data-driven way. Although the exact reward function parameters cannot be obtained directly from common methods, the individual preferences of GENCOs are hidden in their actual bidding behaviors and can be unveiled by analyzing the real data. As power market data have become more open and transparent in recent years, we are able to obtain access to more real market data. Thus, it might be feasible to identify the bidding preferences of GENCOs using a data-driven method that works backward from the actual historical data. For example, the Australian electricity market (AEM), which discloses all its participants' bidding data to the public, provides an opportunity for us to do this work. There have not been any studies to restore the individual reward functions from real market data.

The inverse reinforcement learning (IRL) algorithm is an appropriate data-driven method to accomplish this target. IRL is a branch of machine learning that can be employed to quantify individual reward functions. Based on the Markov decision process (MDP), the goal of IRL is to identify a reward function from expert demonstrations that can explain historical behaviors to the greatest extent [14]. It has been applied in several areas [15], such as the game of go, automatic pilot, robotics, etc. However, IRL methods still have limitations in applications due to their simple reward function structure. With the development of deep learning, deep inverse reinforcement learning (DIRL) [16] has been proposed and applied in numerous areas, such as autonomous vehicles [17] and video game AI [18]. Despite its prevalence, IRL and DIRL have not been utilized in power market studies.

Thus, to identify the reward functions of real GENCOs, this chapter proposes a data-driven framework that customizes DIRL methods to reveal the parameters of reward functions from real market data in power markets. To accomplish the framework, there are three major procedures. The first procedure is decision process formulation, which is set to formulate the individual bidding decision-making process standardly as an MDP, which establishes a platform for further analysis. The second procedure is reward function identification, which applies the deep inverse reinforcement learning method based on maximum entropy [16] to identify the individual reward functions, which are subsequently saved in multilayer perceptions (MLPs). The third procedure is bidding behavior simulation, which is proposed to simulate the individual bidding behaviors based on the obtained MLP-based reward functions. A deep reinforcement learning (DRL) model that is based on a deep Q-network (DQN) [19] is customized to conduct the simulation. The detailed process is shown in Fig. 13.1.

The real individual bidding data in AEM is applied to a numerical example, which is used to show the performance of the proposed data-driven framework.

The main contributions can be summarized as follows:

1. A data-driven framework to identify GENCOs' reward functions during individual bidding decision-making is proposed.
2. The DIRL method, which is based on maximum entropy, is introduced and customized to identify the individual reward functions in power markets.
3. A DRL model is proposed to utilize the MLP-based reward functions with high-dimensional nonlinearity for individual bidding behavior simulation based on the DQN.
4. A numerical example is conducted based on the real bidding data in power markets. The effectiveness and feasibility of the proposed framework are indicated.

The remainder of the chapter is organized as follows: The market assumptions and data-driven framework are given in Sect. 13.2. In Sect. 13.3, the decision process formulation procedure is detailed. In Sect. 13.4, the reward function identification procedure is explained to present the bidding preference identification process based on the DIRL. In Sect. 13.5, the bidding behavior simulation procedure is developed with the obtained MLP-based reward functions. In Sect. 13.6, a case study that is based on the real data from AEM is provided. Conclusions are given in Sect. 13.7.

13.2 Assumptions and Framework

13.2.1 Market Assumptions

For simplicity, this chapter focuses on GENCOs' strategic bidding behaviors in DA wholesale power markets, such as in the Pennsylvania-New Jersey-Maryland (PJM) market and the Midcontinent Independent System Operator (MISO) market in the US and AEM in Australia.

On the day before operation day, an independent system operator (ISO) publishes forecasting information about the system conditions on the operation day. Considering the published information and their conjectures on market situations, GENCOs make their bidding strategies and submit bids to the ISO. All bids are in the stepwise format with multiple blocks. The ISO then clears the market, which generates market results of clearing price, generator dispatch schedule, power flows, etc. Each generator obtains the cleared price and capacities of their bids. The DA market finishes.

13.2.2 Data-Driven Framework

The detailed proposed data-driven framework is shown in Fig. 13.1.

First, a decision process formulation procedure needs to be performed as the framework foundation. To satisfy the requirements of DIRL deployment, an MDP

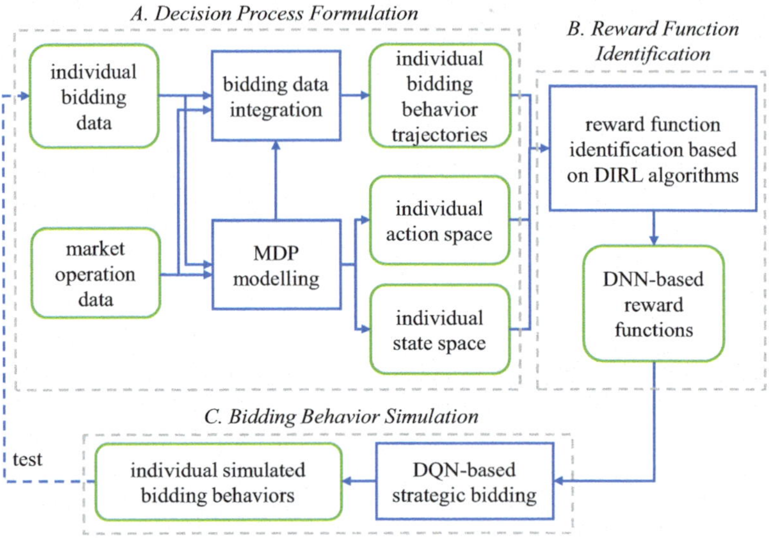

Fig. 13.1 Proposed data-driven framework

framework for the bidding decision-making process of GENCOs is developed. MDP modeling can be considered extraction of the key steps from a generator's decision-making process and an establishment of a universal decision process. MDP has been employed in many studies to formulate the individual bidding decision-making process [6, 12]. The generated MDP works as a platform for subsequently applying the DRL and DIRL algorithms. The individual bidding data are then collected and integrated as demonstrated trajectory data, which serves DIRL training.

Second, a reward function identification procedure is proposed to reveal the reward functions of generators. A DIRL algorithm that is based on maximum entropy is customized according to the power market transaction features. After the determination of the reward function structure, the parameters of the reward function can be obtained. The core idea of this method is to search for the most appropriate parameter set that suits the existing individual bidding behavior trajectories based on maximum likelihood estimation. After the training of DIRL, the high-dimensional nonlinear bidding preferences of various generators will be saved by the diversified parameters of an MLP.

Third, a bidding behavior simulation procedure is given to simulate the bidding behaviors of GENCOs with the utilization of the MLP-based reward functions obtained from their historical bidding data. A DQN model is introduced and customized to restore the individual bidding decision-making process and conduct bidding behavior simulations. By analyzing whether the simulated behaviors have the bidding characteristics of real generators, the effectiveness of the reward function identification can be indicated from the other side.

13.3 Bidding Decision Process Formulation

13.3.1 Markov Decision Process in Wholesale Markets

The MDP is a mathematical framework for modeling the decision-making process in situations where outcomes are partly random and partly under the control of a decision maker [20]. In general, an MDP is a 4-tuple $(\mathcal{S}, \mathcal{A}, \mathcal{P}, \mathcal{R})$, which contains the following elements: (1) a state space, $\mathcal{S}$; (2) an action space, $\mathcal{A}$; and (3) a state conditional transition probability set, $\mathcal{P} = p(s_{t+1}|s_t, a_t)$. The state transition should satisfy the Markov property; (4) a reward function, $\mathcal{R} = r(s_{t+1}|s_t, a_t)$.

The elements in the MDP for wholesale power markets are defined as follows:

(1) Agent: The strategic generator i is considered the agent, which interacts repeatedly with the power market by submitting its bids and obtaining market clearing results. The agent gradually learns from its bidding experience and improves its bidding decision-making ability.

(2) Environment: The power market is considered the environment, which collects bids from the agent and other participants and clears the market while providing market results.

(3) State: The state s represents the status of generator i in the power market. A state is described by a series of market features, including the market clearing results (used as a feedback signal of its bidding behaviors) and market forecasting results (used for future bidding references). The market features in state s_t constitute a vector, which is referred to as the state vector $\mathbf{s}_t$.

(4) Action: The action a_b represents a bid that generator i chooses and submits to the ISO. b is the index of bidding patterns. The bid is in stepwise format.

(5) Reward: The reward r_t represents the individual utility received after transitioning from state s_t to state s_{t+1}. The reward function is designed to map some features of each state, such as the cleared price and capacity, to a reward value. In common studies, reward functions are usually designed as profits of generators. In this chapter, the practical reward functions of generators are extracted from real market data based on the proposed DIRL framework.

Since generators simultaneously submit their bids of each time period for the next day, the time step in the MDP should be separated by days. Due to the load variation, the similarity of the market circumstances at each time period among different days is higher than that at different time periods of one day. Thus, when generators make bidding decisions, they pay more attention to historical bids in the same time period. Considering these bidding characteristics, different MPDs are set for different time periods in this chapter. A time step in an MDP is referred to as the "step", and a time period in a day is referred to as the "time period" in subsequent text.

13.3.2 Reinforcement Learning Process

After modeling the decision-making process in an MDP framework, RL algorithms are usually employed to model and simulate individual learning processes.

At step t, based on the information from state s_t, generator i decides its bid a_t with utilization of stochastic policy π_i, which represents its bidding experience. The ISO then clears the market with bid a_t, provides market results and moves generator i into the new state s_{t+1}. With the market results, which are shown as several features in state vector s_{t+1}, generator i can calculate its reward r_t based on the reward function. With reward r_t, generator i can update its stochastic policy π_i. Generator i continues to make its bidding decision at step $t+1$ until the MDP is over.

The exact form of stochastic policy π_i is π: $\mathcal{S} \times \mathcal{A} \rightarrow \mathbb{R}$. Policy π is a set that contains the probabilities of choosing action a_b at state s_n, which complies with

$$\pi\left(a_b \mid s_n\right) = \Pr\left(a_t = a_b \mid s_t = s_n\right) \tag{13.1}$$

$$\sum_b \pi\left(a_b \mid s_n\right) = 1, \forall n \tag{13.2}$$

The performance evaluation of a stochastic policy depends on the cumulative reward value that is obtained after the execution of the strategy. A goal of RL is to obtain a policy that maximizes the cumulative reward value of an MDP.

$$R_{t=1} = \mathrm{E}\left[\sum_{t=1}^{\infty} \gamma^{t-1} r_t\right] \tag{13.3}$$

where r_t is the discount rate, and r_t represents the reward generator that is obtained at state s_t, which is calculated by the preset reward function.

Based on this analysis, the significance of the reward function is further highlighted. As a preset parametric function, if it deviates from the actual bidding preferences of generators, the evaluation for the stochastic policy will be affected, which will further interfere with the simulation results of the individual bidding behaviors.

13.3.3 Bidding Data Integration

Then, historical bidding data need to be integrated as demonstrated trajectories, which will be utilized as training data of the DIRL algorithm.

According to the analysis in Sect. 13.3.2, a reward function will affect the policy learning process and eventually determine the actions chosen at different states. It is intuitive to understand that the agent's bidding preferences can be reflected by the paths of states and actions. Thus, a path of states and actions taken by the agent in a whole MPD is defined as the demonstrated trajectory $\boldsymbol{\tau}$.

$$\boldsymbol{\tau} = \{s_{t=1}, a_{t=1}, \ldots, s_t, a_t, \ldots, s_{t=T}, a_{t=T}\} \tag{13.4}$$

All individual bidding data of a generator can be assembled into multiple demonstrated trajectories. The index of trajectory $\boldsymbol{\tau}$ is denoted as m. All demonstrated trajectories can constitute a set, which is denoted as demonstration set Ω.

13.4 Reward Function Identification

The basic idea of DIRL is to compute the parameters of the reward function for a given agent's state space, action space, and demonstration set so that the action results of the RL model that is trained with the computed reward function parameters are the most consistent with the demonstrated trajectories. The detailed process is stated in Sect. 13.4.1. In addition, to use the DIRL method, the state space and action space should be discretized. Thus, a discretization method is proposed for states and actions considering the power market characteristics in Sect. 13.4.2.

13.4.1 Deep Inverse Reinforcement Learning Algorithm

In this section, the maximum entropy deep inverse reinforcement learning [16] is customized and used to identify the reward function of a generator.

Since the reward is mapped from the features of states, many artificial neural networks can be utilized for reward function approximation. Here, an MLP is used as an example. Thus, the reward of state s can be calculated as

$$r_s = h\left(\mathbf{f}_s, \boldsymbol{\theta}\right) = \left(g_n\left(\ldots g_2\left(g_1\left(\mathbf{f}_s\right)\right)\right)\right) \tag{13.5}$$

where $\mathbf{f}_s$ is the feature vector, which represents the features that influence the individual rewards in the state vectors; $\boldsymbol{\theta}$ represents the set of parameters in the MLP; and $g(\cdot)$ is the mapping function in each layer of the MLP. The detailed structures are written as follows:

$$\boldsymbol{\theta} = \{\mathbf{W}_1, \mathbf{b}_1, \mathbf{W}_2, \mathbf{b}_2, \ldots, \mathbf{W}_n, \mathbf{b}_n\} \tag{13.6}$$

$$g_n(x) = \sigma\left(\mathbf{W}_n \cdot x + \mathbf{b}_n\right) \tag{13.7}$$

$$\sigma(x) = \frac{1}{1 + \exp(-x)} \tag{13.8}$$

The crucial problem is to determine what parameter set $\boldsymbol{\theta}$ will let the individual model behave similarly with the demonstrated trajectories as much as possible.

First, based on the findings of [15], we can assume that the probability of each trajectory should be exponentially related to its reward value under the maximum entropy hypothesis.

$$P\left(\tau_m \mid \boldsymbol{\theta}\right)=\frac{1}{Z(\boldsymbol{\theta})} e^{\boldsymbol{\theta}^{\mathrm{T}} \mathbf{f}_{\tau m}}=\frac{1}{Z(\boldsymbol{\theta})} e^{\sum_{s_j \in \tau m} \boldsymbol{\theta}^{\mathrm{T}} \mathbf{f}_{s_j}} \tag{13.9}$$

where $Z(\boldsymbol{\theta})$ is a partition function to ensure that the summation of all probabilities equals one, and τ_m represents the mth trajectory. The detailed form of $Z(\boldsymbol{\theta})$ is

$$Z(\boldsymbol{\theta})=\sum_{\tau_m} e^{\boldsymbol{\theta}^{\mathrm{T}} \mathbf{f}_m} \tag{13.10}$$

Presently, we have the probability of each demonstrated trajectory. Thus, the maximum likelihood estimation method is further employed to calculate the exact parameters. The problem can be transformed into an optimization problem:

$$\max_{\boldsymbol{\theta}} L(\boldsymbol{\theta})=\max_{\boldsymbol{\theta}} \sum_{\boldsymbol{\tau}_m \in \Omega} \log \left(P\left(\boldsymbol{\tau}_m \mid \boldsymbol{\theta}\right)\right) \tag{13.11}$$

According to the findings of [16], the gradient of the problem can be derived as

$$\nabla_{\boldsymbol{\theta}} L(\boldsymbol{\theta})=\sum_{\boldsymbol{\tau}_m \in \Omega} \frac{d R_{\boldsymbol{\tau}_m}}{d \boldsymbol{\theta}}-M \frac{1}{\sum_{\boldsymbol{\tau}_m \in \Omega} \exp \left(R_{\boldsymbol{\tau}_m}\right)} \sum_{\boldsymbol{\tau}_m \in \Omega} \exp \left(R_{\boldsymbol{\tau}_m}\right) \frac{d R_{\boldsymbol{\tau}_m}}{d \boldsymbol{\theta}} \tag{13.12}$$

where M is the number of trajectories. Second, the gradient (13.12) can be further derived as

$$\nabla_{\boldsymbol{\theta}} L(\boldsymbol{\theta})=M \sum_s\left[\left(\tilde{D}(s)-D(s)\right) \frac{d}{d \boldsymbol{\theta}} h\left(\mathbf{f}_{s_j}, \boldsymbol{\theta}\right)\right] \tag{13.13}$$

where $\tilde{D}(s)$ represents the proportion of state s that occurs in all historical data, which is named the observed state visitation frequency, and $D(s)$ represents the probability of state s that occurs in the simulated results given the parameter set $\nabla_{\boldsymbol{\theta}}$, which is named the expected state visitation frequency and can be calculated based on the methods proposed in [21].

We obtain the gradient for updating the maximum likelihood estimation optimization problem. Next, the parameters of the MLP-based reward functions can be solved using the common cumulative error backpropagation algorithm.

13.4.2 Discretization Methods for States and Actions

To apply the DIRL algorithm, the state space and action space should be discrete. The key intermediate variables, such as $\tilde{D}(s)$, $D(s)$ and $P^a_{s,s'}$, can then be calculated. Thus, discretization methods for the state space and action space are proposed in this section.

The action space can be considered as the bidding strategy space. According to the findings of [13], the bidding strategy space of generators has the obvious characteristics of discretization and patterning. Thus, each generator's action space can be directly discretized according to its bidding patterns.

The state space includes a series of market features, which are continuous variables. It is inappropriate to discretize these features by clustering methods. Thus, the state space is discretized based on quantiles. The cumulative distribution of each market feature, such as the clearing price, is generated based on the historical demonstrated trajectory data. All data are then discretized by U quantiles with equal intervals. The quantiles that work as boundaries are denoted as q_u.

Although the state space and action space are discretized, they are only utilized as training data for DIRL and do not affect the continuity of the obtained reward function parameters.

13.5 Bidding Behavior Simulation

After obtaining the MLP-based reward functions, we develop an individual bidding decision-making simulation model that is based on DQN, considering the actual operating environment of power markets. The DQN model was introduced in [19] and has been employed for residual load control [22], building energy management [23], EV charging scheduling [24], etc.

13.5.1 DQN-Based Bidding Simulation Model

The proposed DQN-based bidding behavior simulation model is shown in Fig. 13.2. The model can be divided into two parts: the market environmental model and the individual behavior model. The two models are operated separately and sequentially.

The market environmental model can be customized according to different research requirements. In this chapter, considering the accessible data and modeling simplicity, there are three features regarded as states: cleared price λ_t, cleared capacity p_t, and future system demand d_t. At state s_t, the market receives bidding action a_t from the generator. The market is then cleared and moved into state s_{t+1} with market features at time $t + 1$.

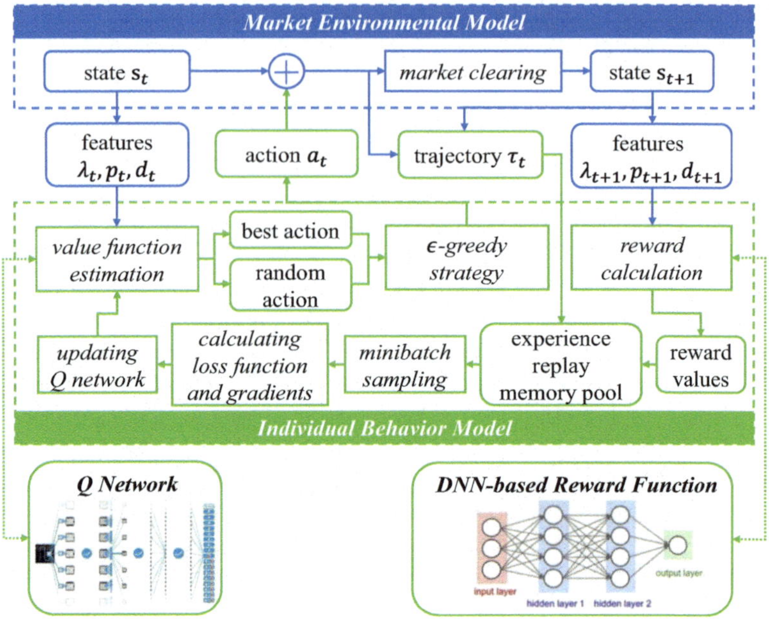

Fig. 13.2 DQN-based bidding behavior simulation model

The *individual behavior model* is operated as follows. First, given s_t, the generator calculates the value functions of all possible actions in its policy π based on the Q-network. The action with the highest value function is chosen as the best action. Second, the final submitted action a_t is decided between the current best action and a random action by the ϵ-greedy algorithm. With action a_t, the market is moved into state s_{t+1}. Third, the generator calculates its reward value r_t based on the reward function obtained from the last section. Fourth, the interaction experience $[s_t, a_t, s_{t+1}]$ is integrated as trajectory τ_t and further saved in the experience replay memory pool. Last, several results are randomly sampled from the pool to calculate the gradients of the Q-network, which are subsequently employed to update the Q-network parameters.

Based on the alternate operation of the market environment model and individual behavior model, the DQN-based bidding simulation model operates and ultimately provides optimal bidding strategies for generators.

13.5.2 Value Function and Q-Network

The value function is introduced to evaluate the accumulated reward of policy π. The function can be categorized as state-action value function $Q^{\pi}(s, a)$ and state value function $V^{\pi}(s)$.

$$Q^{\pi}(s,a) = r(s,a) + \gamma \sum_{s' \in \mathbf{S}} p\left(s' \mid s,a\right) \cdot V^{\pi}\left(s'\right) \tag{13.14}$$

$$V^{\pi}(s) = \sum_{a \in \mathbf{A}} \pi(a \mid s) \cdot Q^{\pi}(s,a) \tag{13.15}$$

The two value functions have recursive relationships:

$$\begin{aligned} Q^{\pi}(s,a) =& r(s,a) \\ &+ \gamma \sum_{s' \in \mathbf{S}} \left[p\left(s' \mid s,a\right) \sum_{a' \in \mathbf{A}} \pi\left(a' \mid s'\right) \cdot Q^{\pi}\left(s',a'\right) \right] \end{aligned} \tag{13.16}$$

The value functions can be iteratively updated by dynamic programming methods.

The state-action value function $Q^{\pi}(s,a)$ in DQN methods is modeled by an MLP, which is denoted as the Q-network. The state value function with parameter $\boldsymbol{\theta}$ of the Q-network can be written as $Q(s,a|\boldsymbol{\theta})$. When the generator submits action a at state s and obtains reward r, state value function $Q(s,a)$ can be updated as $Q'(s,a)$. The algorithm goal is to reduce the deviations between the value function before and after updates. Thus, the deviation can be set as the loss function:

$$L\left(\boldsymbol{\theta}^{(k)}\right) = \mathbf{E_M}\left[\left(\begin{matrix} r + \gamma \max_{a' \in \mathbf{A}} Q\left(s',a' \mid \boldsymbol{\theta}^{(k-1)}\right) \\ -Q\left(s,a \mid \boldsymbol{\theta}^{k}\right) \end{matrix}\right)^2\right] \tag{13.17}$$

where k is the serial number of trainings and $\mathbf{M}$ represents the number of minibatches sampled from the experience replay memory pool.

Then, we can derive the gradient of the loss function (13.17).

$$\nabla_{\boldsymbol{\theta}^{(k)}} L\left(\boldsymbol{\theta}^{(k)}\right) = \mathbf{E_M}\left[\begin{matrix} r + \gamma \max_{a' \in \mathbf{A}} Q\left(s',a' \mid \boldsymbol{\theta}^{(k-1)}\right) \\ -Q\left(s,a \mid \boldsymbol{\theta}^{k}\right) \cdot \nabla_{\boldsymbol{\theta}^{(k)}} Q\left(s,a \mid \boldsymbol{\theta}^{k}\right) \end{matrix}\right] \tag{13.18}$$

With gradient (13.18), the parameters of the Q network can be updated and the DQN model can be trained.

13.6 Case Study

13.6.1 Dataset Description

To demonstrate the effectiveness of the proposed framework, the real market data in the AEM have been utilized. Bidding behavior data and power market operation data from October 19th, 2017 to September 18th, 2018 are employed.

The DA market in the AEM is operated with the following settings. There are 48 time periods in a day. Each time period contains 0.5 h. All generators can submit various bids for different time periods. The bids submitted by the generators are in the stepwise format with a majority of 10 blocks.

The numerical example section can be divided into two parts. The bidding preferences of several different generators are visually demonstrated and compared, as shown in Sect. 13.6.3. The second part simulates individual bidding behaviors based on the obtained MLP-based reward functions, as shown in Sect. 13.6.4.

13.6.2 Parameter Setting

The state space is discretized in three dimensions. The clearing price and cleared capacity are discretized with 10 intervals, and the forecasting system demand is discretized with 5 intervals. Thus, the state space is discretized into 500 discrete states. The action space is discretized based on the pattern extraction method proposed in [13]. The pattern extraction number is set to 30.

As stated in Sect. 13.3, the same time periods on different days are integrated as demonstrated trajectories. The length of each trajectory is 28 days (four weeks). Thus, 12 demonstrated trajectories are generated for each time period. Considering that there are 48 periods in a day, the total number of demonstrated trajectories is 576.

The structures of the MLP, which are utilized to represent individual reward functions, are set as follows. It has one input layer, two hidden layers, and one output layer. The units of two hidden layers are set to 10 and 4. The output layer has 1 unit to generate the reward values.

The parameters of the DIRL algorithms are set as follows. The discount rate r_t is set to 0.5. The maximum iteration number is set to 1000. The learning rate is set to 0.001.

The parameters of the DRL algorithms are set as follows. The discount rate and learning rate are set similar to the rates in the DIRL algorithm. The size of the minibatch is 64. The greedy probability in the ϵ-greedy algorithm is set to 0.1.

The numerical examples are conducted using Python 3.7 and Deep Learning Toolbox in MATLAB on a PC with an Intel i7-8700 CPU that runs at 4.3 GHz and 48 GB of RAM.

13.6.3 Reward Function Identification

After the DIRL calculation, the parameters of the MLP-based reward functions can be obtained. However, since the number of MLP parameters is immense, which hinders their presentation, we attempt to present generators' bidding preferences by drawing individuals' reward values at various states in three dimensions.

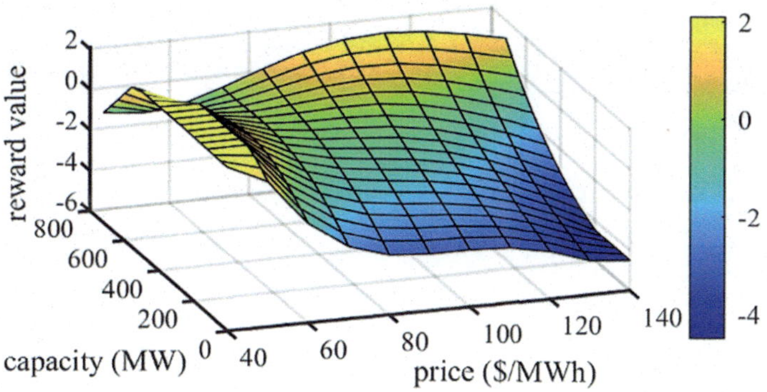

Fig. 13.3 Bidding preferences of ER04

Several thermal generators with different fuel types are chosen for reward function identification. Two generators are chosen for a detailed presentation: coal-fired generator ER04 with 720 MW capacity and gas-fired generator CPSA with 144 MW capacity. Four thermal generators with various capacities are presented to demonstrate the adaptivity of the proposed method, and the diversity of individual bidding preferences is demonstrated.

(1) Coal-fired ER04

The detailed bidding preferences of ER04 are shown in Fig. 13.3. Each point represents a reward function value that corresponds to the price and capacity of its located state. The color of the points corresponds to the reward values.

The bidding preferences of coal-fired generator ER04 among various price-capacity states are different and have complex reflections. The shape of the bidding preference resembles a saddle, where the reward values are higher when the prices and capacities are both higher or both lower but become lower when the prices and capacities are in opposite directions. The bidding preference of ER04 has a clear strategic feature, which is "*prefer larger capacity to be cleared at higher prices and avoid being cleared at lower prices*".

To better demonstrate ER04's bidding preferences, three 2D slices of Fig. 13.3 at 50, 70 and 90 $/MWh prices are presented in Fig. 13.4. It is obvious that the reward value trends with an increase in capacity are different at the three prices.

When the prices are lower, such as 50 $/MWh, a smaller cleared capacity, such as 100 MW to 400 MW, will yield a higher reward value. In this case, ER04 tends to submit capacity-withholding bids to ensure that it is cleared at smaller capacities. When the clearing price increases to 70 $/MWh, the reward value increases with an increase in capacity, and is maintained at approximately 0 when the capacity exceeds 400 MW. When the clearing price reaches 90 $/MWh, the preferences for higher cleared capacities become more prominent, as shown in a steeper reward value curve. Thus, to obtain a higher reward value, ER04 tends to adjust its bids to clear at larger capacities in the last two cases.

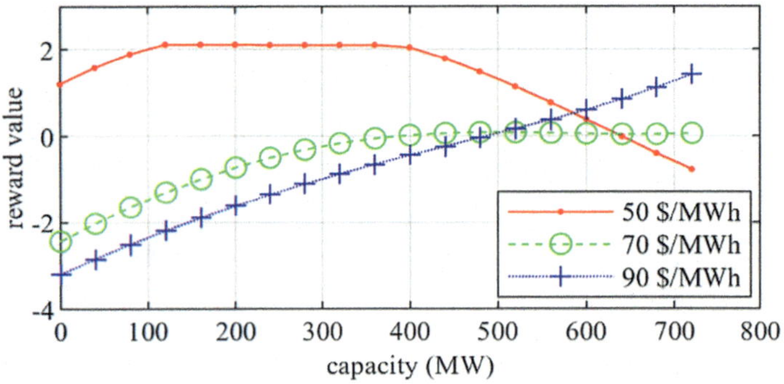

Fig. 13.4 Reward values of ER04 at 50, 70, 90 $/MWh

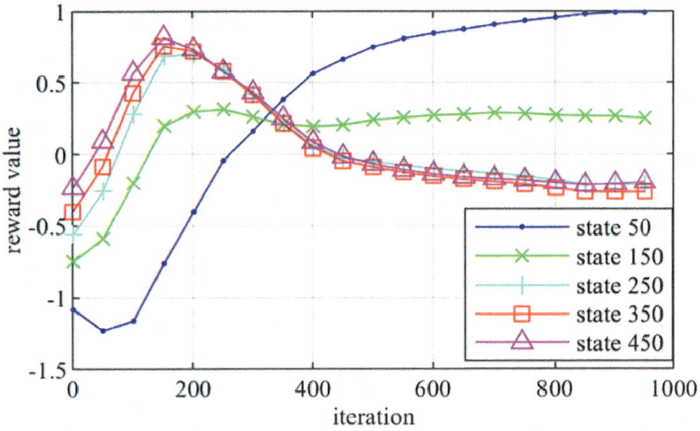

Fig. 13.5 Reward values of ER04's 50, 150, 250, 350, and 450 states during iteration

To demonstrate the convergence of the proposed DIRL method, the iteration process of the algorithm needs to be presented. Since the MLP-based reward functions have numerous parameters, which hinders their presentation, the reward values of several states during iterations are shown in Fig. 13.5 to demonstrate the convergence.

At the beginning of the iteration, the parameters of the reward functions are assigned with random initial values. The reward function parameters are updated during the iteration, which causes fluctuations in the reward values. It can be seen that the reward values fluctuate before 500 iterations and subsequently tend to be stable. When the iteration number reaches 800, most of the reward values stabilize, and the algorithm converges and stops at 948 iterations.

(2) Gas-fired CPSA

The bidding preferences of CPSA are shown in Fig. 13.6. The shape of gas-fired CPSA's bidding preferences resembles a saddle, which has a feature that is similar to

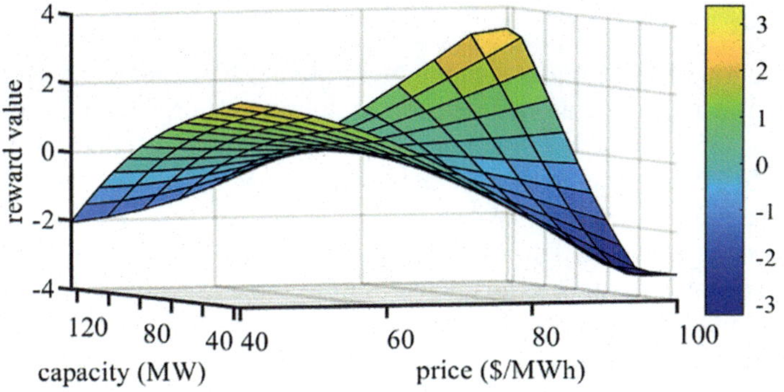

Fig. 13.6 Bidding preferences of CPSA

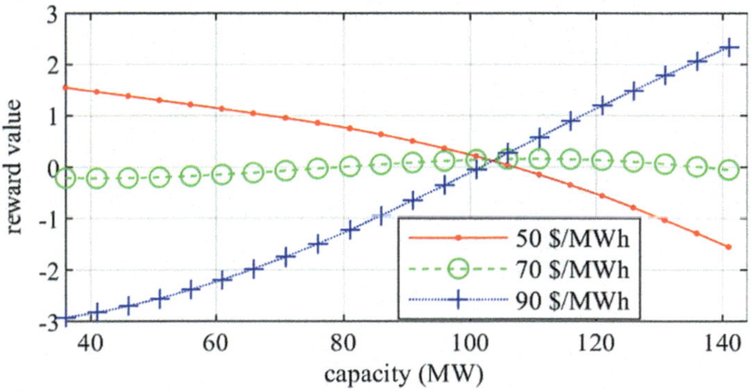

Fig. 13.7 Reward values of CPSA at 50, 70, and 90 $/MWh

the coal-fired ER04's bidding preferences. However, the detailed bidding preferences of the two generators are largely different.

For better presentation, 2D slices of Fig. 13.6 at 50, 70 and 90 $/MWh prices are presented in Fig. 13.7. When the clearing price is 50 $/MWh, CPSA's reward values decrease as cleared capacities increase. When the clearing price is 70 $/MWh, reward values fluctuate at approximately 0. When the clearing price is 90 $/MWh, CPSA's reward values have a clear positive correlation with cleared capacities.

When the bidding preferences of ER04 and CPSA are compared, we determine that the latter gas-fired generator's preferences are more aggressive, which is reflected in the violent variations in the individual bidding preference surfaces. The reason is inferred as follows. Due to the better dynamic performance of gas-fired generators, the adjustment range of CPSA is better than ER04. Thus, CPSA's pursuit of additional cleared capacities at higher prices might be more intense than ER04's, which causes a more aggressive bidding preference.

(3) Bidding preference diversity

In this section, four thermal generators' bidding preferences are presented in Fig. 13.8. The four generators are listed as follows: (1) STAN-1, coal-fired generator with 365 MW capacity, (2) GSTONE4, coal-fired generator with 280 MW capacity, (3) BW03, coal-fired generator with 660 MW capacity, and (4) DDPS1, gas-fired generator with 280 MW.

It is obvious that although the bidding preferences of the four generators have a saddle shape, their details differ. The different shapes demonstrate the diversities among the individual bidding preferences of generators. The proposed reward function identification method is demonstrated with the ability to reveal the diversified bidding preferences of different generators.

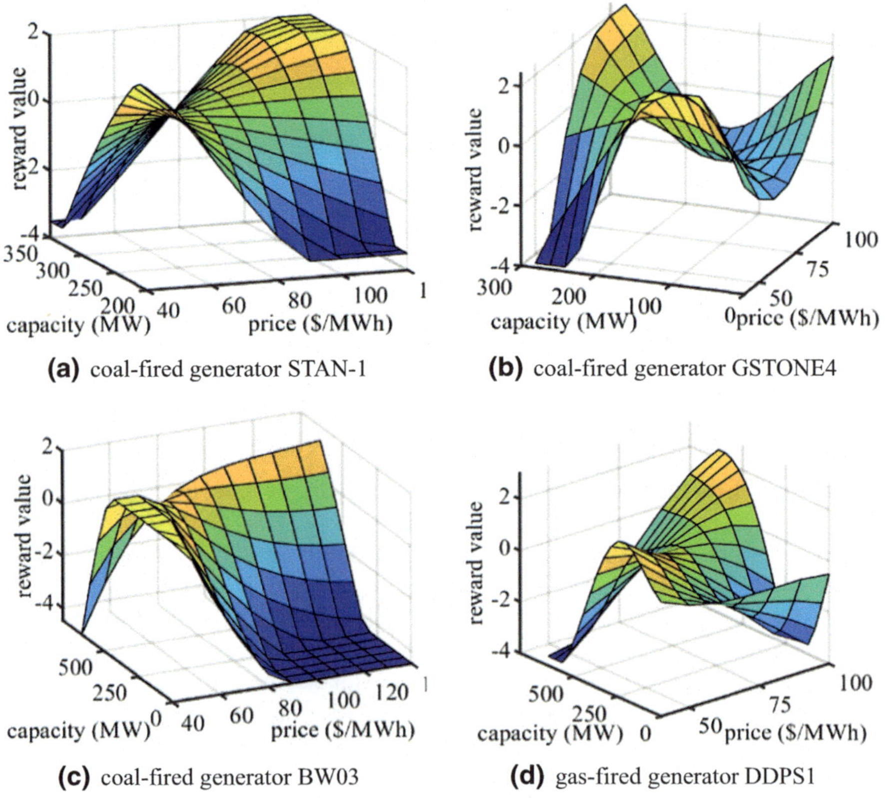

(a) coal-fired generator STAN-1 **(b)** coal-fired generator GSTONE4

(c) coal-fired generator BW03 **(d)** gas-fired generator DDPS1

Fig. 13.8 Bidding preferences of four thermal generators

13.6.4 Bidding Behavior Simulation

(1) Simulation based on identified reward functions
With the obtained MLP-based reward functions, the individual bidding behaviors of generator ER04 in April 2018 are simulated. By analyzing the simulated behaviors, we are able to test the effectiveness of the proposed DIRL method to reveal the individual bidding preferences.

The simulated bidding behaviors of ER04 on April 7th, 2018 are shown in Fig. 13.9, where the x-axis represents the time periods from 1 to 48, the y-axis represents the generator's capacity in bids, and the z-axis represents the generator's price in bids. The red "x" markers are the cleared price-capacity values at each time period.

(2) Simulation based on assumed reward functions

To demonstrate the performance of the proposed method, a similarly structured DQN model with the traditional reward function, which is employed to calculate the individual profit from bidding, is conducted to provide a comparison. Since the cost parameters of the reward function are unknown, they are assumed based on the statistic cost data in AEM [25] as in other studies. The simulated bidding behaviors on the same day are shown in Fig. 13.10.

We observe that the simulated bidding behaviors of ER04 are obviously different from Fig. 13.9. ER04 submits bids with lower prices to obtain additional cleared capacities regardless of the time period. The reason for this situation can be inferred as follows. Since the parameters used for profit calculation are assumed based on statistical data, they might deviate from real situations. If the cost parameters are assumed to be lower than the real costs, the generator bidding simulation model might indicate that the prices in all time periods are profitable. The generator will likely submit bids with lower prices to obtain more cleared capacities and generate unrealistic bidding simulation results.

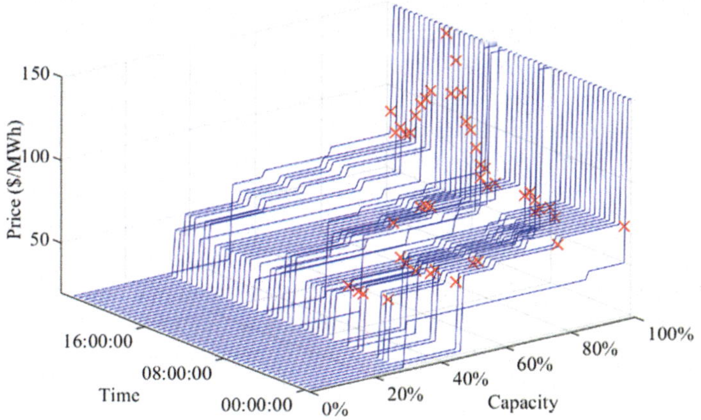

Fig. 13.9 Simulated bidding behaviors of ER04 on April 7th, 2018 by the proposed method

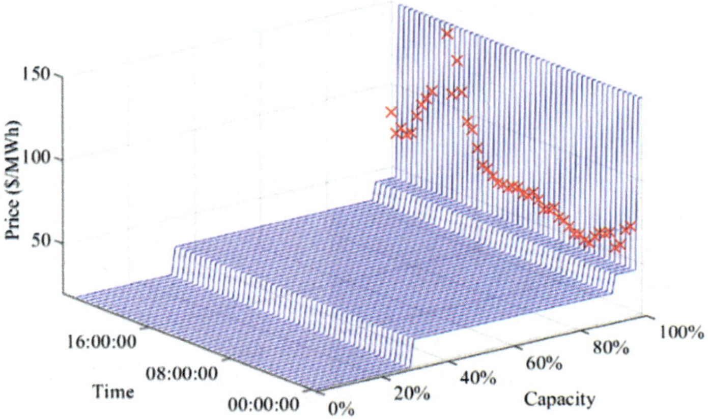

Fig. 13.10 Simulated bidding behaviors of ER04 on April 7th, 2018 by the reference method

The comparison results conclude that the identified reward functions have better performances than traditional assumption-based reward functions. Diversified bidding preferences should be considered in bidding behavior simulation studies for more realistic results.

13.7 Conclusions

To identify the reward functions of generators in power markets, a data-driven reward function identification framework is developed in this chapter. The framework contains three procedures: decision process formulation, reward function identification, and bidding behavior simulation. The first procedure is conducted based on MDP. The second procedure is developed based on the maximum entropy DIRL. The third procedure is realized based on the DQN.

The proposed methods are validated based on real market data from AEM. The reward functions of six thermal generators are identified and presented. The results demonstrate that the extracted reward functions can clearly reflect the diversified bidding preferences of different generators. The iteration of reward values at various states suggests that the proposed method can be convergent. The obtained reward functions are further employed to simulate individual bidding behaviors. The simulated results demonstrate that the proposed DQN model can effectively simulate individual behaviors based on the identified reward functions. Based on a compared example, we can conclude that the reward functions identified in this chapter have better performances than the traditional assumption-based reward function formulation.

References

1. Moiseeva, Ekaterina, and Mohammad Reza Hesamzadeh. 2018. Bayesian and robust Nash equilibria in hydrodominated systems under uncertainty. *IEEE Transactions on Sustainable Energy* 9 (2): 818–830.
2. Pozo, David, Enzo Sauma, and Javier Contreras. 2017. Basic theoretical foundations and insights on bilevel models and their applications to power systems. *Annals of Operations Research* 254 (1–2): 303–334.
3. Nan-Peng, Yu., Chen-Ching. Liu, and James Price. 2010. Evaluation of market rules using a multi-agent system method. *IEEE Transactions on Power Systems* 25 (1): 470–479.
4. Ye, Yujian, Dawei Qiu, Mingyang Sun, Dimitrios Papadaskalopoulos, and Goran Strbac. 2020. Deep reinforcement learning for strategic bidding in electricity markets. *IEEE Transactions on Smart Grid* 11 (2): 1343–1355.
5. Kiani, Arman, and Anuradha Annaswamy. 2014. Equilibrium in wholesale energy markets: Perturbation analysis in the presence of renewables. *IEEE Transactions on Smart Grid* 5 (1): 177–187.
6. Cao, Jun, Dan Harrold, Zhong Fan, Thomas Morstyn, and Kang Li. 2020. Deep reinforcement learning-based energy storage arbitrage with accurate lithium-ion battery degradation model. *IEEE Transactions on Smart Grid* 11 (5): 4513–4521.
7. Chen, Tao, and Su. Wencong. 2018. Local energy trading behavior modeling with deep reinforcement learning. *IEEE Access* 6: 62806–62814.
8. Hanchen, Xu., Hongbo Sun, Daniel Nikovski, Shoichi Kitamura, Kazuyuki Mori, and Hiroyuki Hashimoto. 2019. Deep reinforcement learning for joint bidding and pricing of load serving entity. *IEEE Transactions on Smart Grid* 10 (6): 6366–6375.
9. Ding, Huajie, Pierre Pinson, Hu. Zechun, Jianhui Wang, and Yonghua Song. 2017. Optimal offering and operating strategy for a large wind-storage system as a price maker. *IEEE Transactions on Power Systems* 32 (6): 4904–4913.
10. Ruiz, Carlos, Antonio J Conejo, and Yves Smeers. 2012. Equilibria in an oligopolistic electricity pool with stepwise offer curves. *IEEE Transactions on Power Systems* 27 (2) :752–761.
11. Ye, Yujian, Dawei Qiu, Jing Li, and Goran Strbac. 2019. Multi-period and multi-spatial equilibrium analysis in imperfect electricity markets: A novel multi-agent deep reinforcement learning approach. *IEEE Access* 7: 130515–130529.
12. Liang, Yanchang, Chunlin Guo, Zhaohao Ding, and Huichun Hua. 2020. Agent-based modeling in electricity market using deep deterministic policy gradient algorithm. *IEEE Transactions on Power Systems* 35 (6): 4180–4192.
13. Guo, Hongye, Qixin Chen, Gu. Yuxuan, Mohammad Shahidehpour, Qing Xia, and Chongqing Kang. 2020. A data-driven pattern extraction method for analyzing bidding behaviors in power markets. *IEEE Transactions on Smart Grid* 11 (4): 3509–3521.
14. Piot, Bilal, Matthieu Geist, and Olivier Pietquin. 2017. Bridging the gap between imitation learning and inverse reinforcement learning. *IEEE Transactions on Neural Networks and Learning Systems* 28 (8): 1814–1826.
15. Ziebart, Brian D, Andrew Maas, J. Andrew Bagnell, and Anind K. Dey. 2008. Maximum entropy inverse reinforcement learning. In *Proceedings of the 23rd National Conference on Artificial Intelligence - volume 3*, AAAI'08, 1433–1438. AAAI Press.
16. Wulfmeier, Markus, Peter Ondruska, and Ingmar Posner. 2015. Maximum entropy deep inverse reinforcement learning. arXiv preprint arXiv:1507.04888.
17. You, Changxi, Lu. Jianbo, Dimitar Filev, and Panagiotis Tsiotras. 2019. Advanced planning for autonomous vehicles using reinforcement learning and deep inverse reinforcement learning. *Robotics and Autonomous Systems* 114: 1–18.
18. Lerner, Itamar, Ravi Sojitra, and Mark Gluck. 2018. Inverse reinforcement learning for video games. *Aging* 10 (12).
19. Mnih, Volodymyr, Koray Kavukcuoglu, David Silver, Andrei A Rusu, Joel Veness, Marc G Bellemare, Alex Graves, Martin Riedmiller, Andreas K Fidjeland, Georg Ostrovski, et al. 2015. Human-level control through deep reinforcement learning. *Nature* 518 (7540): 529–533.

20. Haili Song, C-C Liu, Jacques Lawarrée, and Robert W Dahlgren. 2000. Optimal electricity supply bidding by markov decision process. *IEEE Transactions on Power Systems* 15 (2): 618–624.
21. Alger, Matthew. 2016. Deep inverse reinforcement learning. Technical report. https://cs.anu.edu.au/courses/csprojects/15S2/Final_presentations/Presentations
22. Claessens, Bert J, Peter Vrancx, and Frederik Ruelens. Convolutional neural networks for automatic state-time feature extraction in reinforcement learning applied to residential load control. *IEEE Transactions on Smart Grid* 9 (4): 3259–3269.
23. Mocanu, Elena, Decebal Constantin Mocanu, Phuong H Nguyen, Antonio Liotta, Michael E Webber, Madeleine Gibescu, and Johannes G Slootweg. 2019. On-line building energy optimization using deep reinforcement learning. *IEEE Transactions on Smart Grid* 10 (4): 3698–3708.
24. Wan, Zhiqiang, Hepeng Li, Haibo He, and Danil Prokhorov. 2019. Model-free real-time EV charging scheduling based on deep reinforcement learning. *IEEE Transactions on Smart Grid* 10 (5): 5246–5257.
25. Graham, Paul, Jenny Hayward, James Foster, Oliver Story, and Lisa Havas. 2019. Gencost 2018: Updated projections of electricity generation technology costs. https://doi.org/10.25919/5c587da8cafe7.

Printed by Printforce, the Netherlands